D0914472

# ENVIRONMENTAL HYDROGEOLOGY

**Mostafa M. Soliman**
*Professor, Faculty of Engineering*
*Ain Shams University*
*Cairo, Egypt*
**Philip E. LaMoreaux**
*Senior Hydrogeologist*
*P.E. LaMoreaux & Associates, Inc.*
*Tuscaloosa, Alabama*
**Bashir A. Memon**
*Executive Vice President*
*P.E. LaMoreaux & Associates, Inc.*
*Tuscaloosa, Alabama*
**Fakhry A. Assaad**
*Consultant Geologist*
*P.E. LaMoreaux & Associates, Inc.*
*Tuscaloosa, Alabama*
**James W. LaMoreaux**
*President*
*P.E. LaMoreaux & Associates, Inc.*
*Tuscaloosa, Alabama*

LEWIS PUBLISHERS
Boca Raton Boston London New York Washington, D.C.

**Library of Congress Cataloging-in-Publication Data**

Environmental hydrogeology / Mostafa M. Soliman ... [et al.].
p. cm.
Includes bibliographical references and index.
ISBN 0-87371-949-2 (alk. paper)
1. Hydrogeology. 2. Environmental geology. I. Soliman, Mostafa M. (Mostafa Mohammed)
GB1005.E58 1997
628.1'68—dc21 97-5718
CIP


Lewis Publishers is an imprint of CRC Press LLC


International Standard Book Number 0-87371-949-2
Library of Congress Card Number 97-5718
Printed in the United States of America 1 2 3 4 5 6 7 8 9 0
Printed on acid-free paper

# PREFACE

World population is expected to grow in the next two decades with an increased population of 1.7 billion, which will bring the earth's inhabitants to about 7 billion. The increased population must have adequate food, clothing, and shelter with minimum additional impact on the environment.

We have learned that there must be a readily accessible reserve of professionals, mainly geoscientists, for the governmental infrastructure to guide research, regulation, and remediation. We have also learned that environmental problems are complex and not only of local concern, but also of national and global concern. Some of these problems such as water pollution, acid rain, and air pollution extend beyond country boundaries and between continents. Experienced professional geoscientists are needed to implement programs and solve such problems.

Our environmental growing pains resulted in a great flurry of professional acquisitions. Suddenly experienced hydrogeologists are in great demand. Their knowledge is needed regarding the study of the groundwater movement as influenced by depositional geologic environments and geologic structures. Geophysicists and geochemists could help describe "the bucket" containing the water. Hydrologists and engineers can determine the hydrologic characteristics and identify, monitor, and safely remove polluting constituents.

We have learned by experience that there is much waste of financial resources and time without trained professional geoscientists to carry out the task to clean up. One of the biggest problems associated with future environmental programs is directly related to the availability of professional staffing to do the job. As we consider the future, how do we assess this factor? This can be measured in part by the number of new courses, seminars, and training programs pertaining to the environment offered at universities, by professional associations, and by private training organizations. In the next few years these programs will provide a reserve of professionals trained in hydrogeology, environmental geology, environmental engineering, and environmental chemistry.

For the aforementioned reasons, this textbook was prepared to aid geoscientists in their understanding of environmental hydrogeology. Chapter 1 covers an introduction. Chapter 2 is devoted to geological aspects of potential disposal sites. Chapter 3 covers an introduction to surface water hydrology, groundwater hydrology, and the design of wells. Chapter 4 enlightens the professional graduates and undergraduate students about the relation between the environmental impacts and the hydrogeological system. Chapters 5 describes the kinds and sources of wastes and their properties. Chapter 6 focuses on the environmental impacts on water resource systems, and Chapter 7 gives a clear idea about waste management for groundwater protection.

Chapter 8 contains selected problems inside and outside the USA as examples to show some of the environmental impacts on water resource systems and what has to be done to protect those hydrogeological systems.

The final section, includes three appendices: Appendix A, a glossary of important hydrogeological terms; Appendix B, conversion tables; and Appendix C, mathematical modeling of some of the hydrogeologic cases with accompanying software manual and computer diskette containing an executable file and a solved problem with its data file for demonstration.

**Mostafa M. Soliman**
**Philip E. LaMoreaux**

# ACKNOWLEDGMENTS

The authors are grateful to the many people who helped with the preparation of this book and particularly to all reviewers of the manuscript, Mr. William J. Powell, Dr. John Moore, Dr. Nick Schneider, Dr. Pat Madison, and Ms. Caryl Alfaro.

Most of the graphics were prepared by the Graphics and Computer Department of P.E. LaMoreaux & Associates, Inc.

Many thanks to Judy T. Tanner, manuscript manager, who was responsible for organizing and typing the manuscript.

# TABLE OF CONTENTS

# 1 INTRODUCTION

## 1.1 INTRODUCTION

It is not possible to read a daily newspaper or magazine — *The Wall Street Journal, USA Today, Newsweek,* or *Time* — without seeing the word *environmental* or reading about a tangential catastrophic event. We live with this constant reminder of local, statewide, national, and international issues, actions, or politics regarding our environment. The Earth Summit held in Brazil is an example.

On June 1, 1992, *Newsweek* blazoned headlines: "No More Hot Air It's Time to Talk Sense About the Environment."[1] The article described the meeting of world leaders in Rio the following week. Their mission: "To save the ship from its passengers." The feature article was titled "The Future Is Here" and emphasized that: Antarctica suffers an ozone hole; North America takes the lion's share of world's resources; South America is custodian of the world's largest rain forest; Australia is overcultivated; Africa faces population density doubling; and Asia has stressed resources.

Another smaller and less pretentious example: *Newsweek*, July 27, 1992, with a full-page color ad illustrating a relatively complicated geologic cross section at a nuclear waste disposal facility.[2] How many people 10 years ago would have known what a geologic cross section was, let alone understood it! The caption read: "To most people, it's a complex diagram, to Scientists, it's a clear summary of safe nuclear waste disposal." The ad implied to a supposedly rather sophisticated reading public the Madison Avenue concept of a very controversial scientific, social, and political issue: nuclear power and its associated waste disposal. This concept relates to world energy needs and is a major geoscience issue. Two different objectives are described; both, however, illustrate the great need for capable geoscientists.

World population is expected to grow in the next two decades with an increased population of 1.7 billion. This will bring the Earth's total inhabitants to about 7 billion. This article describes the environmental situation with regard to water, air, land, trees, industry, energy, species risk, and climate change. The bottom line: this increased population must have adequate food, clothing, and shelter, with minimum additional impact on the environment. This population increase with the corollary resource development also clearly identifies another substantial need for expertise in geoscience.

Environmental problems are not new. About 2000 years ago, the first written religious documents, the Bible and the Koran, related humans' relationship to their environment and recognized the importance of water to their existence. Springs and wells are the subject of numerous stories of famine, migration, war, hate, greed, and jealousy. In fact, Dr. O.E. Meinzer, the father of "ground water," in Water Supply Paper 489 remarked that parts of the Bible read like a Water Supply Paper.[3]

Since the 1970s, the environmental movement has progressed from an emotional adolescence to maturity. In the beginning there was a great cry of anguish from the general population. "There was pestilence in the land." Symptoms included sores on children who had been playing in abandoned industrial fields, cancer in adults, and pollution in our waters. Problems ranged from minor to major but all were given headlines. Love Canal was the "battle cry" and the beginning of "not in my back yard," or the NIMBY syndrome. Concern and hysteria reigned in many localities. The population was scared and was not sure of "the truth" from anyone — politician, government employee or scientist. Confidence level in these representatives was low. Everyone — individuals, politicians, industry, and government — agreed that "something" had to be done! Politicians responded and the Resource Conservation and Recovery Act (RCRA) and the Comprehensive Emergency Response Compensation and Liability Act (CERCLA) resulted. There were companion bills and rules and regulations for each state. Initially it was thought that money could solve the problem. It was soon learned it could not. Experienced professional geoscientists were needed to implement programs and solve problems. We have learned that there must be a readily accessible reserve of professionals, mainly geoscientists, for the governmental infrastructure to guide research, regulation, and remediation. We have also learned these problems are complex and not only of local, county, and state concern, but national and oceanwide, and include water pollution, acid rain, and air pollution. Some environmental problems extended beyond country boundaries and between continents. Saddam Hussein showed the world what one individual could do to jeopardize the environment that we live in.

In the late 1980s, our environmental growing pains resulted in a great flurry of professional acquisition. Suddenly hydrogeologists were in great demand. Advertisements for hydrologists appeared in the trade and technical magazines. New jobs were created for geoscientists capable of writing and implementing regulations as well as serving in regulatory roles in local, state, and federal government and, subsequently, in remedial roles in business and industry and the consulting fields. Geoscientists were suddenly charged with studies to provide the basis for intelligent remediation. Experienced hydrogeologists were particularly in demand for it was their knowledge regarding the relationship between ground-water recharge, storage, and movement as influenced by geologic depositional environments and geologic structure that was needed. Geophysicists and geologists could help describe "the bucket" containing the water, and hydrologists and engineers could determine how fast water flowed through this complicated system. Polluting constituents had to be identified, monitored, removed, and safely disposed.

A new set of industries associated with environment and environmental clean-up developed at the same time. See the "Guide to Environmental Stocks," published monthly, or refer to lists on the New York Stock Exchange, NASDAQ, or over-the-counter stocks and compare 1960 versus 1990 to learn about the large number of new firms becoming involved in environmental activities. There are old names and new ones. Companies retreaded or new ones formed to include names such as DuPont, Westinghouse, Weston, NUS, Chemical Waste Management, Rollins, Waste Management, Inc., as well as a host of other smaller specialized firms.

To evaluate the greater financial impact from the environmental movement, review the appropriations for environmental investigation and remediation in government. The U.S. Environmental Protection Agency (EPA), U.S. Department of Defense (DOD), and U.S. Department of Interior (DOI) are being appropriated each year to support research and remediation. To this we can add the corollary billions of dollars spent by commercial and industrial firms. This rapid injection of money into governmental and associated remediation has created a whole new set of demands on the geoscience community since the 1980s.

Our concept of the environment in the 1990s is much more comprehensive than in the 1960s. However, even with much progress, there remains much work to accomplish, including at least 20 years of greater emphasis on many complex problems. It will become necessary to quantify certain types of ground-water movement through rocks, geochemical interrelationships between rocks, natural constituents in water as well as pollutants, risk assessment, and one of the biggest problems of all — adequate communication about these factors and their solution with the public.

We have learned by experience that there is much waste of financial resources and time without trained professional geoscientists to carry out the task of clean up and that one of the biggest problems associated with future environmental programs is directly related to the availability of professional staffing to do the job. As we consider the future, how do we assess this factor? This can be measured in part by the numbers of new courses, seminars, and training programs pertaining to the environment offered at universities, by scientific societies, and by a very substantial number of new environmental institutes inaugurated since 1980. Newly developed academic programs and degrees are now available in environmental geology, environmental engineering, and environmental chemistry which in the next few years will provide a reserve of professionals. Another indicator would be the increased number of scientific papers in journals on the subject illustrating a reorientation of thought and emphasis on the environment. A search of the American Geological Institute GEOREF database provided the following:

**Environmental**

Key words: pollution, water quality, ecology, land use, reclamation, conservation, nonengineering aspects of geologic hazards, and nonengineering aspects of waste disposal, plus the general term environmental geology.

| Period | Citations | |
|---|---|---|
| 1785–1979 | 41,000 | over 194 years |
| 1980–1987 | 75,000 | over 7 years |
| 1988–1991 | 60,000 | over 3 years |

Concurrently, within the geoscience societies there are a number of new environmental divisions, activities, and journals, for example, the Institute for Environmental Education of the Geological Society of America (GSA) established in 1991, the newly organized Division of Environmental Geoscience of the American Association of Petroleum Geologists (AAPG) established at Calgary in June 1992, and a major new emphasis by the American Geological Institute (AGI) (see Earth System Science, a current series in *Geotimes*).

In the U.S. Government there is greater environmental awareness, for example, the U.S. Army Corps of Engineers' (COE) allocation of substantial funding during construction of the Tenn-Tom Waterway (TTW) to employ an "Environmental Advisory Board" with the specific assignment to provide guidance that included changes in construction, to minimize soil erosion, loss of wetlands, attention to wild life, protection of groundwater supplies, and many other environmental considerations. One such recommendation changed the course of the waterway to protect a famous old geologic locality at Plymouth Bluff, Mississippi. This large project required the efforts of many geoscientists. One aspect of their work included communication with the public, politicians, and government about what should be considered proper planning and construction and the adequate consideration of environmental impacts. This illustrates the need for the geoscientist to communicate, a responsibility that will become more important in the future.

As we look into the future, environmental activities will exert the greatest demand for geoscientists. Specific identity of four of these activities will illustrate the point. The first two, RCRA and CERCLA, in the 1970s and 1980s provided a whole new body of law with significance to environmental activities that affected all facets of the private, agricultural, commercial, and industrial sectors, as well as to local, state and federal governments.

1. RCRA was created in 1976 and applied to future waste management. It included criteria for location, groundwater monitoring, operations, and contingencies. The law was converted to comprehensive regulations and criteria to be implemented in each state by legislative and legal action.
2. CERCLA was created in 1980 and applied to the clean up of old, abandoned hazardous-waste facilities. It was a massive program of investigation of climate, geology, hydrology, biology, botany, and other environmental risks. SARA (Superfund Amendments and Reauthorization Act), 1986, National Priority List (NPL) found in 40 CFR, Part 300, Appendix B. The last issue of list was in February 1991 (proposed listing as of March 1992) of 1179 NPL sites, and 84 sites removed from the list from 1980 to the present.
3. Environmental Audits: The impact from the environmental laws of the 1970s was even more far reaching as the private sector as well as commercial and industrial activities began to need an environmental audit prior to property transfers. The EPA policy on July 9, 1986, *Federal Register*, recommends the use of environmental audits.[4] Banks and other loaning institutions require audits. Millions of property transfers now require an audit by a certified environmental scientist. Criteria for audits have been established by the Resolution Trust, Small Business Administration, as well as by individual banks. This represents a massive amount of work in the future.
4. LUST: To the uninitiated, LUST does not mean what you think it does. It means Leaking Underground Storage Tanks. In 1984 Congress responded to the problem by adding Subtitle I to RCRA. Subtitle I requires EPA to develop regulations to protect human health and the environment from leaking USTs. Between three and five million underground storage tanks are currently being used in the U.S. to store motor fuels and chemical products. Nearly 80% of these tanks are constructed of bare steel. Not surprisingly, 60% of all leaks results from corrosion.

EPA UST rules are promulgated by 40 CFR Parts 280 and 281. Final rules on technical requirements were published in the *Federal Register* (September 23, 1988).[5] The most significant problem is the sheer size of the regulated community. Nationally, over 700,000 UST facilities account for over 3 million UST systems, an average per state of about 14,000 UST facilities and 40,000 UST systems. Estimates indicate that roughly 79% of existing UST systems are unprotected form corrosion. In addition, because a relatively high proportion of UST facilities (10–30%) already have had a leak, or will soon leak unless measures are taken to upgrade them, the average number of leaking UST systems may range from 1,400 to 4,200 per state in the near future. The

LUST problems must be handled by knowledgeable geologists, hydrologists, and engineers.

Information on the magnitude of the problems relating to waste management, acid rain, water pollution, non-point sources of pollution such as agricultural use of insecticides and pesticides, mining activities, oil and gas activities, and construction of all types, even the acquisition of any property transfers in the future will require an environmental assessment. These issues will require a whole new team of sophisticated scientists over the next 20 years.

In *Geotimes* January 1991, there appeared two excellent articles, "Tomorrow's Geoscientist" by Marilyn Suiter[6] and "Geoscience Careers" by Nick Claudy,[7] which contain appropriate and accurate information about the demand for geoscientists in the future. Suiter makes the point that women and ethnic minorities will make up much of the human resources potential for the work force in the future. Also the demand in science and engineering for qualified workers will grow (Figure 1.1). Claudy concludes that the geosciences offer unparalleled diversity for career opportunities. He identifies by percentage their major employment categories — oil and gas (50%), mining (9%), Federal/State (12%), research institutions (4%), consulting (11%), and academia (14%). Claudy also identifies correctly the need for geoscientists with MS degrees for professional categories and extensive job opportunities as well as for geotechnicians with BA or BS degrees. These articles, however, do not call attention to the major shift to be expected in demand for scientists in the broad environmental activities area. The state and federal government agencies are limited by appropriation constraints; however, the biggest demand will be in the broad area of environmental work. We predict that at least 50% of the new jobs will fall in this category and the need will be critical.

If the solid earth sciences are to meet the demands of society's environmental problems, the profession must recruit, train, and place in the professional work force a sufficient number of well-qualified professionals to carry out the task ahead. According to a recent survey, about one half of earth scientists in the United States (about 120,000), including petroleum and mining engineers, are employed by the petroleum industry. The U.S. Government employs about 14,000 and academia about 9,000. The remainder are employed on environmental work related to waste management, hazardous and toxic radioactive waste permitting litigation, underground storage problems, environmental audits, and environmental impact studies. The supply of and demand for earth scientists over the past 50 years has historically been out of phase. In the early 1980s, because of the dramatic decline in petroleum prices, employment in oil and gas activities decreased by about 30%. This was also a depressed period in the mining industry, and there resulted a loss of thousands of jobs in the earth scientist categories. We are just recovering from this cycle. It was a traumatic period for the geosciences with over 4,000 laid off and geologists a glut on the market. Experienced PhDs were searching for any respectable employment. Qualified geoscientists, especially the younger gen-

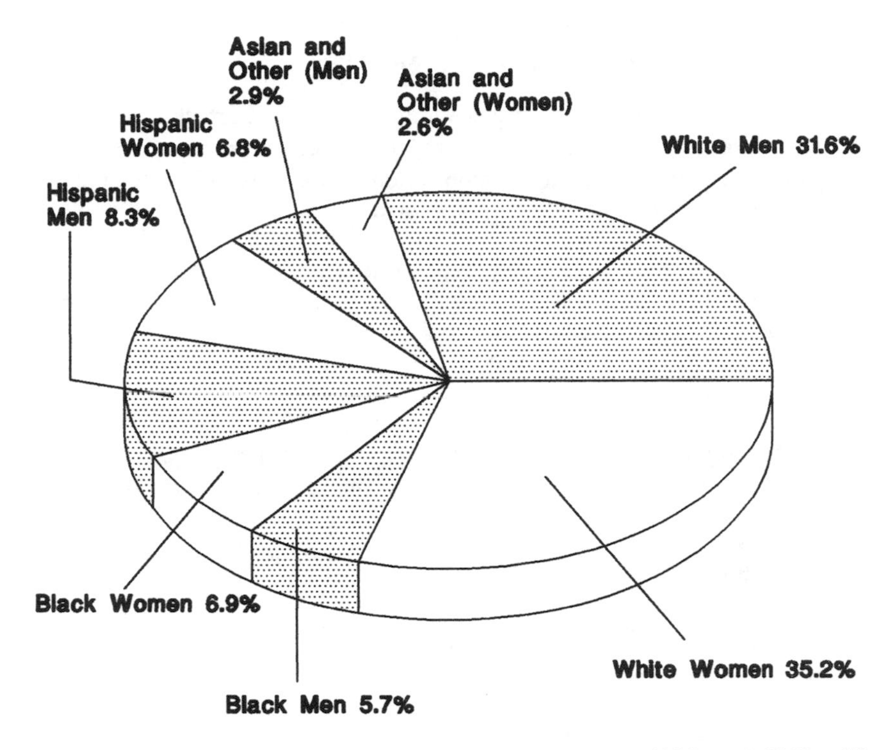


**Figure 1.1 People entering the work force between 1988 and 2000 will be mostly women and minorities, according to the U.S. Bureau of Labor Statistics. (From Suiter, M., Tomorrow's geoscientist, *Geotimes,* January 1991.)**

eration, were unable to find jobs. This had a detrimental affect on the recruitment of geoscience majors and the production of professionals. Environmental legislation dealing with waste disposal was enacted in the early 1980s, and employment projections indicate employment in the earth sciences is growing rapidly, with emphasis on groundwater issues, the siting of waste repositories, and needs for environmental cleanups. The down cycle in the oil industry had its detrimental impact in the decreased number of new scientists entering the field. In the future we must recognize that great opportunities exist in the environmental areas and that there will be critical needs for new vigorous members for the profession. With the recovery of the oil and gas and mining industries to produce resources for a rapidly expanding world population, the demand for earth scientists will become strong. Further, unless these reserves of competent professionals are forthcoming, the nation will face a critical situation. These are reasons for a textbook on *Environmental Hydrogeology* at this time.

## 1.2 SUGGESTIONS AND REFERENCES

Several recent publications provide important reference material for sound environmental geologic, geoscience, and hydrogeological programs. Each emphasizes the need for good communication between the scientific community and political, industrial, and private citizens.

*Solid-Earth Science and Society,* National Academy Press, 368 pp., 1993
*Citizens' Guide to Geologic Hazards,* American Institute of Professional Geologists, 134 pp., 1993
*Societal Value of Geologic Maps,* Circular 1111, U.S. Geological Survey, 53 pp., 1993

A special journal, *Environmental Geology,* is published eight times annually by Springer-Verlag and is available by subscription. It provides many good case histories on the subject.

A modified statement is taken from the Citizens' Guide to Geologic Hazards to illustrate the present:[8]

> *Hazardous geological processes* most familiar to the public are those that occur as rapid events, i.e. over a period of minutes, hours or days. Examples include: *earthquakes* produced by the process of rapid snapping movements along faults; *volcanoes* produced by upward-migrating magma; *landslides* produced by instantaneous failure of rock masses under the stress of gravity; and *floods* produced by a combination of weather events and land use. These events all produce massive fatalities and make overnight headlines. Other geologic processes, such as soil creep (slow downslope movement of soils that often produces disalignment of fence posts or cracked foundations of older buildings), *frost heave* (upheaval of ground due to seasonal freezing of the upper few feet) and land subsidence act more slowly and over wider regions. These slower processes however also take a toll on the economy. Human interaction can be an important factor in triggering or hastening these natural processes (see Table 1.1).
>
> *Lack of awareness* induces human complacency which sometimes proves fatal. It is difficult to perceive of natural dangers in any area where we and preceding generations have spent our lives in security and comforting familiarity. This is because most catastrophic geologic hazards do not occur on a timetable that makes them easily perceived by direct experience in a single lifetime. Yet development within a hazardous area inevitably produces consequences for some inhabitants. Hundreds of thousands of unfortunate people who perished in geological catastrophes such as landslides, floods or volcanic eruptions undoubtedly felt safe up until their final moments. In June 1991, Clark Air Force Base in the Philippines was evacuated when Mount Pinatubo,

**Table 1.1 Economic Costs of Geologic Hazards in the U.S.**

| Geologic hazard | Cost in 1990 dollars* | Source(s) |
|---|---|---|
| | **Hazards from materials** | |
| Swelling soils | $6 to 11 billion annually | Jones and Holtz, 1973, Civil Engrg. v. 43, n. 8, pp. 49–51; Krohn and Slosson, 1980, ASCE Proc. 4th Intnl. Conf. Swelling Soils, pp. 596–608 |
| Reactive aggregates[1] | No estimate | — |
| Acid drainage | $365 million annually to control; $13 to 54 billion cumulative to repair | USBM, 1985, IC 9027; Senate Report, 1977, 95–128 |
| Asbestos | $12 to 75 billion cumulative for remediation of rental and commercial buildings; total well above $100 billion including litigation and enforcement | Croke et al., 1989, The Environmental Professional, v. 11, pp. 256–263. Malcolm Ross, USGS, 1993, personal communication. Costs depend on extent and kind of remediation done; removal is most expensive option. |
| Radon | $100 billion ultimately to bring levels to EPA recommended levels of 4 PCi/L | Estimate based on remediating about 1/3 of American homes at $2500 each plus costs for energy and public buildings |
| | **Hazards from processes** | |
| Earthquakes | $230 million annually decade prior to 1989; over $6 billion in 1989 | USGS, 1978, Prof. Paper 950; Ward and Page, 1990, USGS Pamphlet, "The Loma Prieta Earthquake of October 17, 1989" |
| Volcanoes | $4 billion in 1980; Several million annually in aircraft damage | USGS Circular 1065, 1991, and Circular 1073, 1992 |
| Landslides/ Avalanches | $2 billion/50.5 million annually | Schuster & Fleming, 1986, Bull. Assoc. Engrg. Geols., v. 23, pp. 11–28/Armstrong & Williams, 1986, *The Avalanche Book* |

**Table 1.1 Economic Costs of Geologic Hazards in the U.S. *(continued)***

| Geologic hazard | Cost in 1990 dollars* | Source(s) |
|---|---|---|
| Subsidence[2] and permafrost[3] | At least $125 million annually for human-caused subsidence; $5 million annually from natural karst subsidence | Holzer, 1984, GSA Reviews in Engrg. Geology VI; FEMA, 1980, Subsidence Task Force Report |
| Floods | $3 to 4 billion annually | USGS Prof. Paper 950 |
| Storm surge[4] and coastal hazards | $700 million annually in coastal erosion; over $40 billion in hurricanes and storm surge 1989–early 1993 | Sorensen and Mitchell, 1975 Univ. CO Institute of Behavioral Sci., NSF-RA-E-75-014; Inst. of Behavioral Sci., personal comm. |

* Costs from dates reported in "Source(s)" column have been reported in terms of 1990 dollars. This neglects changes in population and land use practice since the original study was done but gives a reasonable comparative approximation between hazards.

1 Aggregates are substances such as sand, gravel, or crushed stone that are commonly mixed with cement to make concrete.

2 Subsidence is local downward settling of land due to insufficient support in the subsurface.

3 Permafrost consists of normally frozen ground in polar or alpine regions that may thaw briefly due to warm seasons or human activities and flow.

4 Storm surge occurs when meteorological conditions cause a sudden local rise in sea level that results in water piling up along a coast, particularly when strong shoreward winds coincide with periods of high tide. Extensive flooding then occurs over low-lying riverine flood plains and coastal plains.

From Nuhfer, E.B., Proctor, R.J., and Moser, P.H. (Eds.), American Institute of Professional Geologists, *The Citizens' Guide to Geologic Hazards,* Arvada, CO, 134 pp., 1993.

a volcano dormant for over 600 years, began to erupt and put property and lives at risk.

Geologic hazards are not trivial or forgiving; in terms of loss of life, geologic hazards can compare with the most severe catastrophes of contemporary society. Where urban density increases and land is extensively developed, the potential severity of loss of life and property from geologic hazards increases.

We are often faced with the decision about whether we can wisely live in areas where geologic forces may actively oppose otherwise pleasant living conditions. There follows some guidelines:

(1) *Avoid an area where known hazards exist.* Avoidance or abandonment of a large area is usually neither practical nor necessary. The accurate mapping of geologic hazards delineates those very specific areas which should be avoided for particular kinds of development. Otherwise hazardous sites may make excellent green belt space or parks in areas zoned as floodplains, thus avoiding placing expensive structure where flooding will cause damage.

(2) *Evaluate the potential risk for hazards.* Risks can never be entirely eliminated, and the process of reducing risk requires expenditures of effort and money. Assuming, without study, that a hazard will not be serious is insufficient. Life then proceeds as though the hazard were not present at all. "It can't happen here" expresses the view that is responsible for some of the greatest losses. Yet it is equally important not to expend major amounts of society's resources to remedy a hazard for which the risk is actually trivial.

(3) *Minimize the effect of the hazards by engineering design and appropriate zoning.* Civil engineers who have learned to work with geologists as team members can be solid and effective contributors to minimizing effects of geologic hazards. More structures today fail as a result of incorrectly assessing (or ignoring) the geological conditions at the site than fail due to errors in engineering design. This fact has led many jurisdictions to mandate that geological site assessments be performed by a qualified geologist. Taking geological conditions into account when writing building codes can have a profound benefit. The December 1988 earthquake in northwestern Armenia that killed 25,000 people was smaller in magnitude (about 40% smaller) than the October 1988 Loma Prieta earthquake in California. The latter actually occurred in an area of higher population density but produced just 67 fatalities. Good construction and design practice in California was rewarded by preservation of lives and property.

Academic training for civil engineers must include basic courses in geology taught by qualified geologists. A more comprehensive geologic education is needed for civil and environmental engineers. Engineers should be cognizant of the benefits of a geological assessment and be able to communicate with professional geologists.

California, in 1968, became the first state to require professional geological investigations of construction sites and has reaped proven benefits for that decision. Since then many states have enacted legislation to insure that qualified geologists perform critical site evaluations of the geology beneath prospective structures such as housing developments and landfills. Most of these laws were enacted after 1980.

Zoning ordinances and building codes that are based on sound information and that are conscientiously enforced are the most effective legal documents for minimizing destruction from geologic hazards. After a severe flood, citizens have often been relocated back to the same site with funding by a sympathetic government. This is an example of "living with a geologic hazard" in the illogical sense. A less costly alternative might be to zone most floodplains out of residential use and to financially encourage communities or neighborhoods that suffer repeated damage to relocate to more suitable ground. When damage or injuries occur from a geologic hazard in a residential area, the "solution" is often a lawsuit brought against a developer. The problem has not truly been remedied; the costs of the mistake have simply been transferred to a more luckless party — the future purchasers of liability and homeowners' insurance at higher premiums. A solution would be a map that clearly delineates those hazardous areas where residential development is forbidden. A suitable alternative would be a statute requiring site assessment by a qualified geologist before an area can be developed. Sound land use that takes geology into account can prevent unreasonable insurance premiums, litigation, and repeated government disaster assistance payments for the same mistakes.

(4) *Develop a network of insurance and contingency plans to cover potential loss or damage from hazards.* Planners and homeowners need not be geologists, but it is useful to them to be able to recognize the geological conditions of the area in which they live, and to realize when they need the services of a geologist. A major proportion of earthquake damage is not covered by insurance. Despite public awareness about earthquakes in California, the 1987 Whittier quake produced 358 million dollars worth of damage of which only 30 million dollars of this was covered by insurance.

For the property owner, especially the prospective homeowner, a geological site assessment may answer the following:

> *Is the site in an area where landslides, earthquakes, volcanoes or floods have occurred during historic time? Has the area had past underground mining or a history of production from wells? Did the land ever have a previous use that might have utilized underground workings or storage tanks that might now be buried? Does the site rest on fill, and is the quality of the fill and the ground beneath it known? Are there swelling soils in the area? Have geologic hazards damaged structures elsewhere in the same rock and soil formations which underlie the site in question? Has the home ever been checked for radon? IF the home is on a domestic well, has the water quality been recently checked? Is the property on the flood plain of a stream? Is the property adjoining a body of water such as a lake or ocean where there have been severe shoreline erosion problems after infrequent (such as 20-year or 50-year) storms?*

Insurance agents are not always familiar with local geological hazards. After risks have been assessed, the individual can then consult with insurance professionals (agents, brokers, salesperson) to learn which firms offer coverage that would include pertinent risks. Consulting with the state insurance boards and commissions can assist one in finding insurers who provide pertinent coverage.*

Local governments should make plans for zoning and for contingency measures such as evacuations with involvements from a professional geologist. The first line of help for local governments lies in their own state geological surveys. Hydrogeologists are employed for service to the public and can provide much of the available information that is known about the site or region in question and can direct the inquirer to other additional resources. Geologic maps and reports from public and private agencies are most useful in the hands of those trained to interpret them. Significant evidence that reveals a potential geologic hazard may be present in the reports and maps. If significant risks of hazards are thought to exist, then consultation with a professional geologist may be warranted.

Geologic hazards annually take more than 100,000 lives and take billions of dollars from the world's economy. Such hazards can be divided into hazards that result dominantly from particular earth materials or from particular earth processes. Most of these losses are avoidable, provided that the public at large makes use of state-of-the-art geologic knowledge in planning and development. A public ignorant of geology cannot usually perceive the needs for geologists in many environmental, engineering or even domestic projects. The result is a populace prone to making expensive mistakes, particularly in the area of public policy.

Education is one of the most effective ways of preparing to deal successfully with geologic hazards. Every state geological survey produces useful publications, distributes maps, and answers inquiries by the public. Unfortunately, lack of good earth science education leaves many citizens unaware of the resources that their geological surveys provide.

The literature of geologic hazards falls primarily under two indexed subfields of geology: environmental geology and engineering geology. Flood hazards may also be found under the subfield hydrology.

The following list provides suggested sources of references pertaining to environmental hydrogeology and geology:

American Society of Civil Engineers, 1974, *Analysis and Design in Geotechnical Engineering*, New York, Amer. Soc. Civil Engrs.

American Society of Civil Engineers, 1976, *Liquefaction Problems in Geotechnical Engineering,* New York, Amerc. Soc. Civil Engrs.

American Society of Foundation Engineers, 1975 — ongoing, Case History Series: ASFE/The Association of Engineering Firms Practicing in the Geosciences, 811

* Homeowners' policies do not routinely cover geologic hazards such as subsidence or earthquake damage.

Colesville Road, Suite G 106, Silver Spring, MD, 20910. A series of case studies arranged in terms of background, problems and outcomes, and lessons learned in brief one-page, two-sided formats.

Bennison, A.P., et al. (Eds.), 1972, *Tulsa's Physical Environment,* Tulsa Geol. Society Digest, 37. Tulsa Geol. Soc., Suite 116, Midco Bldg., Tulsa, OK, 74103.

Bolt, B.A., Horn, W.L., Macdonald, G.A., and Scott, R. F., 1977, *Geological Hazards*[E], New York, Springer-Verlag.

Bryant, E.A., 1991, *Natural Hazards,* New York, Elsevier.

Coates, D.R., 1981, *Environmental Geology,* New York, John Wiley.

Coates, D.R., 1985, *Geology and Society,* New York, Chapman and Hall.

Dodd, K., Fuller, H.K., and Clarke, P.F., 1989, Guide to Obtaining USGS Information[E], U.S. Geol. Survey Circular 900. Our federal geological survey serves more than just other geologists. This free circular tells how to access their vast storehouse of information and how to order many of the USGS publications. Write Books and Open-File Reports Section, USGS, Federal Center, Box 25425, Denver, CO 80225.

El-Sabk, M.I., and Marty, T.C., Eds., 1988, *Natural and Man-Made Hazards,* Dordrecht, Netherlands, Reidel.

Federal Emergency Management Agency (FEMA), 1991, Are You Ready? Your Guide to Disaster Preparedness[E], FEMA, Publications Dept., P.O. Box 70274, Washington, DC, 20224.

Foster, H.D., 1980, *Disaster Mitigation for Planners: The Preservation of Life and Property,* New York, Springer-Verlag.

Freedman, J.L. (Ed.), 1977, "Lots" of Danger — Property Buyers Guide to Land Hazards of Southwestern Pennsylvania[E], Pittsburgh Geol. Soc., 85 pp. This is a model publication that serves property owners and prospective property owners of southwestern Pennsylvania.

Gerla, P.J., and Jehn-Dellaport, T., 1989, Environmental impact assessment for commercial real estate transfers, *Bull. Assoc. Engrg. Geologists,* v. 26, pp. 531–540.

Griggs, G.B., and Gilchrist, J.A., 1983, *Geologic Hazards, Resources, and Environmental Planning (2nd ed.),* Belmont, CA, Wadsworth.

Haney, D.C., Mankin, C.J., and Kottlowski, 1990, Geologic mapping: a critical need for the nation, Washington Concentrates, Amer. Mining Congress, June 29, 1990.

Hays, W.W. (Ed.), 1981, Facing geologic and hydrologic hazards[E], U.S. Geol. Survey Prof. Paper 1240-B.

Henderson, R., Heath, E.G., and Leighton, F.B., 1973, What land use planners need from geologists, in *Geology, Seismicity, and Environmental Impact*, Assoc. Engrg. Geologists Spec. Pub., Los Angeles, CA, pp. 37–43.

Keller, E.A., 1985, *Environmental Geology (5th ed.)*[E], Columbus, OH, Charles E. Merrill.

Legget, R.F., 1973, *Cities and Geology,* New York, McGraw-Hill.

Legget, R.F., and Hatheway, A.W., 1988, *Geology and Engineering (3rd ed.),* New York, McGraw-Hill.

Legget, R.F., and Karrow, P.F., 1982, *Handbook of Geology in Civil Engineering,* New York, McGraw-Hill.

McAlpin, J., 1985, Engineering geology at the local government level: planning, review, and enforcement, Bull. Assoc. Engrg. Geologists, v. 22, p. 315-327.

Mileti, D.S., 1975, Natural Hazard Warning Systems in the U.S., Natural Hazards Research and Applications Information Center, Univ. Colorado at Boulder.

Montgomery, C.W., 1985, *Environmental Geology*[E], Natl. Geog., May, pp. 638-654.

Palm, R.I., 1990, *Natural Hazards: An Integrative Framework for Research and Planning,* Baltimore, MD, Johns Hopkins University Press.

Peck, D.L., 1991, Natural hazards and public perception: earth scientists can make the difference[E], *Geotimes,* 36 (5), 5.

Rahn, P.H., 1986, *Engineering Geology — An Environmental Approach,* New York, Elsevier.

Scheidegger, A.E., 1975, *Physical Aspects of Natural Catastrophes,* New York, Elsevier.

Slosson, J.E., 1969, The role of engineering geology in urban planning, in Governor's Conference on Environmental Geology, Colorado Geol. Survey Spec. Publ. No. 1, pp. 8–15.

Smith, K., 1992, *Environmental Hazards,* New York, Routledge, Chapman Hall, Inc.

Steinbrugge, K.V., 1982, Earthquakes, Volcanoes, and Tsunamis, Anatomy of Hazards: New York, Skandia America Group.

Tank, R.W. (Ed.), 1983, *Environmental Geology; Text and Readings (3rd ed.)*[E], New York, Oxford University Press.

United States Geological Survey, 1968–present, *Earthquakes and Volcanoes*[E], A magazine that combines news reporting with journal articles. It is published bimonthly and is designed for both generalized and specialized readers, USGS, Denver Federal Center, Bldg. 41, Box 25425, Denver, CO 80225.

Wermund, E.G., 1974, Approaches to Environmental Geology — A Colloquium and Workshop, Austin, TX, Texas Bureau Econ. Geol.

Whittow, J., 1979, *Disasters: The Anatomy of Environmental Hazards,* University of Georgia Press.

Wiggins, J.H., Slosson, J.E., and Krohn, J., 1978, Natural hazards: earthquake, landslide, expansive soil loss models, Natl. Sci. Foundation, NTIS, PB-294686/AS.

## REFERENCES

1. No More Hot Air It's Time to Talk Sense About the Environment, *Newsweek*, June 1, 1992.
2. Full page color illustration showing a complicated geologic cross section at a nuclear waste disposal facility, *Newsweek*, July 27, 1992.
3. Meinzer, O.E., The occurrence of ground water in the United States, with a discussion of principles, Water Supply Paper 489, 321 pp., 1923.
4. *Federal Register,* EPA policy on use of environmental audits, July 9, 1986.
5. *Federal Register,* Final rules on technical requirements, 53 (185), September 23, 1988.
6. Suiter, M., Tomorrow's geoscientist, *Geotimes*, January 1991.
7. Claudy, N., Geoscience careers, *Geotimes*, January 1991.
8. Nuhfer, E.B., Proctor, R.J., and Moser, P.H. (Eds.), American Institute of Professional Geologists, *The Citizens' Guide to Geologic Hazards,* Arvada, CO, 134 pp., 1993.

# 2 GEOLOGICAL ASPECTS FOR ASSESSMENT, CLEAN-UP AND SITING OF WASTE DISPOSAL SITES

## 2.1 INTRODUCTION[1]

Geology and hydrogeology are broad-based multidisciplines developed from many different sciences. The origin of hydrogeology required multidiscipline concepts from mathematics, physics, chemistry, hydrology, geology, and the processes of evaporation, transpiration, and condensation. Meinzer noted that the science of hydrogeology could not be undertaken until the basic concepts of geology were understood.

A knowledge of rock type, stratigraphy, and structure is imperative as a basis of understanding groundwater, recharge, storage, and discharge characteristics. An understanding of geology is a prerequisite to the development of an understanding of the source, occurrence, availability, and movement of groundwater. The application of quantitative methods for groundwater requires an accurate description of the container (aquifer) or geologic framework.

State and Federal regulations have established restrictions for location of hazardous waste and municipal solid waste landfills. Regulations require owners/operators to demonstrate that the hydrogeology has been completely characterized at proposed landfills and that locations for monitoring wells have been properly selected. Owners/operators are also required to demonstrate that engineering measures have been incorporated in the design of both hazardous and municipal solid waste landfills, so that the site is not subject to destabilizing events, as a result of location in unsuitable or unstable areas.[2]

Proposed, new, or existing landfills are subjects of controversy and sources of continuing debate as to whether an area can provide suitable sites for construction of landfills for hazardous or non-hazardous waste. Issues of concern are the potential threats to human health and the environment which could result from a) collapse or subsidence, with the associated loss of structural integrity of the landfill; b) release of contaminants through collapse, subsidence, or leakage from the landfill; and c) contamination of groundwater and/or surface water, which may result from a release.

Conceptually, the selection of waste management sites involves collection of information necessary to answer a few simple questions, including those which follow. Will the natural hydrogeologic system provide for isolation of wastes, so that disposal will not cause potential harm to human health or the environment? Is the site potentially susceptible to destabilizing events, such as collapse or subsidence, which will provide a sudden and catastrophic release from the facility and be rapidly and irrevocably transmitted to important aquifers or bodies of surface water? Are the monitoring wells in proper positions to intercept groundwater flow from the facility? If minor releases (leakages) occur, will contaminants be readily detected in monitoring wells? If a release is detected, is knowledge of the hydrogeologic setting sufficient to allow rapid and complete remediation of release? Is the hydrogeologic system sufficiently simple to allow interception and remediation of contaminated groundwater?

Answers to the above questions depend upon the thoroughness of geologic and hydrogeologic studies by which each site was assessed and evaluated prior to construction of a land disposal facility. In the experience of the authors, most significant environmental problems, resulting from releases from land disposal facilities, occur from facilities for which preliminary hydrogeologic studies were inadequate to answer the above questions. In many such cases, studies designed to gain understanding of the hydrogeologic system did not begin until after a release was detected. Compliance monitoring and remediation of groundwater are costly processes, all of which can be avoided by assiduous care in selection of proper sites for land disposal.

Specific regulations for siting landfills in all geologic settings have not been promulgated. However, regulations for protection of groundwater and monitoring as well as other regulations[3,4] require characterization of the hydrogeologic system and proper location of monitoring wells at landfills. Figure 2.1 is a conceptual hydrogeological model showing the gradient flow, geologic setting, and monitoring wells.

Consideration of candidate sites for land disposal facilities is a process that requires careful screening of many potential sites, rejection of unsuitable sites, avoidance of questionable sites, and demonstration that the selected site is hydrogeologically suitable for disposal of waste.

The screening process typically includes a) selection of a large number of candidate sites within the geographic area of interest; b) ranking of the candidate sites in order of apparent suitability for disposal of wastes; c) rejection of areas or sites that are obviously not suitable for disposal of wastes; and d) selection of one or more of the sites for further evaluation.

Tasks during screening typically involve review of published and unpublished engineering, geologic, and hydrologic literature, discussions with appro-

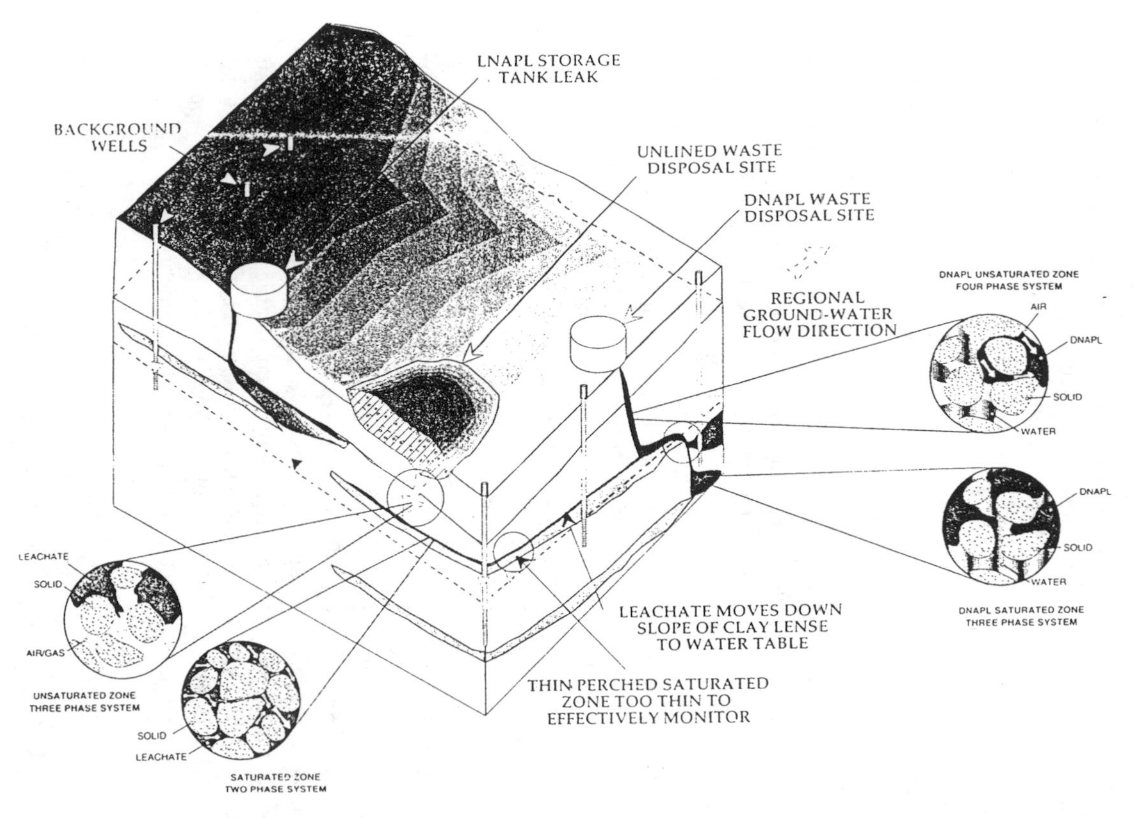


**Figure 2.1** **Conceptual flow models for NAPL sites. (From Sara, M.N., *Standard Handbook for Solid and Hazardous Waste Facility Assessments*, Lewis Publishers, Boca Raton, FL, 1994, pp. 10–68.)**

priate state or federal personnel, study of topographic maps, interpretation of sequential aerial photographs, and verification of studies by field reconnaissance. Most of the preliminary screening can be rapidly accomplished in the office at low cost. The stratigraphic intervals and structural anomalies, along with other geologic features, are defined in published geologic literature. General geologic maps are often available, and this knowledge can be extended to site-specific locations through use of aerial photographs, topographic maps, and field work. The locations, depths, water levels, producing horizons, and rates of pumping for wells in the vicinity of the site are often available in the files of state or federal agencies.

A thorough review of published and unpublished literature must be done during the preliminary investigation. Older literature often contains more complete descriptions of the geology than more recent publications and should not be overlooked.

Table 2.1 illustrates concepts presented by many authors, including Hughes,[6] LeGrand,[7] LeGrand and Stringfield,[8,9] LeGrand and LaMoreaux,[10] LaMoreaux et al.,[11,12] Newton,[13,14] Parizek et al.,[15] Sweeting,[16] and White.[17-19] Sara provides a comprehensive review of the site assessment process.[5] His guide, Standard Handbook for Solid and Hazardous Waste Facility Assessments, is a comprehensive manual to assist in the planning, performing, and interpretation of investigations for the facility's suitability for disposal of solid or hazardous waste. The manual also provides appropriate locations for groundwater monitoring and effectiveness of the landfill's leachate collection and contamination system. This guide was the result of a substantial team effort that included professionals from government, industry, and academia. It is a comprehensive compilation of information on solid and hazardous waste management that all professionals involved with environmental hydrogeology should be aware of and use.

Sequential aerial photographs have been available in many parts of the United States since the late 1930s. Sets of aerial photographs, as stereo pairs, can be used to determine if karst features have changed, if collapse features have locally occurred, or if depressions have been enlarged with time and can serve as a means to catalog changes in land use over a period of about 50 years. In addition, the aerial photographs can be used for preliminary mapping of stratigraphic contacts, structural features, and lineaments.

## 2.2 GEOLOGICAL ASPECTS

### 2.2.1 Rock Types

Rocks are classified according to their origin as igneous, metamorphic, and sedimentary, and according to their lithology which describes the rock composition and texture.

Unfractured metamorphic and plutonic igneous rocks have maximum porosities of 2% with minute intercrystalline voids that are not interconnected.

## Table 2.1 Important Characteristics of Different Geologic Terrain

**Stratigraphy**
(Regional and Local)
- Stratigraphic Column
- Thickness of Each Carbonate Unit
- Thickness of Noncarbonate Interbeds
- Type of Bedding
  - Thin
  - Medium
  - Thick
- Purity of Each Carbonate Unit
  - Limestone or Dolomite
    - Pure
    - Sandy
    - Silty
    - Clayey
    - Siliceous
    - Interbeds

**Overburden**
(Soils and Subsoils)
- Distribution
- Origin
  - Transported
    - Glacial
    - Alluvial
    - Colluvial
- Residual
- Other

Characteristics and Variability
- Thickness
- Physical Properties
- Hydrologic Properties

**Hydrology**
- Surface Water
  - Discharge
    - Variability
      - Seasonal
      - Gaining
      - Losing
- Groundwater
  - Diffuse Flow
  - Conduit Flow
  - Fissure Flow
  - Recharge
  - Storage
  - Discharge

Fluctuation of Water Levels
Relationships of Surface-Water and Groundwater Flow

**Geologic Structure**
(Regional and Local)
- Nearly Horizontal Bedding
- Tilted Beds
  - Homoclines
  - Monoclines
- Folded Beds
  - Anticlines
  - Synclines
  - Monoclines
  - Domes
  - Basins
  - Other
- Fractures
  - Lineaments
  - Locations
  - Relationships with
    - Geomorphic Features
    - Karst Features
    - Stratigraphy
    - Structural Features
- Joint System
  - Joint Sets
    - Orientation
    - Spacing
    - Continuity
    - Open
    - Closed
    - Filled
  - Faults
    - Orientation
    - Frequency
    - Continuity
    - Type
      - Normal
      - Reverse
      - Thrust
      - Other
    - Age of Faults
      - Holocene
      - Pre-Holocene

**Activities of Man**
- Construction
  - Excavation
  - Blasting
  - Vibration
  - Loading
    - Fill
    - Buildings

**Geomorphology**
(Regional and Local
- Relief-Slopes

Density of Drainage Network
- Characteristics of Streams
  - Drainage Pattern(s)
    - Dendritic
    - Trellis
    - Rectangular
    - Other
  - Periennial
  - Intermittent
- Terraces
- Springs and/or Seeps
- Lakes and Ponds
- Flood Plains and Wetlands
- Karst Features — Active, Historic
  - Karst Plains
  - Poljes
  - Dry Valleys, Blind Valleys, Sinking Creeks
  - Depressions and General Subsidence
  - Subsidence Cones, in Overburden
  - Sinkholes
  - Roof-Collapse
  - Uvalas
  - Caverns, Caves and Cavities
  - Rise Pits
  - Swallow Holes
  - Estavelles
  - Karren
  - Other
- Paleo-Karst

**Climate**
- Precipitation (Rain and Snow)
  - Seasonal
  - Annual
  - Long-Term
- Temperature
  - Daily
  - Seasonal
  - Annual
  - Long Term
- Evapotranspiration
- Vegetation

**Table 2.1 Important Characteristics of Different Geologic Terrain *(continued)***

| | |
|---|---|
| | Changes in Drainage |
| | Dams and Lakes |
| | Withdrawal of Groundwater |
| | Wells |
| | Dewatering |
| | Irrigation |

From Hughes, T.H., Memon, B.A., and LaMoreaux, P.E., Landfills in karst terrains, Bulletin of the Association of Engineering Geologists, Vol. 31, 2, 1994, p. 203–208.

The primary permeabilities of these rocks therefore are low because of the lack of void interconnection.[20]

Fractured plutonic igneous and crystalline metamorphic rocks have secondary (fracture) permeability developed along the fracture openings that are generally more common at a depth of less than 100 feet with some occurring at a maximum depth of 200 feet. The permeability of fracture zones in the crystalline rocks decreases with depth where the fractures tend to close because of vertical and horizontal stresses imposed by overburden loads.

Volcanic rocks are formed from the solidification of magma which when discharged at land surface flows out as lava. The rocks that form on cooling are generally very permeable. Columnar joints and bubble-like pore spaces are formed due to rapid cooling and escape of gases. The blocky rock masses and associated gravel deposits that are interbedded in recent basalts create a very high permeability.

Sedimentary rocks such as sandstones, carbonate rocks, and coal beds form aquifers to store and transmit groundwater. Sandstones constitute 25% of the sedimentary rocks of the world, and the permeable zones in these types of rocks form regional aquifers which contain large quantities of potable water. Friable sandstones generally have a high porosity (30–50%) which diminishes greatly with depth due to compaction and the intergranular cementing materials, mainly quartz, calcite, iron, and clay minerals. The latter are precipitated from hydrothermal solution circulating into the sandstone aquifers at depths where temperature and pressure are high.[20]

Carbonate-type rocks such as limestone and dolomite consist mostly of calcite and dolomite minerals with minor inclusions of clay. Dolomitic rocks, or dolostones, are secondary in origin, formed by geochemical alternation of calcite which creates an increase in porosity and permeability as the crystal lattice feature of dolomite occupies about 13% less space than that of calcite. Geologically young carbonate rocks commonly have porosities that range from 20% for coarse, blocky limestone to more than 50% for poorly indurated chalk.[21] At depth, the soft minerals that constitute the matrix of the carbonate rock are normally compressed and recrystallized into a more dense, less porous rock. Fractures or openings along bedding planes of carbonate beds create appreciable secondary permeability, whereas secondary openings due to stress conditions may be enlarged as a result of dissolution of calcite or dolomite by circulating groundwater.

Karst terrains have specific hydrologic characteristics and are composed of limestone, dolomites, gypsum, halite, or other soluble rocks. Karst landscapes which exhibit irregularities of the land surface are caused by surface and subsurface removal of rock by dissolution of limestone, calcite, or dolomite by circulating groundwater and erosion. Figure 2.2 shows the complex physical and geochemical processes involved in forming karst and the phenomenon of karst and karstification.[22]

Table 2.1 includes a list of generic categories of rock types and other relevant information that should be considered during evaluation of potential sites for land disposal. It includes broader categories, an understanding of which is necessary during the preliminary screening of sites, and it provides a basis for formulation of a conceptual hydrogeologic model.

Coal beds are lithologic units within sequences of sedimentary rocks formed in floodplain or deltaic environments.

Shale beds constitute the thickest semi-pervious units in most sedimentary basins. Shale beds originate as mud laid down in the gentle-water areas of deltas, on ocean bottoms, or in the backswamp environments of floodplains. Clay is transformed to shale by digenetic processes related to compaction and tectonic activity. In outcrop areas, shale is commonly brittle and fractured with appreciable amounts of permeability, whereas, at depth, it is less fractured and permeability is generally very low. Unfractured shale, clay, anhydride, gypsum, and salt usually provide good seals against upward or downward flow of fluids.

Alluvial deposits are unconsolidated materials deposited by streams in river channels or on flood plains. They consist of particles of clay, silt, and/or sand and gravel. Alluvial deposits include alluvial cones which consist of loose material washed down the mountain slopes by ephemeral streams and deposited at the mouth of gorges, alluvial fans which are formed by a tributary of high declivity in the valley of a stream, or those deposits built by rivers issuing from mountains upon lowland. Numerous other alluvial features are associated with alluvial processes.

## 2.2.2 Candidate Sites

Based on knowledge gained during the preliminary assessment of the geologic framework, a rank-ordered list of candidate sites can be prepared. One or more of the candidate sites, which have the highest rankings, are typically selected for site-specific hydrogeologic and geotechnical studies. The additional studies may be performed on each of two or three selected sites or for one site with the highest ranking. The hydrogeologic studies should be designed to seek and discover fatal flaws, if they are present. In the absence of fatal flaws, detailed studies provide a means of completely characterizing

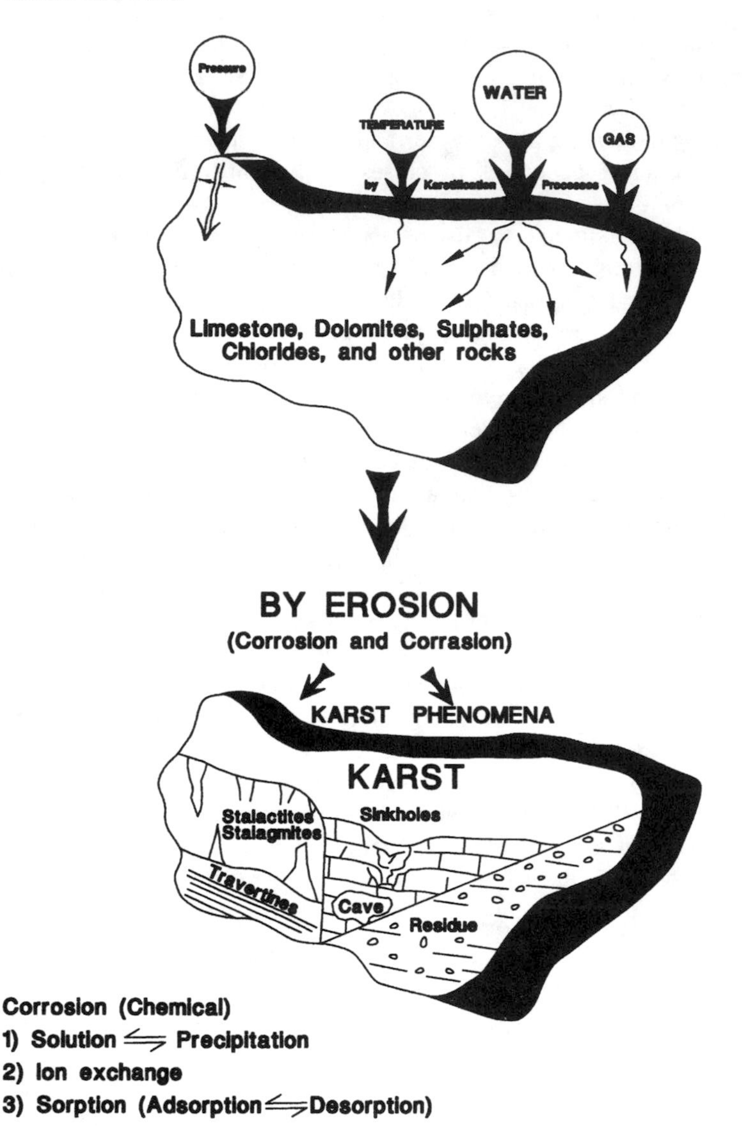


**Figure 2.2 A flow diagram showing karstification and karst. (Modified from F. Assaad and H. Jordan, 1994.[22])**

the hydrogeology of the site, demonstrating the suitability of the site for a potential land disposal facility, and defining locations for monitoring the site.

Some conditions that may, but do not always, lead to rejection of a candidate site include: areas that contain well-developed karst features and recent karst activity; recharge areas for aquifers (i.e., particulate stratigraphic intervals); specific geologic structures (e.g., some folds, faults, and lineaments); areas that contain thin or geotechnically unsuitable soil; areas of wellhead protection for public water supplies; and areas of significant pumping (e.g., quarries, mines, and industrial wells).

The U.S. Environmental Protection Agency[3] (US EPA) has established definite landfill siting requirements and some restrictions on location of municipal solid waste landfills, which locally may also constitute cause of rejection of proposed sites. The restrictions include proximity to airports, floodplains, wetlands, Holocene faults, seismic impact zones, and unstable areas. When hydrogeologic conditions are unfavorable, or when costs of overcoming deficiencies of the site are too high, the site should be rejected from further consideration as a potential site for disposal of wastes to land.[2]

The EPA and state governments have established the following landfill siting requirements for waste disposal facilities and practices:

### *Location Restrictions*

1. *Fault Areas.* Landfills should not be located within 200 feet of the active fault zones that have had displacement in Holocene time.
2. *Airport Safety.* Landfills should be located at least 10,000 feet from airports handling turbojets and 5,000 feet from airports handling piston-type aircraft to avoid bird hazard to aircraft.

   EPA requires that landfills should not be located in the 100-year flood plain. Landfills shall not restrict the flow of the 100-year flood, reduce the temporary water storage capacity of the floodplain, or result in the washout of solid waste and pose a hazard to human health and to the environment. However, new MSWLFs or existing landfills, if located in a 100-year floodplain, should be designated and operated to mitigate and/or minimize adverse impacts on the flow of the 100-year flood and water storage capacity of the floodplain.
3. *Wetlands.* New landfill units cannot be placed in wetlands unless the owner or operator makes specific demonstrations to the state for assuring that the facility will not result in "significant degradation" of the wetlands.

Dredged or fill material should not be deposited or discharged into the aquatic ecosystem unless it can be demonstrated that such actions will not have an unacceptable adverse impact either individually or in combination with known or probable impacts of other activities affecting the ecosystems of concern. The degradation or destruction of special aquatic sites, such as filling operations in wetlands, is considered to be among the most severe environmental impacts.

### *Operating Criteria*

EPA has established operating requirements for landfills such as application of daily cover and post-closure care; random inspections of incoming waste loads; and record keeping of inspection results.

1. *Explosive gases control.* The concentration of methane generated by landfills should not exceed 7.5% of the lower explosive limit (LEL) in facility structures and at the property boundary.
2. *Air criteria.* Air criteria prohibit the open burning of waste but allow infrequent burning of agricultural wastes, silvi culture wastes, land cleaning debris, diseased trees, and debris from emergency cleanup operations. Any of these infrequent burnings should be conducted in areas dedicated for that purpose and at a distance from the landfill unit so as to preclude the accidental burning of other wastes.

## 2.2.3 Stratigraphy

Stratigraphy is the study of the thickness, age, lithology, and chronological sequence of rocks. The lithostratigraphic column, a graphic representation of the rock units, is the basic display of data used in stratigraphic studies. Figure 2.3 is a generalized columnar section for northeast Illinois, showing a variety of rock types typical of the east-central states.

Cross sections show the sequence of stratigraphic data constructed from the material penetrated in several deep wells. The stratigraphic correlation aids in understanding the depositional environment of subsurface material. Understanding of depositional environment of subsurface material beneath the potential site for land disposal is important as the flow of groundwater, recharge, storage, and discharge of the groundwater is controlled by various lithologies of stratigraphic section of the geologic column.

## 2.2.4 Structural Geology[23]

Structural geology is related to folding and faulting and the geographic distribution of these features which greatly affect the fluid flows, the physical properties of rocks, and the localization of mineral deposits and earthquakes.

Faults are fractures in the rock sequence along which displacement of two blocks took place. Such fractures may range from inches to miles in length, and displacements are of comparable magnitudes. Faults may act either as barriers to or as channels for fluid movement. Generally, geologists should consider any significant fault to be a potential flow path for purposes of preliminary evaluation of its importance. Accordingly, the fault would be an environmental hazard according to this assumption. If further investigation indicates this assumption to be true, it would be necessary to abandon potential disposal sites. Fractures that occur without any movement lead to cracks or

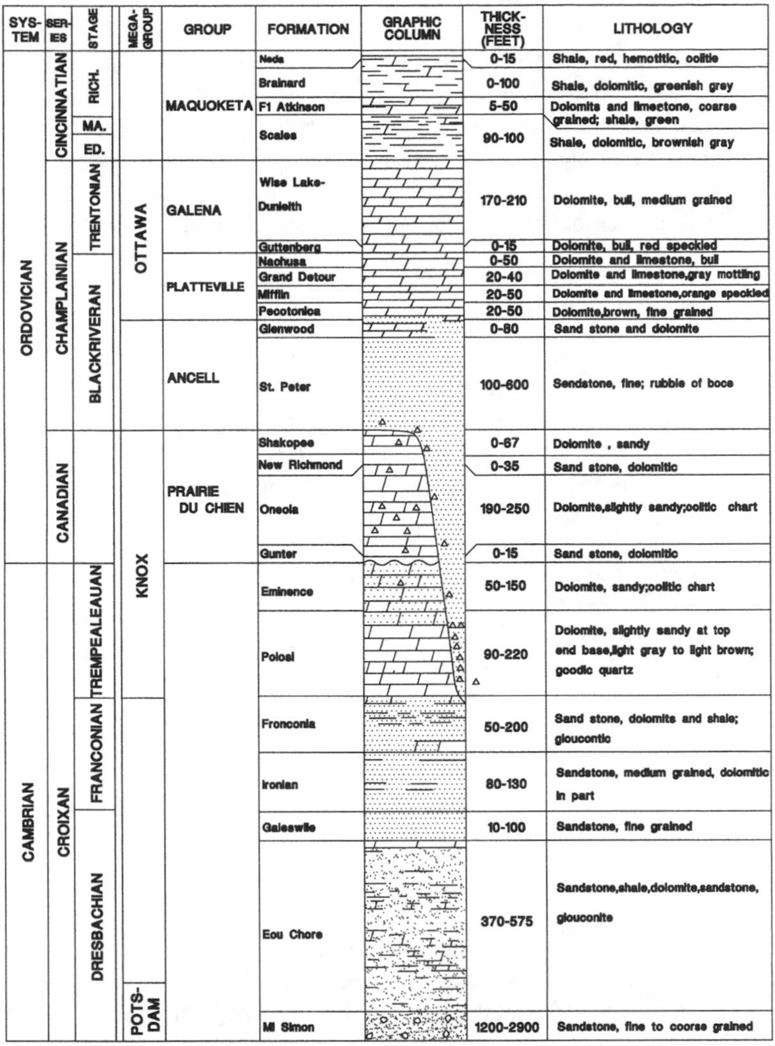

| SYSTEM | SERIES | STAGE | MEGAGROUP | GROUP | FORMATION | GRAPHIC COLUMN | THICKNESS (FEET) | LITHOLOGY |
|---|---|---|---|---|---|---|---|---|
| ORDOVICIAN | CINCINNATIAN | RICH. | | MAQUOKETA | Neda | | 0-15 | Shale, red, hemotitic, oolitie |
| | | | | | Brainard | | 0-100 | Shale, dolomitic, greenish grey |
| | | MA. | | | F1 Atkinson | | 5-50 | Dolomite and limestone, coarse grained; shale, green |
| | | ED. | | | Scales | | 90-100 | Shale, dolomitic, brownish gray |
| | CHAMPLAINIAN | TRENTONIAN | OTTAWA | GALENA | Wise Lake-Dunleith | | 170-210 | Dolomite, buff, medium grained |
| | | | | | Guttenberg | | 0-15 | Dolomite, buff, red speckled |
| | | BLACKRIVERAN | | PLATTEVILLE | Nachusa | | 0-50 | Dolomite and limestone, buff |
| | | | | | Grand Detour | | 20-40 | Dolomite and limestone,gray mottling |
| | | | | | Mifflin | | 20-50 | Dolomite and limestone,orange speckled |
| | | | | | Pecotonica | | 20-50 | Dolomite,brown, fine grained |
| | | | | ANCELL | Glenwood | | 0-80 | Sand stone and dolomite |
| | | | | | St. Peter | | 100-600 | Sandstone, fine; rubble of boce |
| | CANADIAN | | KNOX | PRAIRIE DU CHIEN | Shakopee | | 0-67 | Dolomite , sandy |
| | | | | | New Richmond | | 0-35 | Sand stone, dolomitic |
| | | | | | Oneola | | 190-250 | Dolomite,slightly sandy;oolitic chart |
| | | | | | Gunter | | 0-15 | Sand stone, dolomitic |
| CAMBRIAN | CROIXAN | TREMPEALEAUAN | | | Eminence | | 50-150 | Dolomite, sandy;oolitic chart |
| | | | | | Potosi | | 90-220 | Dolomite, slightly sandy at top and base,light gray to light brown; geodic quartz |
| | | FRANCONIAN | | | Franconia | | 50-200 | Sand stone, dolomite and shale; glouconitic |
| | | | | | Ironian | | 80-130 | Sandstone, medium grained, dolomitic in part |
| | | DRESBACHIAN | | | Galesville | | 10-100 | Sandstone, fine grained |
| | | | | | Eau Claire | | 370-575 | Sandstone,shale,dolomite,sandstone, glouconite |
| | | | POTSDAM | | Mt Simon | | 1200-2900 | Sandstone, fine to coorse grained |

**Figure 2.3 Generalized columnar section of Cambrian and Ordovician strata in northeastern Illinois. (From USEPA 600/2-77-240, An introduction to the technology of subsurface wastewater injection, in *Environmental Protection Technology Series*, USEPA, Cincinnati, OH, 1977, pp. 21–47, 64–91, 329–344.)**

joints, which are important to the development of porosity and permeability in some aquifers, but can be undesirable when they could be potential for draining fluids rapidly away from the disposal site. Joints can be examined from core samples obtained through drilling, by well logging, and by testing methods and evaluated based on experience with other deep wells drilled in the same region.

Structural geologic data are commonly shown on maps and cross sections. Structure contour maps show the elevation of a particular stratigraphic horizon relative to a selected datum. These maps can be used to estimate the depth to a mapped rock unit, the direction and magnitude of dip, and location of faults and folds that may influence decisions concerning the location of proposed disposal site and associated network of monitoring wells.

### 2.2.5 Physical Properties

Physical properties[1] of fluid and porous media that describe the hydraulic aspects of saturated groundwater flow and include density, ρ, viscosity, μ, and compressibility, β, for the fluid, whereas for the media (aquifers), it is described by porosity, Ø, permeability, *k*, and compressibility, α. These parameters are essential to quantitatively evaluate the hydrogeologic conditions of the potential sites for land disposal.

#### *Porosity*

Porosity is basically a grain formation dependent on grain size and degree of roundness.

$$\text{Quantitatively, Porosity } (\text{Ø}) = \frac{Vv}{Vt}$$

where Ø = Porosity (expressed in percentage)
Vv = Volume of voids
Vt = Total volume of soil sample

**Total porosity** is a measure of all void space whereas the effective porosity is defined as the hydraulic properties of a rock unit which considers the volume of interconnected voids available only to fluids flowing through the rock.[23]

A distinction must be made between primary and secondary porosity.[24] **Primary porosity** is intergranular or intercrystalline. Intergranular porosity in a sandstone depends on the size distribution, shape, angularity, packing arrangements, mineral composition, and cementation.[23] **Secondary porosity** results from fractures, solution channels, cavities or space (particularly in karst), and recrystallization processes and dolomitization.

#### *Permeability*

Permeability is expressed as the coefficient of permeability (k in $cm^2$). It is a formation property which allows the flow of liquids within the rock under applied potential gradient, and it is a rock parameter that influences the flow velocity. In general, the permeability is usually much lower in vertical directions than in horizontal.[24]

Permeability depends on the grain size. The smaller the grains, the larger will be the surface area exposed to the flowing fluid. As the frictional resistance of the surface area lowers the flow rate, the smaller the grain size, the lower the permeability. Shales, which are formed from extremely small grains, have very small permeability and are classified as confining intervals. The fracture permeability, due to fracturing, as well as the secondary permeability caused by the creation of karst in limestone and dolomite, may be significant for fracture flow.[24] Intrinsic permeability is expressed as follows:

$$k = \frac{Q\mu}{A\rho g} \cdot \frac{dh}{dL} (cm^2)$$

where k = coefficient of permeability
Q = flow rate through porous medium
A = cross sectional area through which flow occurs
$\mu$ = fluid viscosity
$\rho$ = fluid density
L = length of porous medium through which flow occurs
h = fluid head loss along L
g = acceleration of gravity

A simple form of Darcy's Law used in shallow groundwater is:

$$K = \frac{Q}{A} \Big/ \frac{dh}{dL}$$

where K is hydraulic conductivity (cm/sec).

Transmissivity (or transmissibility), T can be interpreted as the rate in which fluid of a certain viscosity and density is transmitted through a unit width of an aquifer at a unit hydraulic gradient. It is measured as the product of the thickness of the aquifer (b) and its hydraulic conductivity (K). Its unit is generally gallon per day per foot$^2$ (gpd/ft$^2$) or m/day.

### *Compressibility*

The compressibility of an aquifer, $\alpha$, encompasses not only the formation or the skeleton of the aquifer but also the contained fluids. Compressibility and the coefficient of storage are combined as a function of the aquifer thickness.

Quantitatively, compressibility of an elastic medium is defined as

$$\alpha = \frac{\delta v}{V \delta p}$$

where $\alpha$ = Compressibility of aquifer ($psi^{-1}$)
$\delta v$ = differential volume V
$\delta p$ = differential pressure P

The compressibility of the aquifer ranges from $5 \times 10^{-6}$ to $10 \times 10^{-6}$ $psi^{-1}$ as compared with that of water alone which is about $3 \times 10^{-6}$ $psi^{-1}$.

### *Storativity*

Storage coefficient for a confined aquifer, which is a parameter related to compressibility, indicates the capacity of the formation to accept water and is defined as the volume of water that an aquifer releases from or takes into storage per unit surface area of the aquifer per unit change in the hydraulic head normal to the surface and is quantitatively expressed as follows:[23]

$$S = \O \gamma b \left( \beta \frac{\alpha}{\O} \right)$$

where S = storage coefficient
$\O$ = Porosity
$\gamma$ = pg = specific weight of water per unit area
b = aquifer thickness
$\beta$ = compressibility of water
$\alpha$ = compressibility of aquifer formation

Values of S are dimensionless and normally range from $5 \times 10^{-5}$ to $5 \times 10^{-3}$ for confined aquifers and $10^{-1}$ to $10^{-3}$ for unconfined aquifers.

### *Viscosity*

The viscosity of the formation water in a porous rock influences the velocity of flow. As temperature increases, viscosity decreases and the velocity of flow increases.

### *Hydrodynamic Dispersion*

Hydrodynamic dispersion is a mixing process by which a liquid diffuses with another liquid on the condition that both are miscible. The coefficient of dispersion is inversely proportional to temperature, porosity, and grain form; whereas an increase in grain size, grain roundness, and the degree of irregularity promotes dispersion.[20-24]

## 2.2.6 Hydrogeologic Considerations

The subsurface rocks are subdivided into groups, formations, and members in descending order. These terms imply mappable rock subdivisions based on mineralogy, fossil contents, or other geological characteristics. However, such subdivisions may or may not be applicable to subsurface flow systems, as the geologic boundaries are not related to the physical properties (porosity and permeability). The following hydrogeologic terms are used to describe rock subdivisions according to their capacity to keep and/or transmit water. An *aquifer* is a saturated permeable geological unit (i.e., a formation or part of it or a group of formations) that can transmit significant quantities of water under ordinary hydraulic gradients to wells and springs. An *aquiclude*, on the other hand, stores water but does not transmit significant amounts. An *aquitard*, which has been used to describe the less permeable beds in a stratigraphic sequence, is in between an aquifer and an aquiclude and transmits enough water to be regionally significant but not enough to supply a well.[23] A *confined aquifer* is confined between two aquicludes and occurs at depth. It may be under artesian conditions when the water level in a well rises above the ground surface. The water level elevations in wells that are tapping a confined aquifer are plotted and contoured to construct a potentiometric surface map that shows the hydraulic head in the aquifer and provides an indication of the direction of groundwater flow. An *unconfined aquifer*, or water-table aquifer, is one in which the water table forms the upper boundary and occurs near the ground surface. A *perched aquifer* is a saturated lens of a formation that is bounded by a perched water table at the top and lense of relatively low permeable material at the bottom and is a special case of an unconfined aquifer. An *aquifuge* will not transmit water. The basement rock which is igneous or metamorphic lies beneath the sedimentary mantle and is generally nonporous and impermeable.[3] The groundwater flow theory and well hydraulics are discussed in Chapter 3.

## 2.3 DATA ACQUISITION OF ROCK AND FORMATION FLUID TESTINGS

### 2.3.1 Data Obtained Prior to Drilling Potential Disposal Sites

Geologic data should be obtained for evaluation of the site selected for drilling a well. Surface geophysical methods including seismic, gravity, magnetic, and electrical surveys may provide considerable subsurface geological information, but because of high costs, surface geophysical surveys are not widely used for water-well site studies. Literature and logs on the basic geological formations are available through national and state geological surveys, state oil and gas agencies, state water resources agencies, and some universities. The available geologic information should be collected and evaluated prior to field activities, including drilling wells. This exercise would help to economize the cost and efforts to study the proposed site.

### 2.3.2 Well Logs[23]

A drilling time log is a record of the rate of penetration of the rocks by the bit. Drilling time may be recorded directly on a log strip by a machine or plotted manually from the record sheets of the driller. It is usually expressed as penetration (feet) for each unit of time (minutes or hours). The rate of penetration changes with cementing material and lithologic rock type. A drilling time log may serve to indicate when the bit needs to be changed and when drill stem tests should be made and is most useful if no electric or radioactivity test is available.

Lithologic logs are prepared by the well-site geologist after examining the cuttings recovered from drilling fluids in a normal rotary-drilling process and the core samples recovered from the reservoir rocks and the confining beds. All cuttings and cores should be retained for future reference.

Logging tools are used for recording the geophysical properties of the formations penetrated and their fluids. Some of the properties measured are electrical resistivity, conductivity, ability to transmit and reflect sonic energy, natural radioactivity, hydrogen ion content, temperature, and density. These geophysical properties are then interpreted in terms of lithology, porosity, and fluid content.

Table 2.2 lists current geological well logging methods, their properties, and their practical applications.[23]

Miscellaneous logs include: (1) caliper logs which measure bore hole diameter needed for quantitative analyses of many geophysical logs and used for the lithologic interpretation; (2) dip-meter logs which measure the angle of dip of beds penetrated by the well and aid in the interpretation of geologic structure; (3) deviation logs which measure the degree of deviation of the well from the vertical; deviation of bore holes from the vertical is undesirable and periodic surveys are made during drilling to check bore hole orientation; and (4) production (or injection) logs which run through tubing or casing after the well is completed and are mainly used to determine the physical condition of subsurface facilities such as the interval of production (or injection) zones, the quantity of fluid produced from (or injected into) a particular zone, and the results of well-bore stimulation treatment.[23]

## 2.4 SUMMARY SITE SELECTION[2]

Selection of a site requires thorough knowledge and understanding of the regional and local stratigraphy; how that specific stratigraphic section has been affected by structural events; and how stratigraphy, structure, climate, and time have interacted to provide the present hydrogeologic system. Through such knowledge one can develop a conceptual hydrogeologic model of the site. The model can include complex computer models of the hydrology or relatively simple hydrologic models, such as flow nets or contours of the potentiometric surface. Hydrogeologic models can be inaccurate representations of the system, if they are derived without inclusion of an intimate knowledge of the stratigraphy and geologic structure.

In practice, the conceptual hydrogeologic model will be modified and improved as studies progress at the selected site. The final model should

**Table 2.2 Geophysical Well Logging Methods and Practical Applications**

| | Method | Property | Application |
|---|---|---|---|
| 1 | Spontaneous Potential (SP) log | Electrochemical and electrokinetic potentials | Formation water resistivity; (Rw) shaliness (sand or shale) |
| 2 | Non-focused electric log | Resistivity | Water and gas/oil saturation; porosity of water zones; Rw in zones of known porosity; formation resistivity (Rt); resistivity of invaded zone (Ri) |
| 3 | Focused and micro-focused micro resistivity logs | Resistivity | Resistivity of the flushed zone (Rxo); porosity; bed thickness |
| 4 | Sonic log | Travel time of sound | Rock permeability |
| 5 | Caliper log | Diameter of borehole | Without casing |
| 6 | Gamma Ray | Natural radioactivity | Lithology (shales and sands) |
| 7 | Gamma-Gamma | Bulk density | Porosity, lithology |
| 8 | Neutron-Gamma | Hydrogen content | Porosity with the aid of hydrogen content |

From USEPA 600/2-77-240, An introduction to the technology of subsurface wastewater injection, in *Environmental Protection Technology Series*, USEPA, Cincinnati, OH, 1977, pp. 21–47, 64–91, 329–344.

provide an accurate integration of the geologic, hydrologic, and geotechnical characteristics of the site that has been tested by installation of borings, piezometers, and monitoring wells; measurement of water levels; and determination of the direction and rate of groundwater flow. The ultimate goal of the siting process is to aid in meeting the need for disposal at locations that will assure protection of the environment.

## REFERENCES

1. Meinzer, O.E., Outline of groundwater hydrology with definitions, U.S. Geol. Survey Water-Supply Paper 494, 1923, 71 pp.
2. Hughes, T.H., Memon, B.A., and LaMoreaux, P.E., Landfills in karst terrains, *Bull. Assoc. Eng. Geol.*, Vol. 31, No. 2, 1994, p. 203.

3. U.S. Environmental Protection Agency, Solid Waste Disposal Facility Criteria, Final Rule, *Federal Register,* Vols. 53 & 56, 40 CFR Parts 257 and 258: U.S. Government Printing Office, Washington, DC, 1991a.
4. U.S. Environmental Protection Agency, *Code of Federal Regulations*, Vol. 40, Parts 260 and 299, U.S. Government Printing Office, Washington, DC, 1075, 1991b.
5. Sara, M.N., *Standard Handbook for Solid and Hazardous Waste Facility Assessments*, Lewis Publishers, Boca Raton, FL, 1994, pp. 10–68.
6. Hughes, T.H., Structure, in *Guide to the Hydrology of Carbonate Rocks*, LaMoreaux, P.E., Wilson, B.M., and Memon, B.A. (Eds.), UNESCO, Paris, 1984, p. 36.
7. LeGrand, H.E., Karst hydrology related to environmental sensitivity, in *Hydrologic Problems in Karst Regions,* Dilamarter, R.R. and Csallany, S.C. (Eds.), Western Kentucky University, Bowling Green, 1977, p. 10.
8. LeGrand, H.E. and Stringfield, V.T., Development and distribution of permeability in carbonate aquifers, *Water Resources Research*, 7, 1284, 1971.
9. LeGrand, H.E. and Stringfield, V.T., Karst hydrology — a review, *Journal of Hydrology*, 20, 97, 1973.
10. LeGrand, H.E. and LaMoreaux, P.E., Hydrogeology and hydrology of karst, in *Hydrology of Karstic Terrains*, International Association of Hydrogeologists, International Union of Geology Sciences, Series B, No. 3, 1975, chap. 1.
11. LaMoreaux, P.E., Wilson, B.M., and Memon, B.A. (Eds.), *Guide to the Hydrology of Carbonate Rocks*, UNESCO, Paris, 1984, 343 pp.
12. LaMoreaux, P.E., Hughes, T.H., Memon, B.A., and Lineback, N., Hydrogeologic assessment — Figeh Spring, Damascus, Syria, *Environmental Geology and Water Sciences*, Vol. 13, No. 2, 1989, p. 77.
13. Newton, J.G., Induced sinkholes — A continuing problem along Alabama highways, in *Proceedings, International Associates of Hydrological Science*, Anaheim Symposium, No. 21, Anaheim, CA, 1976, p. 453.
14. Newton, J.G., *Development of Sinkholes Resulting from Man's Activities in the Eastern United States*, U.S. Geological Survey, Circular 968, U.S. Geological Survey, Denver, CO, 1987, 54 pp.
15. Parizek, R.R., White, W.B., and Langmuir, D., *Hydrogeology and Geochemistry of Folded and Faulted Carbonate Rocks of the Central Appalachian Type and Related Land Use Problems*, Geological Society of America Guidebook, 1971 Annual Meeting, Geological Society of America, Boulder, CO, 1971, 181 pp.
16. Sweeting, M.M., *Karst Landforms*, Columbia University Press, New York, 1973, 355 pp.
17. White, W.B., Conceptual models for carbonate aquifers, *Groundwater*, 7, 15, 1969.
18. White, W.B., Conceptual models for carbonate aquifers: Revisited, in *Hydrologic Problems in Karst Regions*, Dilamarter, R.R. and Csallany, S.C. (Eds.), Western Kentucky University, Bowling Green, 1988, p. 176.
19. White, W.B., *Geomorphology and Hydrology of Karst Terrains*, Oxford University Press, New York, 464 pp. 1988.
20. Freeze, R.A. and Cherry, J.A., *Groundwater*, Prentice Hall, Inc., Englewood Cliffs, New Jersey, 1979, pp. 45–50, 58–61, 152–163.

21. Davis, S.N., *Porosity and Permeability of Natural Materials, Flow Through Porous Media,* DeWiest, R.J.M. (Ed.), Academic Press, New York, 1969, pp. 54–89.
22. Assaad, F.A. and Jordan, P., Karst Terranes and Environmental Aspects, *J. Env. Geol.,* Vol. 23, 228–237, 1994.
23. U.S. Environmental Protection Agency 600/2-77-240, An introduction to the technology of subsurface wastewater injection, in *Environmental Protection Technology Series*, USEPA, Cincinnati, OH, 1977, pp. 21–47, 64–91, 329–344.

# 3 HYDROGEOLOGY

## 3.1 INTRODUCTION

Hydrology is that part of water science that deals with water resources and their development. It has been described by several authors and scientists to include hydrogeology, scientific hydrology, applied hydrology, operating hydrology, stochastic hydrology, and catchment hydrology and its modeling.

The general science of hydrology is considered an integral part of the hydrologic cycle. Numerous sciences relevant to the hydrologic cycle include astronomy, solar physics, cloud physics, meteorology, climatology, environmental sciences, geography, engineering, agriculture, biology, economics, surface water hydrology, limnology, oceanography, soil physics, groundwater hydrology, and geology.

### 3.1.1 Historical Background

The earlier history of hydrogeology is well documented by O.E. Meinzer.[1-3] Historical references to hydrogeology require a search of the early literature, for example, writings by A.C. Veatch (1906);[4] W.H. Norton (1912);[5] C.F. Tolman (1937);[6] C.V. Givan (1934);[7] W.N. White (1941);[8] D.G. Thompson (1928);[9] M. Muskat (1934a);[10] M. Muskat (1934b);[11] M. Muskat (1936);[12] M. Muskat (1937a);[13] M. Muskat (1937b);[14] W. Meyer (1975);[15] H.J. Morel-Seytoux (1975);[16] A.K. Biswas (1970);[17] G. Meyer (1988);[18] A.M. Piper (1940);[19] L.K. Wenzel (1940);[20] R.J. Dewiest (1965);[21] M.M. Soliman (1990);[22] W.C. Walton (1970);[23] and A.J.E.J. Dupuit (1863).[24] The oldest archaeologically dated well, constructed 4000 B.C. using dressed field stone, was recently discovered in the shallow submerged coastal zone in Israel. "Kurker stone" had been used for casings. These early people apparently had a practical knowledge of the occurrence of fresh groundwater over salt water in a coastal zone. Some of these wells date back to before Abraham.

The origin and early use of terms in hydrogeology are interesting, even more so as we still have controversy over terminology. The term *hydrogeology* was used in 1802 by J.B. Lamarck to mean "the study of aqueous erosion and sedimentation." The first published use of the term with its present meaning was by Frenchman A. Daubree in 1887 in his 3rd Volume, *Textbook on Hydrogeology.*[25] It was first used in the U.S. by M.L. Fuller in 1906 in Water Supply Paper 160.[26]

The term *aquifer* was first used in the U.S. in 1896 by William Harmon Norton, a state geologist in Iowa, in a Report on the Underground Water.[5] In this report he revived a European term first used by Frenchman Arago in 1835.

*Artesian well* takes its name from the Province of Artois in France. Within the walls of an old Carthusian convent at Lillers, there is a well that was drilled in 1126 A.D. which still flows today. From France came a term *artesium,* the Latin equivalent of Artois; however, there were artesian-flowing wells long before in Egypt, China, and Persia. Probably the oldest of the big artesian wells were drilled in Kharga Oases in the Western Desert of the Sahara using dome-palm casings to depths of over 900 feet. Remnants of some of these wells remain today as they have been covered and preserved by mounds of wind-blown sand covered over with moisture and vegetation. These wells were used to irrigate large tracts of land and were the oases described by Herodotus as "Islands in the Desert."

Early work on groundwater research in the U.S. was being done at State Geological Surveys and universities. For example, the New York State Geological Survey had been continually operative since 1836 — over 40 years before the founding of the U.S. Geological Survey. By the 1830 and 1840s, there were a number of State Geological Surveys that had published geologic maps beginning with Pennsylvania in 1817. Some examples include the report by S.W. McCallie, *Artesian Well System of Georgia* (1898); and E.A. Smith, *Undergroundwaters of Alabama* (1907). Also, concurrently at universities, there was developing an interest in groundwater and some outstanding contributions resulted, for example, the work of Thomas Crower Chamberlain, *The Relevant and Qualifying Conditions of Artesian Wells* (1885); Franklin H. King of the University of Wisconsin, *Principles of Movement of Groundwater* (1889); and *Law of Undergroundwaters* (1905) by Douglas Johnson (*Water Supply Paper 122*).

Dr. O.E. Meinzer joined the USGS in 1906. He became the Chief of the Groundwater Branch in 1913. It was at this time that "Groundwater," or hydrogeology, became recognized as a discipline in the United States. But, it is best to hear about this phenomena from one of Europe's well-known early hydrogeologists from Ireland, Dr. David Burdon, who, upon receiving the "Aberconway Medal" in 1982 presented an interesting history of hydrogeology published in the *British Geologist.* Burdon recognized three episodes in the genesis of hydrogeology: the first episode (1846–1858) with publications by Darcy, Dupuit, Hagen, and Poiseuille; the second episode (1901–1906) with publications by Forchheimer, Theim, and Veatch; and the third

(1930–1940) the Meinzer Era, with publications by Meinzer,[1-3] Jacob,[27] Piper,[19] Theis,[28] and Wenzel.[20]

It was O.E. Meinzer[20] who first subdivided the science of hydrology, which according to his definition dealt specifically with water completing the hydrologic cycle from the time it is precipitated upon the land until it is discharged into the sea or returned to the atmosphere, into surface hydrology and subterranean hydrology or geohydrology. Groundwater hydrology is more or less following the same concepts of geohydrology or hydrogeology.

This chapter covers, in general, the hydrological aspects in brief whereas groundwater hydrology is covered in more detail. The following chapters are mainly concerned about the industrial, municipal, and agricultural wastes as sources of pollution to the groundwater resource systems and how these valuable sources can be recovered and managed to reduce, as much as possible, the hazardous effects of wastes on water resource systems.

## 3.2 HYDROLOGIC CYCLE

It is important to know how the hydrologic cycle is completed and how its various components are correlated in nature. The hydrologic cycle and its components are illustrated in Figure 3.1 which shows that water, in its three phases (gas, liquid, and solid), starting from the ocean, land, or living matter moves into the atmosphere by evaporation and transpiration. It passes through complicated atmospheric phenomena, generalized as the precipitation process, back to the earth's surface, upon and within which it moves in a variety of ways and is incorporated into nearly all compounds and organisms. This cycle, as demonstrated in Figure 3.1, shows the main hydrologic field of study.

One can conclude that the oceans are the immense reservoirs from which all water originates and to which all water returns. This simple statement may be further explained as, water evaporates from the ocean, forms clouds which move inland, condenses, and falls to the earth as precipitation. From the earth, through rivers and underground, water runs off into the ocean. So far, there is no evidence that water decreases in quantity at a global level. No water is depleted but none is generated either, and that is according to law of conservation of matter. For human usage, however, the physical state of water is important and so its quality. While its available quantity is limited, the need for water is ever increasing, and consumption is bound to exceed the ceiling of supply. Therefore, water conservation and pollution abatement have become very important in today's economic life. For this reason, environmental hydrologists or hydrogeologists should be familiar with the science of hydrology.

## 3.3 MAIN COMPONENTS OF HYDROLOGY

The study of the hydrologic cycle in its wider sense is usually divided into three separate disciplines: meteorology, surface hydrology, and geohydrology[21] or groundwater hydrology.

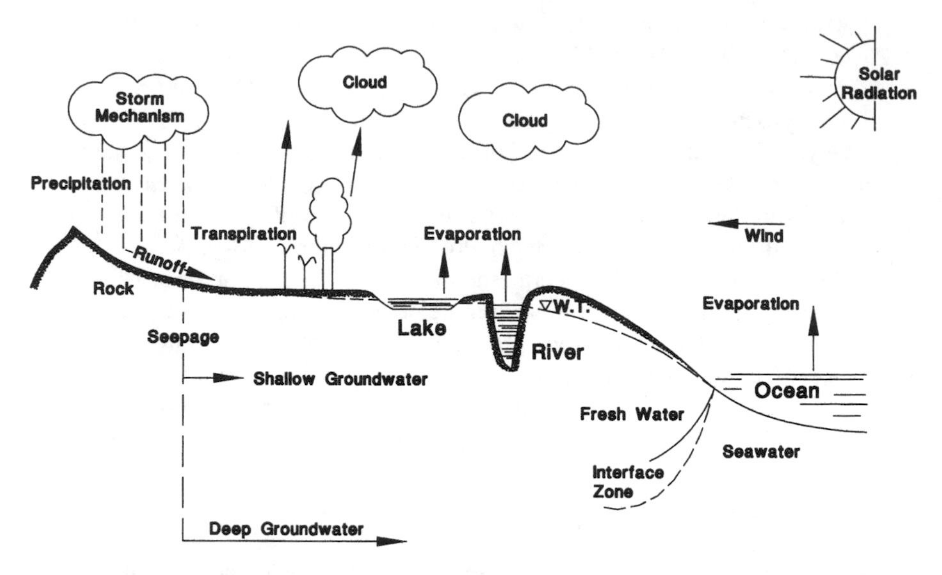


**Figure 3.1 The hydrologic cycle.**

Meteorology or climatology comes first in the study of the hydrologic cycle. It has several aspects: composition and general circulation of the atmosphere; energy balance of the atmosphere; precipitation, rainfall and snow, and snowmelt; and evaporation and evapotranspiration. The random nature of the climate results in a great variability, on different levels of time and space, of the precipitation which is the first link in the chain of the hydrologic cycle.

Surface hydrology is concerned with flow in the hydrographic network. It may be studied with several aims in mind:

1. Evaluation of available resources, either in their natural state or after development as building dams, and the calculation of the reservoir volume necessary to ensure a given flow.
2. Forecasting of flood risks and the works required to control them as detention dams and channel improvement works. Very often those works have to fulfill several simultaneous and often contradictory needs: a reservoir to control floods must be emptied as fast as possible, and this is directly antagonistic to the objective of a reservoir meant to increase flow at low water; hence, the difficult management problems attached to multipurpose installations.

In hydrology, two methods are commonly used:

1. The stochastic method: Because of the variability of rainfall, streamflow is studied as a random variable.

2. The deterministic method: The process of runoff and infiltration is studied from a physical deterministic view point as flow equations based on an impulse assumed to be known.

The basin may be represented as a black box in which components are lumped together and may be analyzed according to the theory of system analysis. On the other hand, one may study the watershed from a physical point of view by considering all the physiographic parameters of the medium.

Because surface water hydrology is interrelated to the environmental hydrogeology, this chapter includes the study of environmental impacts of the hydrologic processing on watersheds, as the mechanism of erosion and deposition of sediments, and its effects on the aquifer upper boundaries.

## 3.4 WATERSHED HYDROLOGY

Watershed is the drainage basin of a watercourse, which is the entire area contributing to the runoff and sustaining part or all of the flow of the mainstream and its tributaries. Strictly speaking, however, a watershed is the divide separating one drainage basin from another. Catchment is another term synonymous to watershed or drainage basin. However, any of these terms may be used to denote the area where the surface runoff travels over the ground surface and through channels to reach the basin outlet.

Runoff is that part of precipitation, as well as any other flow contributions, which appears in surface streams of either perennial or intermittent form. This is the flow collected from a drainage basin or watershed, and it appears at an outlet of the basin. According to the source from which the flow is derived, runoff may consist of surface runoff, interflow, and groundwater discharge. The surface runoff is that part of runoff which flows over the ground surface and through streams to reach the catchment outlet. The part of surface runoff that flows over the land surface toward stream channels is called overland flow. After the flow enters a stream, it joins with other components of flow to form total runoff which is called streamflow.

The interflow, also known as subsurface flow, subsurface storm flow, or storm seepage, is that part of precipitation which infiltrates the soil surface and moves laterally through the upper soil horizons toward the streams as ephemeral, shallow, perched groundwater above the main groundwater level. A part of the subsurface flow may enter the stream promptly while the remaining part may take a long time before joining the streamflow.

The groundwater runoff, or groundwater flow, is that part of the runoff due to deep percolation of the infiltrated water which has passed into the ground, become groundwater, and been discharged into the stream.

For practical purposes of runoff analysis, total runoff in stream channels is generally classified as direct runoff and base flow. The direct runoff, direct surface runoff, or storm runoff is that part of runoff which enters the stream promptly after the rainfall or snow melting. It is equal to the sum of the surface runoff and the prompt subsurface runoff, plus channel precipitation. In the arid and semi-arid regions, direct runoff contains only overland flow in small watersheds, since the subsurface flow percolates deeply inside the ground and does not reach the stream channel where the channel is small. This is also true in rocky areas where seepage loss is very small.

The base flow, or base runoff, is defined as the sustained runoff. It is composed of groundwater runoff and delayed subsurface runoff. However, the base flow is completely excluded in arid zones[22] for the same reason as stated earlier with respect to the subsurface flow.

During a runoff-producing storm, the total precipitation may be considered to consist of precipitation excess and abstractions. The precipitation excess is that part of the total precipitation that contributes directly to the surface runoff. When the precipitation is rainfall, the precipitation excess is known as rainfall excess. The abstractions are that part of precipitation which does not contribute to surface runoff, such as interception, evaporation, transpiration, depression storage, and infiltration.

The part of precipitation that contributes entirely to the direct runoff may be called the effective precipitation, or effective rainfall if only rainfall is involved. Figure 3.2 demonstrates a flow chart identifying the various items from the total precipitation to the total runoff.

From the hydrologic point of view, the runoff from a drainage basin may be considered as a component in the hydrologic cycle, which is influenced by two major groups of factors — climatic factors and physiographic factors.

Climatic factors include mainly the effects of various forms and types of precipitation, interception, evaporation, and transpiration, all of which exhibit seasonal variations in accordance with the climatic environment. Physiographic factors may be classified into two kinds: basin characteristics and channel characteristics. Basin characteristics include such factors as size, shape, and slope of the catchment area, hydraulic conductivity and recharge of groundwater, presence of lakes and swamps, and land use. Channel characteristics are related mostly to hydraulic properties of the channel which govern the movement of streamflows and determine channel storage capacity. It should be noted, however, that the above classification of factors is by no means exact because many factors, to a certain extent, are interdependent. The following two sections list the major factors.

### 3.4.1 Climatic Factors

- Precipitation is classified as rain, snow, frost, etc. and according to type, intensity, duration, time distribution, areal distribution, frequency of occurrence, direction of storm movement, antecedent precipitation, and soil moisture.

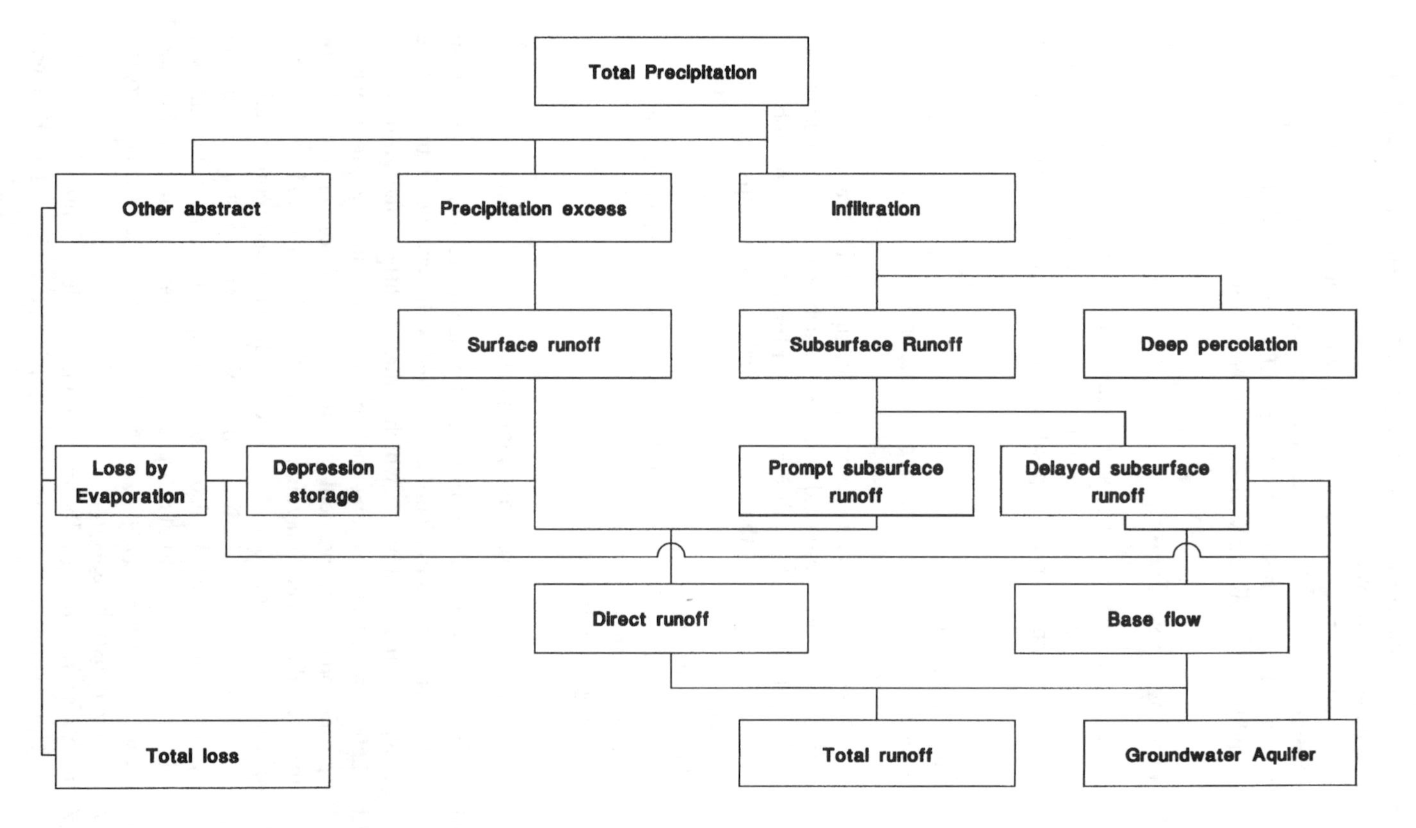


Figure 3.2 Various items from total precipitation to total runoff.

- Interception includes vegetation species, composition, age, density, and season of the year.
- Evaporation varies with temperature, wind, humidity, atmospheric pressure, soluble solids, nature and shape of evaporative surface.
- Transpiration is affected by temperature, solar radiation, wind, humidity, soil moisture, and kinds of vegetation.

### 3.4.2 Physiographic Factors

These are basin characteristics and channel characteristics. Basin characteristics include geometric factors and physical factors, whereas channel characteristics vary with the carrying capacity and storage capacity. Those factors may be itemized as:

1. Basin characteristics
   - Geometric factors: size, shape, slope, orientation, elevation, and stream density.
   - Physical factors: land use and cover, surface infiltration condition, soil type, geological conditions, such as the hydraulic conductivity and capacity of groundwater formations, topographical conditions, such as the presence of lakes and swamps, artificial drainage.
2. Channel Characteristics
   - Carrying capacity: size and shape of cross section, slope, roughness, length, tributaries.
   - Storage capacity: backwater effect.

### 3.4.3 Mechanism of Erosional Deposition

The main environmental effect of runoff on any catchment is the erosion and sedimentation processing. This also includes the change of the surface water quality in some catchments which indirectly affects the groundwater quality where there is an interchange between surface water and groundwater.

The prevention or improvement of soil erosion requires understanding of factors that affect the rate and magnitude of erosion. Soil erosion is responsible for irreversible degradation of vast tracts of arable land and sedimentation of reservoirs and harbors, which diminishs their capacity to store water and supply hydroelectric power. It also causes irreparable damage to transport systems. Even with the support of nearly a century of research data, researchers are unable to combat the erosion menace effectively. In fact, the soil erosion hazard is continuously increasing.

Although far from being fully understood, considerable progress is being attempted. Upland erosion begins with raindrop impact and its interaction with overland flow. Erosion occurs if the combined power of the rainfall energy and overland flow exceed the resistance of soil to detachment. Each of these processes is a complex phenomenon.

Transport and delivery of sediment from catchments are important processes for both downstream and within-catchment considerations. Sediment transported from a catchment presents an environmental problem and a potential hazard downstream in terms of sediment concentration and volumes.

In addition, concentrations and loadings of chemicals adsorbed by the sediment particles may also cause problems due to their toxicity and hazardous impacts. At some locations however, agriculture is enhanced by nutrient-laden sediments deposited in flood plains. In either case, unexpected sediment yield changes have the potential for large economic losses for the affected areas.

Knowledge of the distribution of sediment sources and sinks within a catchment is useful for recommending erosion and sedimentation controls. Sediment sources include agricultural lands, construction sites, disturbed lands, and roadway embankments. Sediment sinks include strips of vegetation, reservoirs, the bases of concave slopes, and areas of diverging flow. Such sinks occur when the load of a stream exceeds its transport capacity.

For a catchment system in the state of quasi-equilibrium, sediment yield at the outlet is directly related to sediment production on upland areas. Control of upland erosion does not always reduce the sediment load immediately because decreasing the upland sediment load increases the erosivity of the channel flow. If sediment yield continues at a higher level until the system readjusts to upland controls, a process which may take several years, an accurate estimate of catchment sediment yield that is undergoing cultural change must consider the entire catchment drainage continuum in addition to its erosion-sedimentation history.

## 3.5 HYDROGEOLOGY

Groundwater and surface water commonly form a linked system. Flow can be in either direction, and the rate of flow varies geographically and seasonally. The interchange is not significant for some aquifers.[23] But it has been estimated that about 30% of total flow in surface streams is supplied from groundwater, and seepage from streams is known to be a principal source of flow to some aquifers. Water withdrawn from wells along a bank of an alluvial stream can effect an appreciable reduction in surface flow, and the diversion of surface flow can reduce groundwater recharge. Supply for groundwater and surface water cannot be evaluated independently unless it is established that the interchange is minimal. There are two distinct types of circumstances concerning the development and management of groundwater supplies. Groundwater is considered a renewable source with optimal use restricted to the average rate of recharge; mining of groundwater, however, is sometimes carried out with fixed term objectives. Average annual recharge can in extreme cases be relatively insignificant, as in the major regional Nubian aquifers in northeast Africa.

The environmental impacts of groundwater withdrawal without a good management plan are land subsidence, sea water intrusion in coastal aquifers during overdraft situations, or water logging of an area during an excess recharge situation.

### 3.5.1. Distribution of Subsurface Water

Water occurs underground in two zones (aeration and saturation) separated by the water table. The occurrence and movements in these two zones are markedly different. The water table exists only in water-bearing formations which contain openings of sufficient size to permit appreciable movement of water. It is generally considered to be the lower surface of the zone of suspended water at which the pressure is atmospheric. The saturated zone extends down as far as there are interconnected openings. The lower boundary may be an impervious layer. The upper saturated zone is called an unconfined aquifer. Sometimes the saturated zone is bounded on top by another impervious layer and is known as the confined aquifer. If one of the impervious layers leaks inward or outward, this aquifer is named a leaky aquifer (Figure 3.3).

Subsoil water is limited to the soil belt and reached by roots. Pellicular water adheres to rock surface throughout zone of aeration and is not moved by gravity but may be abstracted by evaporation and transpiration. Gravity or vadose water, moves downward by force of gravity throughout saturated zone. Perched water occurs locally above an impervious barrier. Capillary water occurs only in the capillary fringe above the water table. Free water is known as the water which moves by gravity in the unconfined aquifer (see Figures 3.3a–e).

The vadose zone can be more difficult to characterize due to the complex localized flow conditions than are found in the saturated zone below the water table. On the other hand, because it is nearer to the land surface, remedial actions may not require complete characterization of the vadose zone flow system for certain site conditions and contaminants if the majority of the affected soils will be treated in place or removed.

### 3.5.2 Groundwater Flow Theories

Groundwater flow is treated in a general way as the flow of fluid in porous media. The classical hydrodynamics of viscous flow are applied. It has been seen that in addition to the equation of continuity and the equation of state, the equation of motion should be considered in any hydrodynamic problem. However for the fluid flow through porous media, the equation of motion is

**Figure 3.3** Distribution of subsurface water.

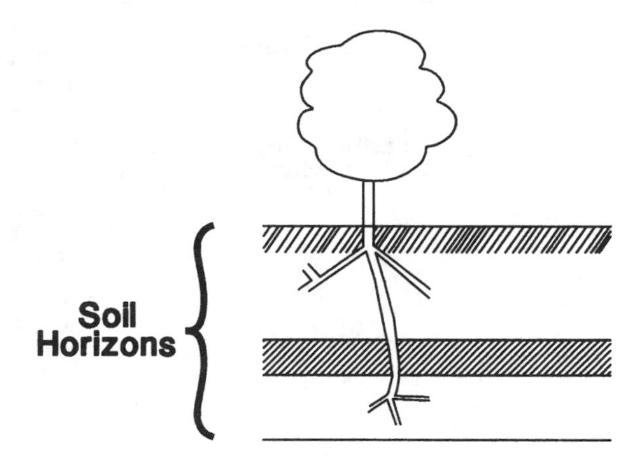


**Figure 3.3a** Soil horizons.

**Figure 3.3b** Pellicular water in granular material (i), and rock fractures (ii).

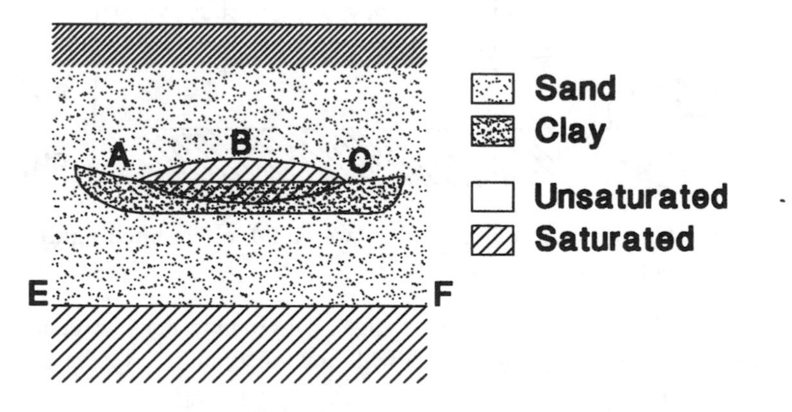


**Figure 3.3c** Perched water table ABC, inverted water table ADC, and true water table EF. (From Freeze, R.A. and Cherry, J.A., *Groundwater,* Prentice-Hall, Englewood Cliffs, NJ, 1979. With permission.)

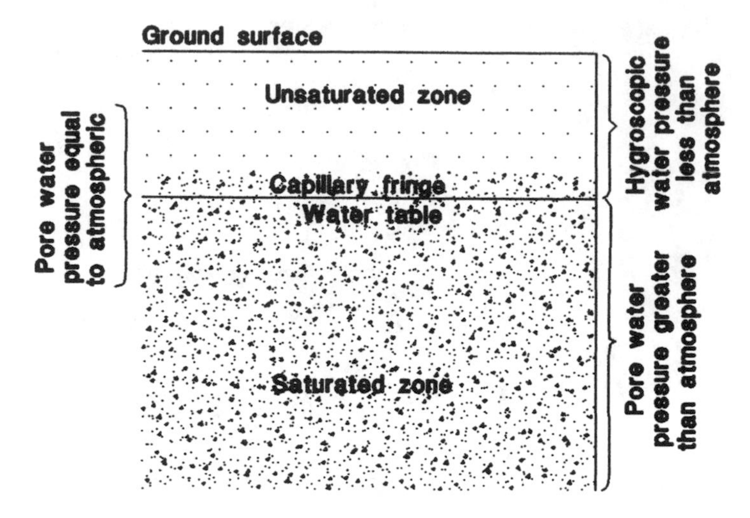


**Figure 3.3d Distribution of fluid pressures in the ground with respect to the water.**

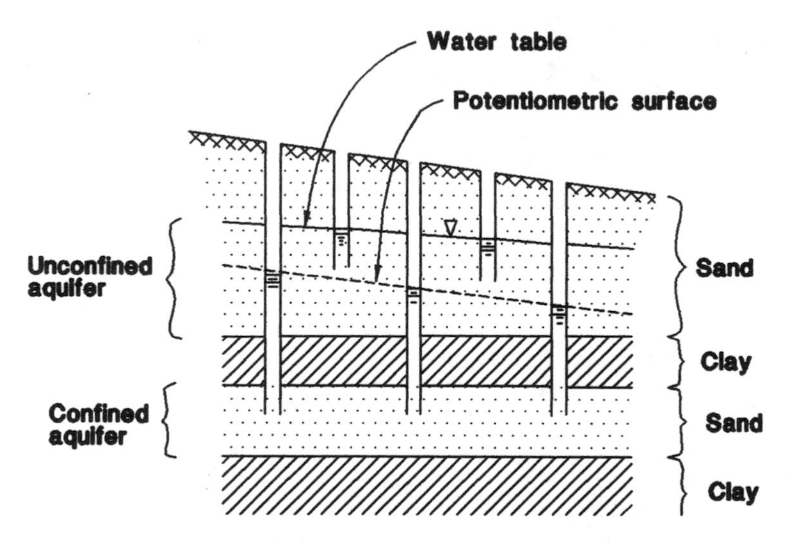


**Figure 3.3e Unconfined aquifer and its water table; confined aquifer and its potentiometric surface.**

replaced by Darcy's Law in order to obtain the groundwater flow equation in a simple manner. Darcy's law for flow through medium is given as:

$$V = -K\frac{dh}{dl} \tag{3.1}$$

where V is the groundwater flow and K is the hydraulic conductivity. However, the actual fluid motion can be subdivided on the basis of the components of flow parallel to the three principal axes:

$$\left.\begin{aligned} u &= -K_x \frac{\delta h}{\delta x} \\ v &= -K_y \frac{\delta h}{\delta y} \\ w &= -K_z \frac{\delta h}{\delta z} \end{aligned}\right] \tag{3.2}$$

where h is the hydraulic head and is defined as:

$h = h_p + z$, where $h_p$ is the pressure head and z is the potential head;

$K_x$, $K_y$, $K_z$ are the hydraulic conductivities with components in the x, y and z directions, respectively

–ve The negative sign means that water is flowing in the direction opposite to increasing hydraulic potentials

u, v, w the velocity of flow in x, y and z directions, respectively

$\frac{\partial h}{\partial x_i}$ hydraulic gradient ($x_i$ = x, y, and z directions)

Combining Eq. 3.2 with both the continuity equation and the equation of state, the groundwater flow equation, in general form, can be given as:

$$\frac{\partial}{\partial x}\left(K_x \frac{\partial h}{\partial x}\right) + \frac{\partial}{\partial y}\left(K_y \frac{\partial h}{\partial y}\right) + \frac{\partial}{\partial z}\left(K_z \frac{\partial h}{\partial z}\right) = S_s \frac{\partial h}{\partial t} \tag{3.3}$$

or

$$K_x \frac{\partial^2 h}{\partial x^2} + K_y \frac{\partial^2 h}{\partial y} + K_z \frac{\partial^2 h}{\partial z^2} = S_s \frac{\partial h}{\partial t} \tag{3.3}$$

where $S_s$ is a specific storage, defined as the volume of water a unit volume of saturated aquifer releases from storage for a unit decline in hydraulic head, per unit depth. Three dimensional flow in aquifers of essential uniform thickness B,

$$T_x \frac{\partial^2 h}{\partial x^2} + T_y \frac{\partial^2 h}{\partial y^2} + T_z \frac{\partial^2 h}{\partial z^2} = S \frac{\partial h}{\partial t} \tag{3.4}$$

where $T_x$, $T_y$, $T_z$ are the transmissivities in x, y, and z direction and T = BK ($m^2$/sec) where B is the aquifer thickness. S is equal to $S_s \times B$ which is the storage coefficient which is defined as the volume of water released from or taken into storage per unit cross sectional area of the vertical column of aquifer per unit change in head. For the unconfined aquifer, S is taken as the specific yield (Sy).

For well problems with radial flow, Eq. 3.4 in polar coordinates becomes:

$$\frac{\partial^2 h}{\partial r^2} + \frac{1}{r}\frac{\partial h}{\partial r} = \frac{S}{T}\frac{\partial h}{\partial t} \tag{3.5}$$

The coefficients S and T may be regarded as empirical values to be determined principally by pumping test technique.

Two cases are considered for groundwater flow toward wells drilled in aquifers: the steady state and the unsteady state flow.

### 3.5.3 Steady State Groundwater Flow in Aquifers

Dupuit (1863)[24] was the first one who combined Darcy's law with continuity equation to derive an equation for well discharge. Dupuit assumed complete axial symmetry, steady flow through an infinitely extending aquifer.[30] He deduced for confined aquifer (Figure 3.4a):

$$Q = \frac{2\pi Km(h_2 - h_w)}{l_n\ r_2 / r_w} \tag{3.6}$$

and for unconfined aquifer (Figure 3.4b):

$$Q = \frac{\pi K(h_2^2 - h_w^2)}{l_n\ r_2 / r_w} \tag{3.7}$$

where, Q is the discharge, K is the hydraulic conductivity, and $h_2$ and $h_w$ are the head levels above the impervious bed at radial distances $r_2$ and $r_w$, respectively.

### 3.5.4 Unsteady State Groundwater Flow in Confined Aquifers

If there is no replenishment, the area of influence of a well increases and potentiometric head declines in such a manner that the water released from storage equals the well discharge. The differential equation governing such unsteady flow to an axially symmetrical well is shown in Eq. 3.5.

For the artesian case, this water is released by consolidation and compression effects associated with release of pressure. For the water table case,

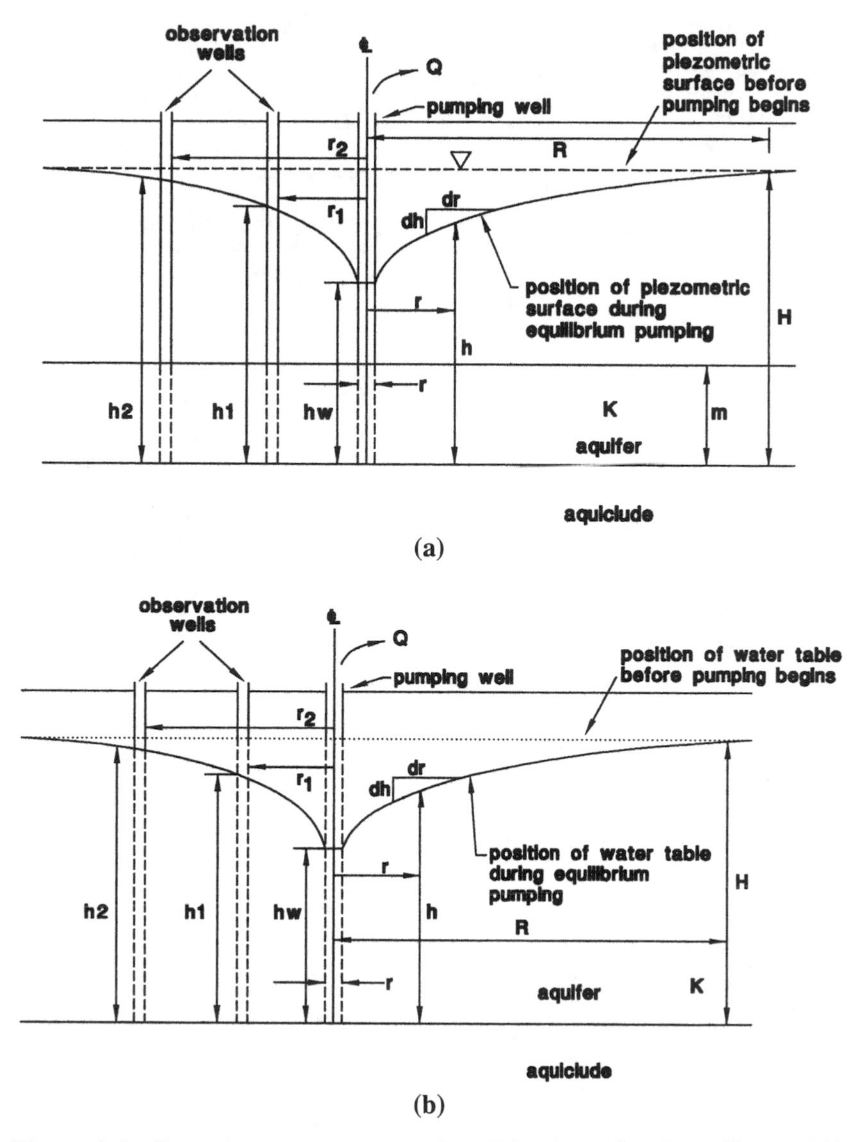


**Figure 3.4 Drawdown curve around well in a) confined aquifer and b) unconfined aquifer.**

the water originates as a result of recession of the water table, and S equals the specific yield. Theis[28] applied a solution of Eq. 3.5 to the case of constant discharge from an infinitely extending artesian aquifer. For this case, the drawdown D and $h_o$, ... h, at any radial distance r and time of pumping t is given by (Figure 3.5):

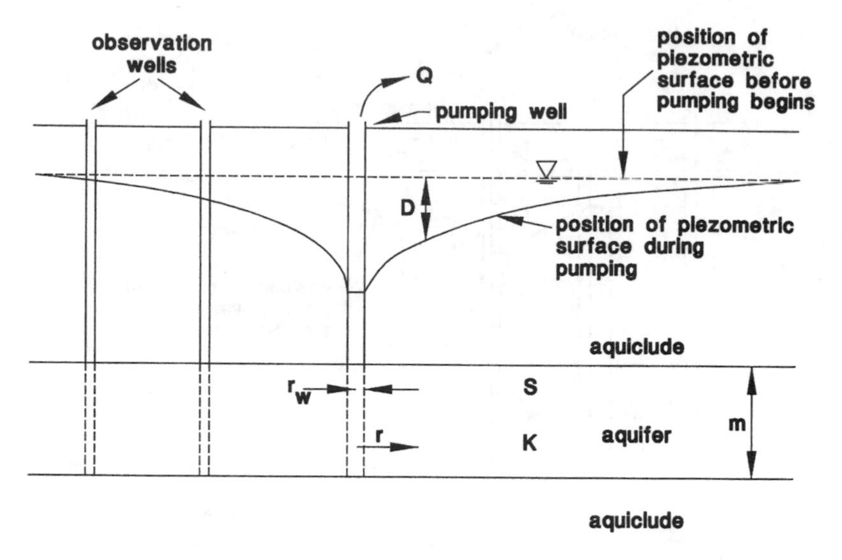


**Figure 3.5 Transient flow toward a well in a confined aquifer.**

$$D = \frac{Q}{4\pi T}\int_{u}^{\infty}\frac{e^{-u}}{u}du \tag{3.8}$$

where:

$$u = \frac{r^2 S}{4Tt}$$

$$\text{putting } w(u) = \int_{u}^{\infty}\frac{e^{-u}}{u}du \tag{3.9}$$

therefore:

$$D = \frac{Q}{4\pi T}W(u) \tag{3.10}$$

Graphical solution could be used to solve for Eq. 3.10. This procedure is called matching procedure. A curve such as shown in Figure 3.6 can be used. This figure shows W (u) plotted as a function of u.

From Eq. 3.8 and 3.9, for any specific well test, u is proportional to $r^2/t$ and D to W(u). By plotting D as ordinate and $r^2/t$ as the abscissa on transparent paper to the same scale as the Theis type curve (Figure 3.6). A field data curve similar to the type curve will be obtained. A portion of the field data curve

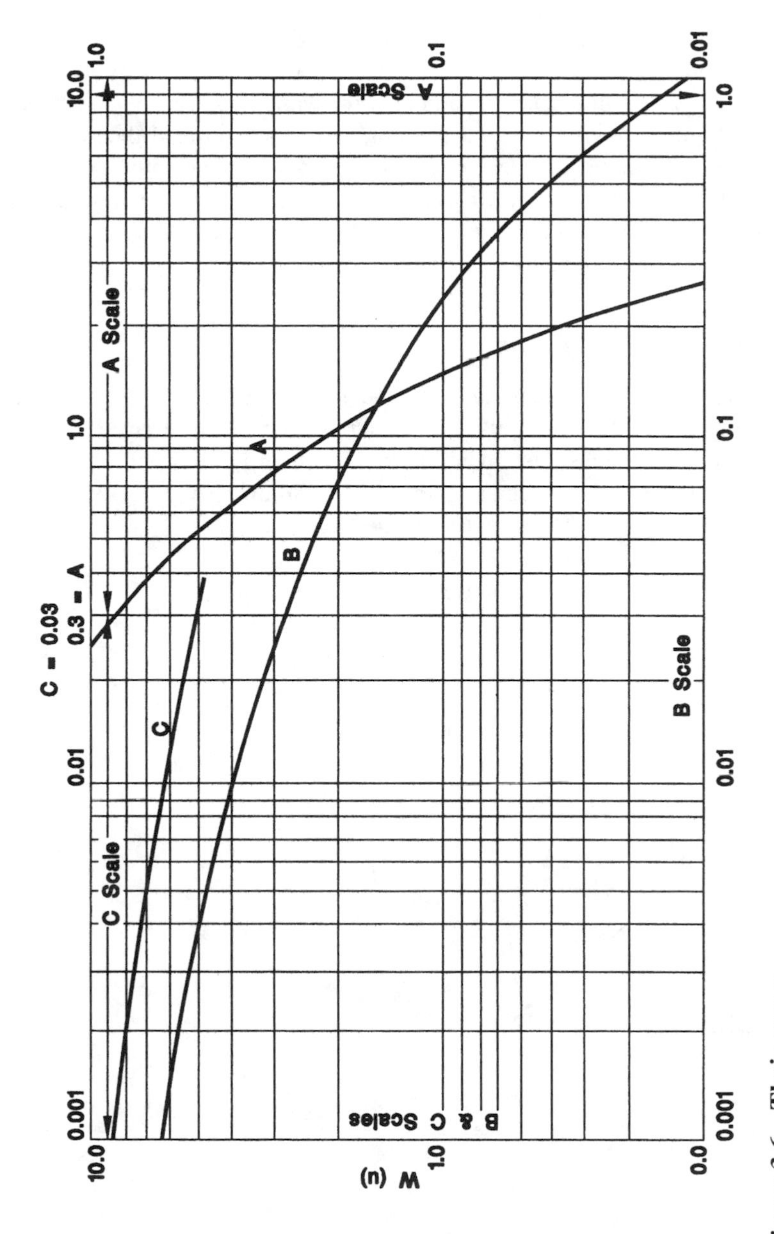


Figure 3.6 Theis curve.

will be superimposed and matched to the type curve keeping coordinates parallel when matching. Choose a specific match point on the matching portion of the curve and record values of u, W(u), D, and $r^2/t$ for this point. Substitute these values for D and W(u) into Eq. 3.8 and solve for T. Using this value of T and substituting values for u and $r^2/t$ into Eq. 3.9, solve for S.

Theis' nonequilibrium equation is of fairly general applicability for artesian wells tapping confined aquifer and also for wells tapping the water table aquifer, if the drawdown is a small percentage of the saturated thickness of the aquifer. For a number of special cases, however, it can be greatly simplified with negligible loss in accuracy. Probably the most important of the modified forms of Theis' nonequilibrium equation are the basic modified equation, the rate of drawdown equation, and the recovery equation.

## *The Basic Modified Equation*

The basic modified equation[27,31] is the least modified. It is the basic modification in that through it all other equations are nearly derived. For its derivation and limitations, consider Figure 3.7 which is the same as Figure 3.6 except that the ordinate scale is linear. The solution of Eq. 3.10 is an infinite series which can be expressed in the form

$$W(u) = (-1.00)(\ln u) + a = (-2.30)(\log u) + a \tag{3.11}$$

where a is nearly constant for small values of u, approaching (–0.5772) as u approaches zero. The type curve is therefore asymptotic to the straight line.

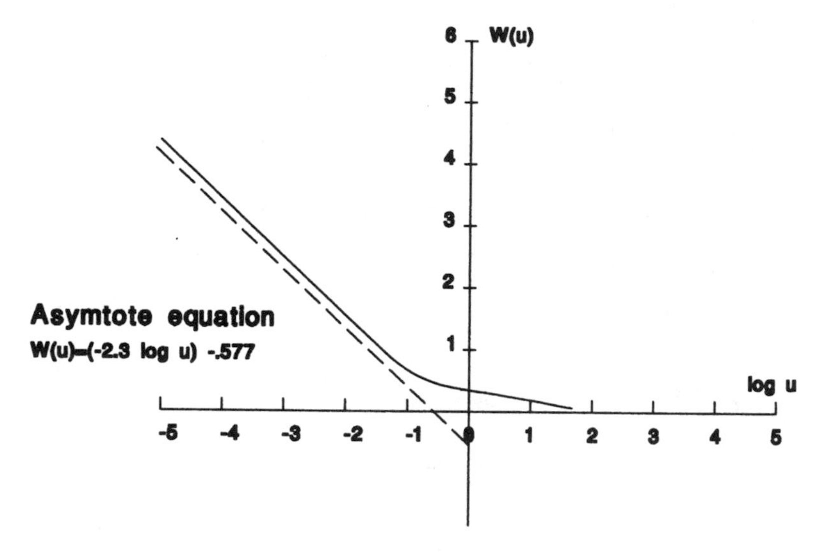


**Figure 3.7 Semi-log type curve.**

$$W(u) = (-2.30\ [\log u]) - 0.57722 \qquad (3.12)$$

For small values of u the type curve nearly parallels the asymptote with slope = –2.30. Thus, the type-curve equation, written in two points, becomes:

$$(y_2 - y_1) = m(x_2 - x_1)$$

which becomes

$$W(u_2) - W(u_1) = -2.3(\log u_2 - \log u_1) = 2.3 \log \frac{u_1}{u_2}$$

Replacing u and W(u) with their equivalents according to Eq. 3.8 and Eq. 3.9 assuming $S_1 = S_2$ and simplifying:

$$D_2 - D_1 = \frac{2.3Q}{4\pi T} \log \frac{r_1^2 / t_1}{r_2^2 / t_2} \qquad (3.13)$$

Eq. 3.11 is the basic modified equation. The choice between the two numbering systems is a matter of convenience.

At a constant radial distance, r, the rate-of-drawdown is given by the following equation:

$$\Delta D = D_2 - D_1 = \frac{2.3Q}{4\pi T} \log \cdot \frac{t_2}{t_1} \qquad (3.14)$$

The values of drawdown ($D_1$ and $D_2$) are taken per log-cycle of time, t

$$\log \frac{t_2}{t_1} = 1$$

thus Eq. 3.14 becomes:

$$\Delta D = \frac{2.3Q}{4\pi T} \qquad (3.15)$$

where $\Delta D$ = drawdown per log-cycle of time (such as $t_1 = 2$, $t_2 = 20$) when D is plotted vs. t on semi-log paper forming a straight line relation and T can be calculated from Eq. 3.15.

By projecting this line to meet the horizontal axis where D = o and t = $t_o$, S can be calculated using Eq. 3.12 as:

$$S \approx \frac{2.25Tt_o}{r^2} \tag{3.16}$$

### *Adjustment of the Modified Equations for Free-Aquifer Conditions*[32]

In the case of free-water table conditions the saturated thickness B is reduced by the drawdown D, so that T = KB is replaced by

$$K(B - D_{ave}) = Ky_{ave} = K(y_1 + y_2)/2$$

where $y_1$ and $y_2$ are the two drawdown curve ordinates corresponding to the two times and/or radial distances and $D_{ave}$ = average drawdown. Then, since $D_2 - D_1 = y_1 - y_2$ the quantity $(D_2 - D_1)$ T is replaced by:

$$(y_1 - y_2)Kx(y_1 + y_2)/2 = ((y_1^2 - y_2^2)/2)K$$

For example, the basic modified Eq. 3.12 then becomes:

$$y_1^2 - y_2^2 = \frac{2.3Q}{2\pi K} \log \frac{r_1^2 / t_1}{r_2^2 / t_2} \tag{3.17}$$

Care must be used in applying a free-aquifer equation which has been derived from artesian well equations by replacing KB by $K(y_{ave})$ as illustrated above. The artesian equations are derived on the basis that all streamlines are horizontal, so that the hydraulic gradient in the Darcy equation is equal to dy/dL = dy/dx; and equi-potential surfaces representing area in the Darcy equation are vertical cylinders. In the case of free-aquifer flow, the upper streamlines slope downward toward the well; so that their head losses vary with the sloping flow-distance, and hydraulic gradient is not equal to dy/dx. Furthermore, in the case of free-aquifer well, the equi-potential surface representing areas in the Darcy equation are not vertical cylinders but semi-cylindrical surfaces which curve inward at the top as indicated by the curvature of the equi-potential lines. No adjustment is made for these factors, so that the resulting free-aquifer equations are applicable only to computations where y represents the height of the hydraulic grade line (P/w) + Z along nearly horizontal streamlines representative of the main flow. This includes all bottom streamlines, a large part of the flow at intermediate elevations where the flow is nearly horizontal, and water table streamlines at such radial distances that the water table is nearly horizontal. Such equations are applicable also in terms of the water level in the pumping well, since the point $(r_e, d_w)$ [$r_e$ = effective radius or radius of the bore hole; $d_w$ = drawdown in the well] is a point on the hydraulic grade line for all streamlines except those which enter the well along the seepage surface.

If the pumping drawdown in the pumping wells is large, the free aquifer equations may be inapplicable in terms of the free-surface drawdown curve for radial distances as great as 1.5B or 2B. In such cases values of water table drawdown closer to the well can be computed on the basis of empirical relationships.

Since Q is proportional to T:

$$\frac{Q\ \text{artesian}}{Q\ \text{free}} = \frac{KB}{K(B - D_{ave})} = \frac{B}{B - D_{ave}}$$

where B is the saturated thickness.

Thus, failure to adjust such modified equations as Eq. 3.14 and 3.8 for free-aquifer conditions by replacing KB with $K(B - D_{ave})$ gives:

$$Q\ \text{computed} = \frac{B}{B - D_{av}} Q\ \text{true}$$

Computed values of K and $(D_2 - D_1)$ are in error to a similar degree.

### *The Recovery Equation*

If, as shown in Figure 3.8, a pumping rate Q is suddenly changed to a new rate Q′. The additional drawdown Z caused by the additional discharge (Q′ – Q) may be expressed in terms of Theis nonequilibrium equation as:

$$Z = \frac{(Q' - Q)}{4\pi T} W(u') \qquad (3.18)$$

where

$$u' = \frac{r^2 S'}{4Tt'}$$

$S'$ = new storage coefficient (3.19)

$t'$ = time with reference to the start of Q′.

However, in the recovery condition the drawdown is called the residual drawdown and Q′ = o, therefore Eq. 3.18 becomes

$$Z' = \frac{Q}{4\pi T} W(u') \qquad (3.20)$$

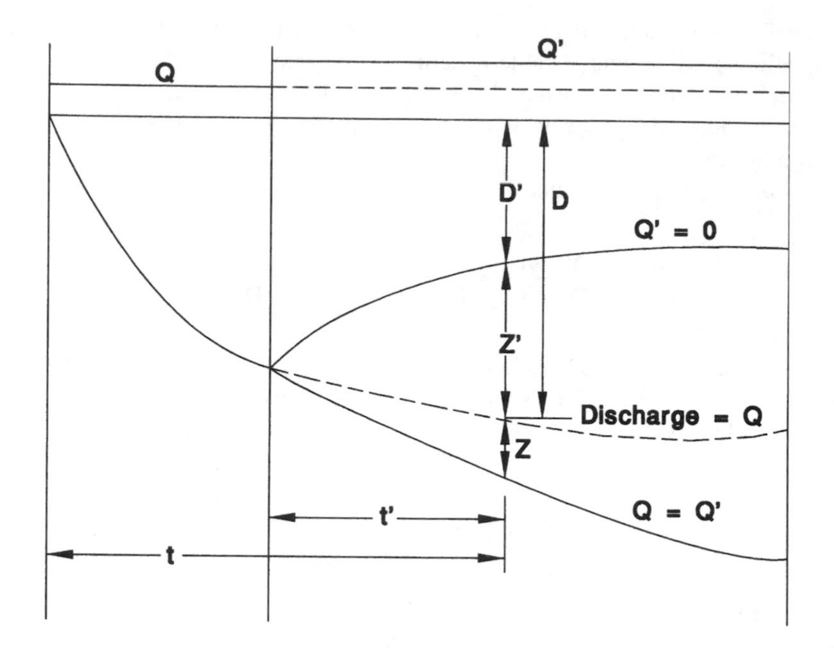


**Figure 3.8 Hydrographs illustrating drawdowns during change in rate and recovery conditions.**

and the residual drawdown D′ becomes

$$D' = D - Z = \frac{Q}{4\pi T}(W(u) - W(u')) \tag{3.21}$$

For small values of u and u′, modified condition can be used as

$$D' = \frac{Q}{4\pi T}(-2.3 \log u + a) - (-2.3 \log u' + a) \tag{3.22}$$

or

$$D' = \frac{2.3}{4\pi T}\left(\log t/t' - \log \frac{S}{S'}\right) \tag{3.23}$$

If S/S′ is taken constant, D′ and t/t′ can be plotted on semi-log paper giving straight line relation. The line need not pass through the origin where t/t′ = 1, as it will do so only when S′ = S. Often S′ << S owing to imperfect elastic recovery and, in the case of free aquifers, owing to air pockets and capillary log or land subsidence.

### *Drawdown Equation for Water-Table Conditions*

Boulton[33-36] was able to derive an equation to solve for the hydrologic factors for water table aquifer with fully penetrating well. It was assumed in his solution that the gravity drainage to water table due to lowering the water level (dz) between the times $T$ and $T + dT$ since pumping commenced, consisted of two parts:

a. A volume (S dz) of water instantaneously released from storage per unit horizontal area, at any time from the start of pumping
b. A delayed yield from storage, at $t(t \geq T)$ from the start of pumping

$$dz \propto S' e^{-\alpha(t-T)} \quad \text{and} \quad n = 1 + S'/S$$

where $\alpha$ is an empirical constant and $S'$ is the total volume of delayed yield from a storage per unit drawdown per unit horizontal area.

Using the last two assumptions, Boulton derived the following equations:

$$D = \frac{Q}{4\pi T} W(U_{a,b}, r/\beta) \tag{3.24}$$

$$U_a = \frac{Sr^2}{4T_t} = \frac{1}{\text{Ø}} \tag{3.25}$$

$$u_b = \frac{S'r^2}{4Tt} = \frac{(r/\beta)^2}{4\alpha t} = \frac{1}{\text{Ø}'} \tag{3.26}$$

The values of $W(u_{ab}r/\beta)$ are plotted against values of $1/U_a$ and $1/u_b$ on logarithmic paper to construct type curves as shown in Figure 3.9. The type curves which lie to the left of the values $r/\beta$ are termed type A curves. The type curves which are shown to the right of the values are termed type B curves.

The method of using the type curves is briefly described as follows: The observed values of the drawdown D at a given distance r from the pumped well are plotted against values of time t on the same logarithmic scale as that used for the type curves to prepare a graph designated as time-drawdown field-data curve. Placing the time-drawdown curve (on transparent paper) over first the type A curves and then the type B curves, and keeping the respective coordinate axes parallel, a value of $r/\beta$ is determined from the type curve which gives the best fit.

Two cases may arise:

1. If the time-drawdown curve becomes horizontal after the early-time drawdown, a "match point" is chosen on this segment and, with the

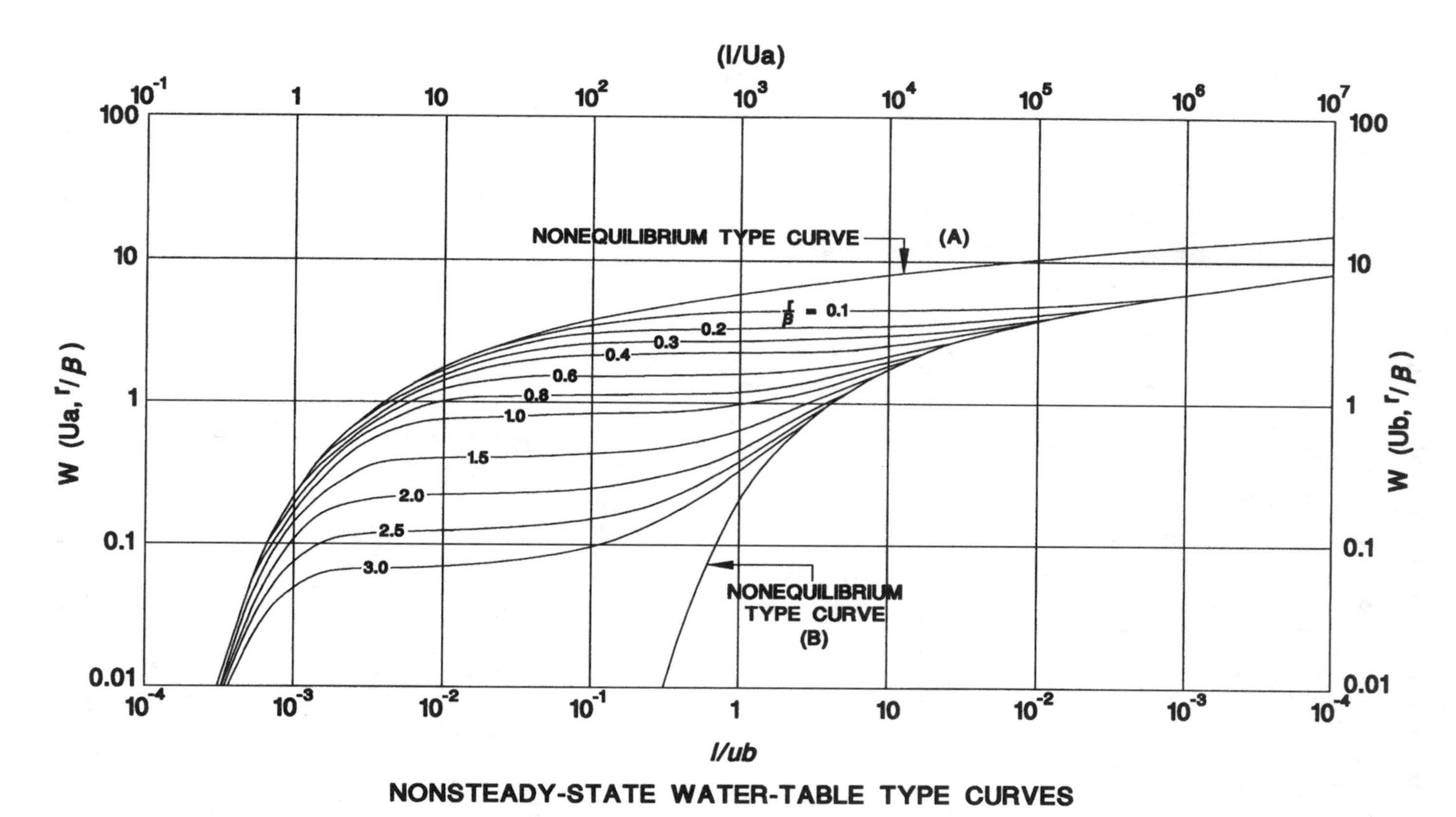


Figure 3.9 Delayed yield type curves.

time drawdown curve fitted to the appropriate type A curve, corresponding values of D, W(u), t, and $1/u_a$ are read off at the match point. The time drawdown curve is then fitted to the appropriate type B curve and the value of $1/u_b$ noted for the match point. (Being on the horizontal segment of the type curve, the match point will give the same value of W $[u_a, u_b]$ as before.) The formation constants, (T, S, S′, t, and α) are calculated from Equations 3.24, 3.25, and 3.26, respectively.

2. If the early time drawdown curve never becomes horizontal, the early time drawdown time segments of the time drawdown curve may be fitted respectively to the type A curve and type B curve having an r/β which gives the best fit. Choosing a match point in each of these segments, values are read from curves and substituted into equations 3.24 to 3.25 to compute hydrogeologic characteristics (T, S, α).

For case (1) above, it is evident that Type A and Type B curves, which are strictly for n equals infinity, are applicable when n has a large finite value since they both have zero slope at their intersection. For case (2), however, type curve B for finite n is generally required. But, if the intermediate slope of the time drawdown curve is not large, the complete type curve is obtained with sufficient accuracy by joining the appropriate Type A and Type B curves (plotted on $1/u_a$ and $1/u_b$ base) by a straight line tangential to both curves. In this case the match points on the Type A and Type B curves must be chosen so as to lie on segments of these curves which are clear for the sloping tangent. If the value of S is not required, the constants T, S′, and α may be directly obtained from the Type B curves, in which case the Type A curves are not needed.

After a relatively long period of pumping, the effects of delayed yield are negligible and aquifer characteristics may be computed using the Theis solution or its approximation as mentioned before. Boulton[10] developed a curve that can be used to estimate the time, $t_o$, when the effects of delayed yield become negligible (Figure 3.10). The figure gives $\alpha\, t_o$ as a function of r/β where α is an emperical coefficient, $T^{-1}$, $\alpha = T/\beta^2 S'$ and r is the radial distance from the pumping well.

Neuman[37-41] gave another solution for the average drawdown $D_{av}$ in an observation well at a distance r at time t after pumping from a fully penetrating well in an unconfined aquifer with saturated thickness m.

$$D_{av} = \frac{Q}{4\pi T}(Wt_s, \sigma, \beta) \tag{3.27}$$

where

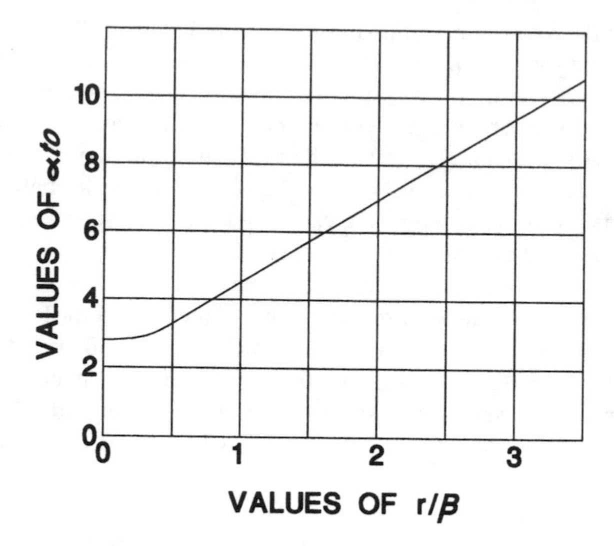


**Figure 3.10 Curve for estimating time, $t_o$, when delayed yield ceases to influence drawdown.**

$W(t_s, \sigma, \beta)$ is the new well function,

$$t_s = \frac{Tt}{Sr^2} \tag{3.28}$$

$$\sigma = S/S' \tag{3.29}$$

$$\beta = \frac{r^2 K_z}{m^2 K_r} \tag{3.30}$$

$K_z$ and $K_r$ are the vertical and horizontal hydraulic conductivity, respectively.

Based on the above equations, the aquifer properties $K_r$, $S'$, and $K_z$ can be obtained simply by plotting the drawdown D vs. t. The procedure is as follows:

1. Plot D vs. log t.
2. Fit a straight line to the last portion of the data. The intersection of this line with the horizontal axis where D = o is denoted by $t_o$. The slope of this line is the change in drawdown over one log cycle, denoted by $\Delta D$.
3. Equation 3.27 can be approximated assuming $S << S'$ and $\sigma \approx o$,

$$D = \frac{Q}{4\pi T} W(t_y, \beta) = \frac{Q}{4\pi T}(2.3 \log 2.25\, t_y) \tag{3.31}$$

and

$$T = \frac{2.3Q}{4\pi\Delta D} \tag{3.32}$$

using this equation to solve for transmissivity (T), then compute horizontal hydraulic conductivity as $K_r = T/m$.

4. The specific yield S′ can be computed using

$$S' = \frac{2.25Tt_o}{r^2} \tag{3.33}$$

5. Using the computed values of T and S′, solve for the dimensionless time $t_y$ from the equation

$$t_y = \frac{Tt}{S'r^2} \tag{3.34}$$

The following equation[36] can give β provided that $4.0 \le t_y \le 100$, otherwise Figure 3.11 may be used

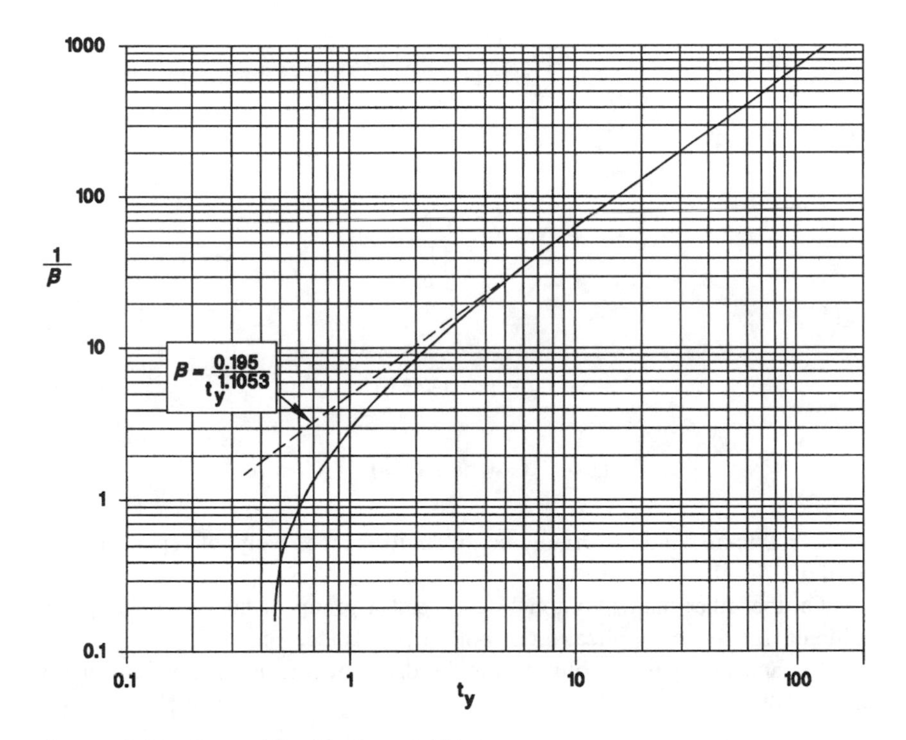


**Figure 3.11 Logarithmic plot of 1/β vs. $t_y$.**

$$\beta = 0.195/(t_y)^{1.1053} \tag{3.35}$$

Eq. 3.28 can be used to solve for $K_Z$ as

$$K_Z = \frac{\beta K r m^2}{r^2}$$

Since the value of S which is the storativity of the early time of pumping is of no importance, only $S'$, $K_r$, and $K_Z$ will be considered.

## *Unsteady State Flow in Semi-Confined Aquifer*[33-34]

The drawdown in a semi-confined aquifer can be described by Huntush and Jacob formula as follows:

$$D = \frac{Q}{4\pi T}\int_u^\infty \frac{1}{u} e^{-(y)} du \tag{3.36}$$

or

$$D = \frac{Q}{4\pi T} W(u, r/L)$$

where

$$u = \frac{r^2 S}{4Tt} \tag{3.37}$$

and

$$y = u + r^2/4L^2u \tag{3.38}$$

$$L = \sqrt{TC}, \text{ where } C = D'/K' \tag{3.39}$$

$D'$, $K'$ = thickness and hydraulic conductivity, respectively, of semi-pervious aquifer.

On the other hand, Walton[23] developed a group of curves defining the value of r/L, for the application of both methods (see Figure 3.12).

Lai and Su[42] gave a solution for the drawdown in the leaky aquifer for large wells:

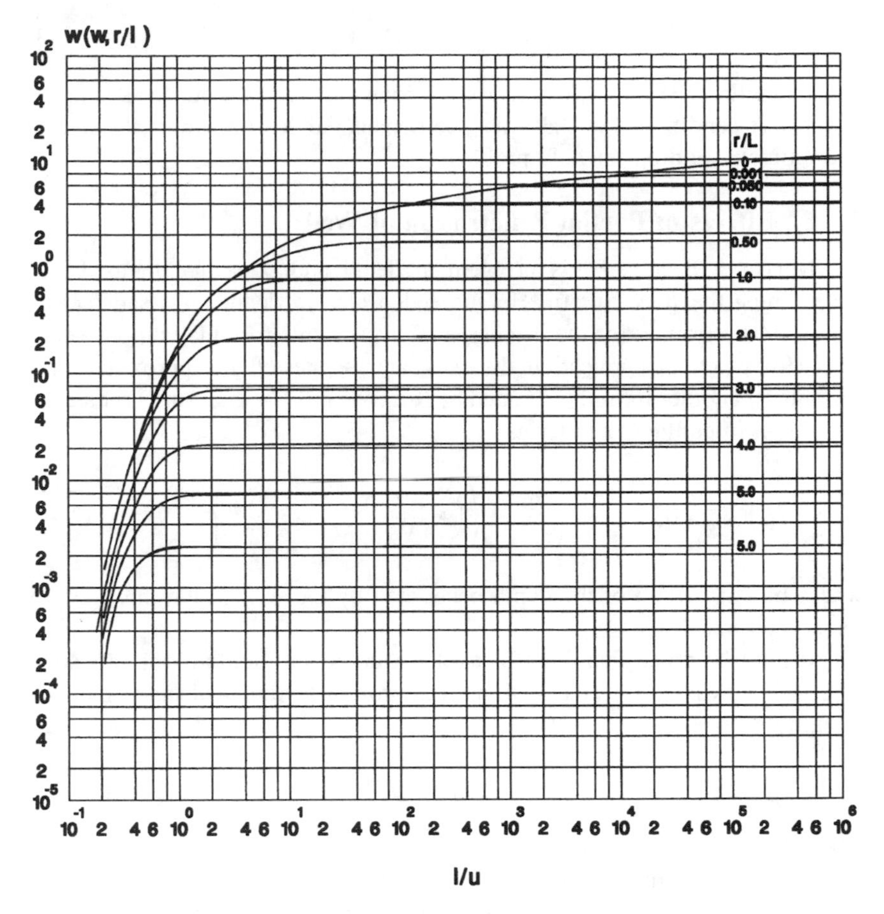


**Figure 3.12 Family of Walton's type curves W(u, r/L) vs. I/u and different values of r/L.**

$$D = \frac{Q}{4\pi T} F(u, \alpha, r_w / \beta, p) \tag{3.40}$$

where

$$u = \frac{r^2 S}{4Tt} \tag{3.41}$$

$$\alpha = r_w^2 S / r_c^2 \tag{3.42}$$

where $r_w$ = effective radius of the well bore or open hole and $r_c$ = radius of the pumping well casing within the range of the water level fluctuation

$$r_w\beta = r_w / \sqrt{T/D'/K'} \tag{3.43}$$

Since Eq. 3.38 needs lots of assumptions to be carried out for a solution in the last condition, Eq. 3.36 may be considered for its simplicity and its approximate solution for field problems.

### 3.5.5 Effects of Partial Penetration of Well

Muskat[43] discussed this problem in detail and presented methods for determining the flow pattern. He succeeded also in deducing a satisfactory approximate formula for the discharge. However, this formula is too complicated for practical application. Muskat later suggested to approximate his formula by another formula obtained by Kozenys (p. 274, Muskat[43]) for steady conditions in confined aquifer as

$$Q = \frac{2\pi D}{l_n} \frac{kmm'}{(r_e / r_w)} \left(1 + 7\sqrt{\frac{r_w}{2mm'}} \cdot \cos\frac{\pi m'}{2}\right) \tag{3.44}$$

where $m'$ is the ratio of the depth penetrated by the well to the thickness of the aquifer m.

It must be noted that the effects of partial penetration are only apparent in drawdown data collected within an approximate radial distance r of the pumping well

$$r < 1.5m\sqrt{K_r / K_z}$$

where m = thickness of aquifer

$K_r$ = horizontal hydraulic conductivity

$K_z$ = vertical hydraulic conductivity (beyond this distance groundwater flow is essentially horizontal)

In case the flow is unsteady, Huntush[44,45] gives an equation for the drawdown (D) at any point in an observation well as

$$D = \frac{Q}{4\pi K_r m}[W(u) + f(u, x, d/m, l/m, z/m)] \tag{3.45}$$

where

$$u = \frac{r^2 S}{4K_r mt} \tag{3.46}$$

W(u) = Theis well function and

$$f = \left[\frac{2m}{\pi(l - d)}\right]\sum_{n=1}^{\infty}\left(\frac{1}{n}\right)\left[\sin\left(\frac{n\pi l}{m}\right) - \sin\left(\frac{n\pi d}{m}\right)\right]\cos\left(\frac{n\pi z}{m}\right)W(u, x) \tag{3.47}$$

where

$$W(u, x) = \int_u^\infty \frac{e^c}{y} dy \tag{3.48}$$

$$c = \left( \frac{-y - x^2}{4y} \right) \tag{3.49}$$

$$x = \frac{r}{m} \sqrt{K_z / K_r} \tag{3.50}$$

The rest of variables are as shown in Figure 3.13.

### 3.5.6 Hydraulics of Well and its Design

The foregoing discussion has treated the flow of fluid through the aquifer under an energy gradient created by a well. The water must also be transferred through the screen and casing or pump column to the point of discharge. Under some circumstances the energy expended in moving the water through the well structure may exceed that used in moving it through the aquifer. Better understanding of hydraulic principles involved in this latter mechanism should lead to improved well design. Some of the considerations are outlined in the following discussions.

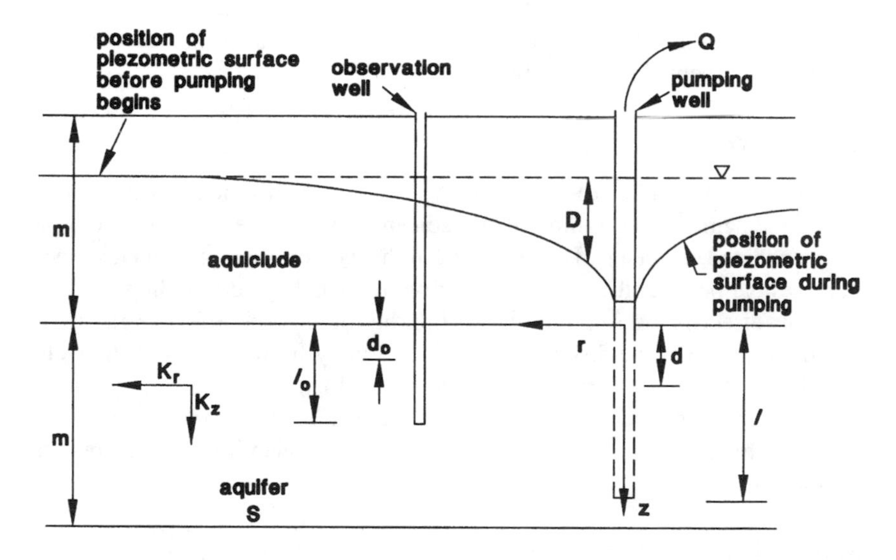


**Figure 3.13 Partially penetrating well in a confined aquifer.**

### *Specific Capacity*

Engineers have designated specific capacity to the ratio of discharge to drawdown. If the hydraulic head losses through the screen and casing were zero and the time effect of storage depletion were ignored, the discharge of an artesian well could be expected to be directly proportional to drawdown. This would lead to a constant value of specific capacity corresponding to all values of discharge of an artesian well — a condition usually assumed. For wells in unconfined aquifers, an increase in drawdown decreases the effective thickness of the aquifer. Thus, even discounting energy losses at the well, the specific capacity would decrease with discharge for the water-table case.

The hydraulic losses through the well cause further non-linearity for the relationship of discharge to drawdown. As mentioned by Jacob,[27] flow through the screen and casing usually occurs in the turbulent regime and resulting head losses are thus proportional to $Q^2$. Aquifer losses under conditions of laminar flow should be proportional to Q for an artesian well.

Thus one may write:

$$D = BQ + CQ^2 \tag{3.51}$$

where D is the total drawdown in the well and B and C are constants.

Equation 3.51 can be evaluated approximately by pumping test at two different discharge rates, $Q_1$ and $Q_2$, and measuring the respective values of drawdown, $D_1$ and $D_2$; substitution successively into Eq. 3.51 provides simultaneous equation in B and C. The difficulty with this procedure is that it is based on the assumption of steady flow and does not take into account the effect of time depletion of storage.

### *Effective Radius*

The effective radius of a well, as it is used in the formulas of flow, may not be the same as the radius of the screen or hole especially for wells in unconsolidated sediments. Development of the well or use of gravel envelopes increases the permeability of the formation immediately surrounding the casing. This effect is the same as increasing the radius. The effective radius is defined by Jacob as the distance, measured radially from the axis of the well, at which the theoretical drawdown based on the logarithmic head distribution equals the actual drawdown just outside the screen. In the reference previously cited, Jacob[46] gives a procedure for determining this quantity using the results of field tests.

### *Well Screens*

Frequently a large part of the energy imparted through a well is expended in transferring the water through the screen and pump. For this reason, attention should be given to the hydraulic performance of the well structure. While considerable progress has been made, collection of data, especially in the field, is difficult because of rapidly changing flow conditions near the well. Laboratory experiments which simulate field conditions are expensive and arduous. Nevertheless, more attention should be given to this important aspect of the problem of well hydraulics.

An important art of the well structure is the screen. Screens of some kind are always required except in consolidated sediments. They may range from rough, haphazard, perforation in a steel casing to highly engineered and carefully manufactured screens of specially selected material. The function of a screen is to exclude the natural sediments while allowing the greatest possible flow of water into the well. The factor of longevity influences the choice of screen.

The hydraulic performance of well screen was ably treated by Peterson et al. in the text edited by Luthin.[47] Water enters the interior of a screen in the form of radial jet at relatively high velocities. The energy of these jets is dissipated and the flow accelerated in the axial direction. From a theoretical consideration of the mechanic involved, these investigators deduced:

$$\frac{\Delta h}{v^2/2g} = \frac{\cosh\,(CL/D+1)}{\cosh\,(CL/D-1)} \tag{3.52}$$

where $\Delta h$ is the hydraulic head loss involved in the screen and v is the first average velocity along the screen axis (Q/A where Q is the well discharge and A the cross-sectional area of the screen). L is the axial length of the screen and D is the screen diameter. C is defined as the screen coefficient.

$$C = 11.31 C_c A_p \tag{3.53}$$

In Eq. 3.53 $C_c$ is the orifice coefficient of discharge applying to the screen opening and $A_p$ is the fractional ratio of screen opening to total screen surface.

The loss coefficient $\frac{\Delta h}{v^2/2g}$ in Eq. 3.52 approaches 2 for values of CL/D exceeding about 6.

### *Velocity Distribution*

The velocity distribution around the well screen was taken constant along its length. This was proved later by Soliman[48] to be curvilinear having the following relations:

$$U = U_o e^{KL/D} \tag{3.53}$$

where K is a constant depending on the well screen and the rest of values are shown in Figure 3.14.

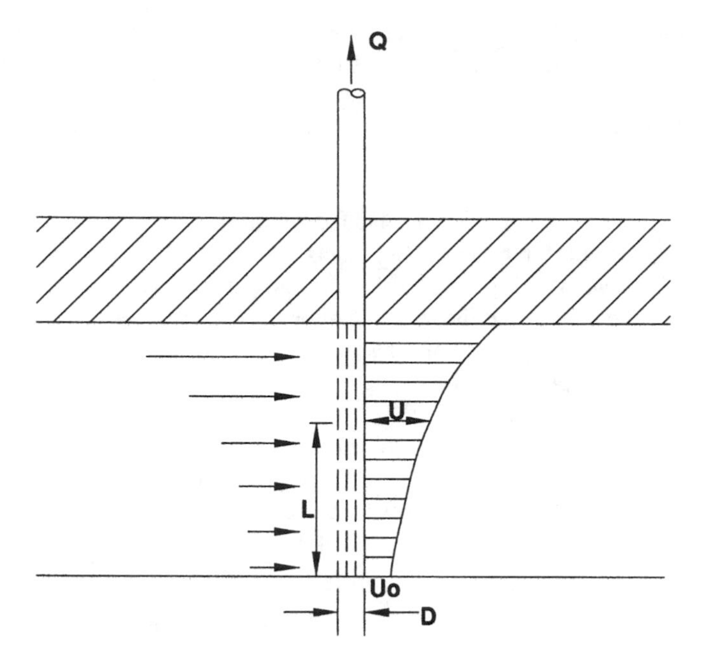


**Figure 3.14 Velocity distribution around well screen.**

These relations should be considered in the designing of screen length.

## 3.5.7 Slug Tests[49]

The slug test which is considered in this event is the test for determining the hydraulic conductivity (K) of unconfined or leaky aquifers confined with completely or partially penetrating wells (Figure 3.15). For any other cases of confined aquifers, data collected from the pumping tests can be used to determine the hydraulic conductivity and storage coefficient.

The equations describing flow are based on a modified form of Thiem (Dupuit) equation (Eq. 3.6). The rate of groundwater flow, Q, from a well screen between the depths d and l for a specified water level in the well, Hw is

$$Q = \frac{2\pi K(l - d)H_w}{\ln(R / r_w)} \tag{3.54}$$

where R is the radius of influence of the injection well and $r_w$ is the effective radius of the well bore; also

$$r_w = \sqrt{r_i^2(1 - n) + nr_o^2}$$

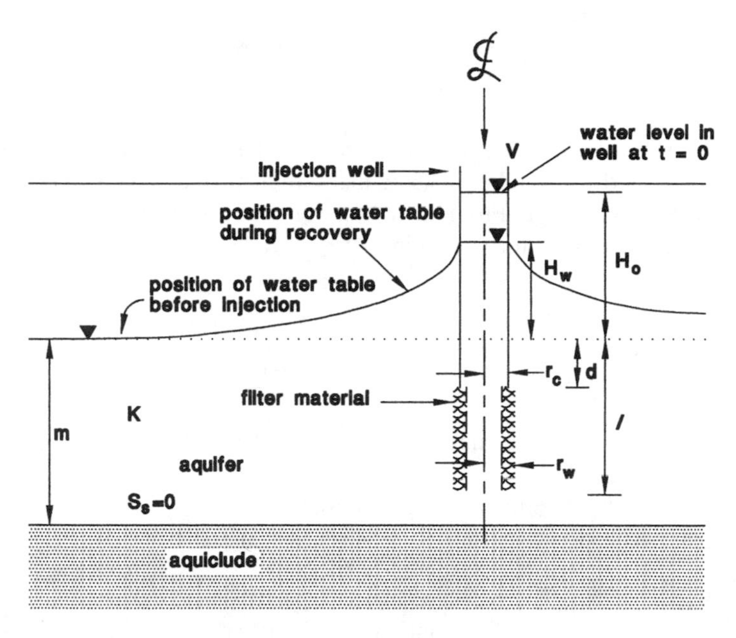


**Figure 3.15 Slug test in unconfined aquifer.**

if the water level is falling within the screen length of the well and the hydraulic conductivity of the filter material or developed zone is much larger than the hydraulic conductivity of the aquifer, and n is the porosity of the filter:

$r_i$ = inside radius of well screen
$r_o$ = outside radius of filter material
$r_c$ = effective radius of the well casing over which the water level in the well changes

In order to develop a simple equation the following assumptions are given:

1. The aquifer is homogeneous and isotropic;
2. A volume of water, V, is injected instantaneously at time t = o;
3. Head losses through the well screen, filter material, and developed zone (if present) are negligible.

The rate of fall of the water level in the well is equal to the flow rate divided by the effective cross-sectional area of the well casing

$$\frac{dHw}{dt} = -\frac{Q}{\pi r_c^2} \tag{3.55}$$

Combining Eq. 3.54 and 3.55 and integrating, K can be given as (refer to Figure 3.15 for limits),

$$K = \frac{r_c^2 \ln(R/r_w)\ln(Ho/Hw)}{2(l-d)t} \tag{3.56}$$

Bouwer and Rice[49] determined the radius of influence, R, for different values of $r_w$, $(l-d)$, $H_w$, and m in using measurements made with an electrical resistance analog model. From their experiments, the following empirical equation was developed for estimating R:

$$\ln(R/r_w) = \left[\frac{1.1}{\ln(l/r_w)} + \frac{A + B\ln[(m-l)/r_w]}{2(l-d)/r_w}\right]^{-1} \tag{3.57}$$

where A and B are dimensionless coefficients which are functions of $(l-d)/r_w$ as shown in Figure 3.16.

If $(\ln(m-l)/r_w) > 6$ then Eq. 3.57 becomes:

$$\ln(R/r_w) = \left[\frac{1-1}{\ln(l/r_w)} + \frac{A + 6B}{(l-d)/r_w}\right]^{-1} \tag{3.58}$$

Also, if the injection well fully penetrates the aquifer, the following equation is used:

$$\ln(R/r_w) = \left[\frac{1-1}{\ln(l/r_w)} + \frac{C}{(l-d)/r_w}\right]^{-1} \tag{3.59}$$

where C can be interpolated from Figure 3.16.

### 3.5.8 Groundwater Recharge

Groundwater recharge may be obtained by artificial or natural means. Artificial groundwater recharge is a planned operation of transferring water from ground surface into aquifers. Natural groundwater recharge is a phenomenon of water reaching aquifers, without man-made activities, from surface sources such as streams, natural lakes, or ponds.

The factors affecting natural groundwater recharge are thickness and properties of soil formation and stratification, surface topography, vegetative cover, land use, soil moisture content, depth to water table, duration, intensity and seasonal distribution of rainfall, air temperature and other meteorological factors (humidity, wind, etc.), and influent and effluent streams.

Groundwater recharge[50] may occur by infiltration, by injection, or by induction. The infiltration process is the entry of water into the saturated zone at the water table surface (Figure 3.17). The injection method is the entry of

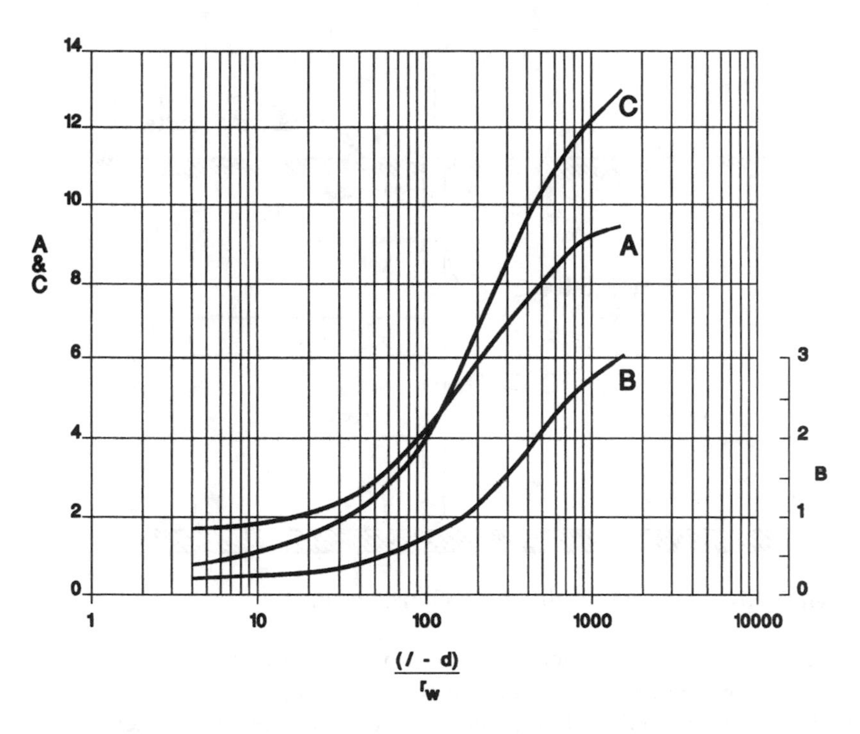


**Figure 3.16** Values of the coefficients A, B, and C for use in estimating the radius of influence, R. (From Bouwer, H. and Rice, R.C., *Water Resources Research,* 1976, p. 426.)[49]

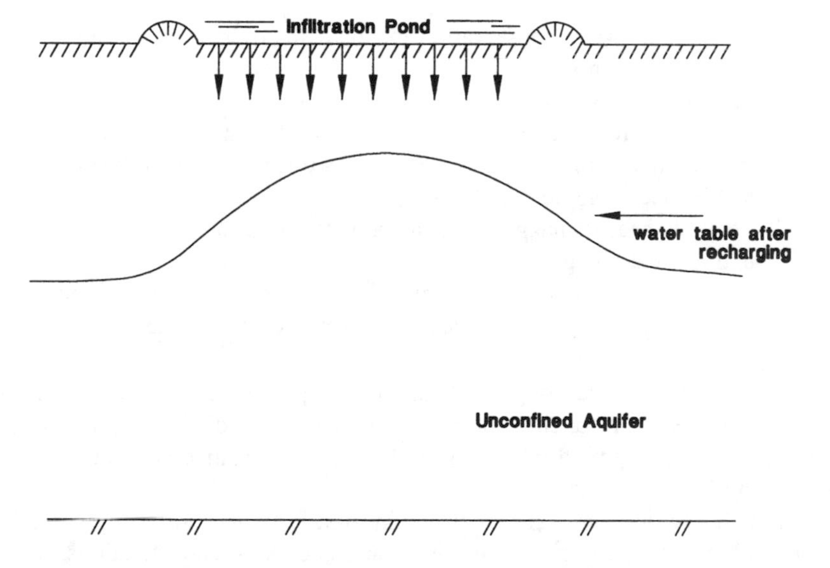


**Figure 3.17** Infiltration from ponds.

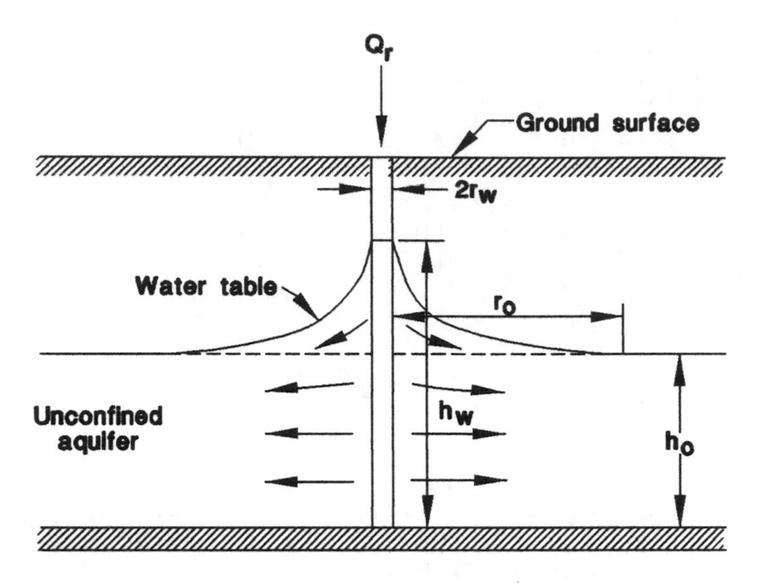


**Figure 3.18 Water injection by recharge well.**

water into confined or unconfined aquifers by the injecting wells (Figure 3.18). Recharge by induction is the entry of water into aquifers from surface water bodies due to extraction of groundwater (Figure 3.19). Groundwater recharge by infiltration could be natural or artificial, while recharge by injection or induction is artificial.

The objectives of artificial groundwater recharge may be given as:

1. To serve as water-conservation mechanisms by subsurface storage for local or imported surface waters, supplement the quantity of groundwater available, and reduce the cost of pumping.
2. To prevent, reduce, and correct adverse conditions such as sea water intrusion, lowering of water table, land subsidence, and unfavorable salt balance (Figure 3.20).
3. To allow heat exchange by diffusion through ground to conserve or extract heat energy.
4. To obtain suspended solid removal of infiltration through ground and storage of reclaimed wastewater for subsequent use.

The sources of water for groundwater recharge may be storm runoff which could be collected in ditches, basins, or reservoirs; a distant surface water which might be imported into a region by pipeline or aqueduct; and treated wastewater.

Depending on source and quality of water, type of aquifer, type of soil, topographical and geological conditions, and economic considerations, there are various artificial groundwater recharge methods.

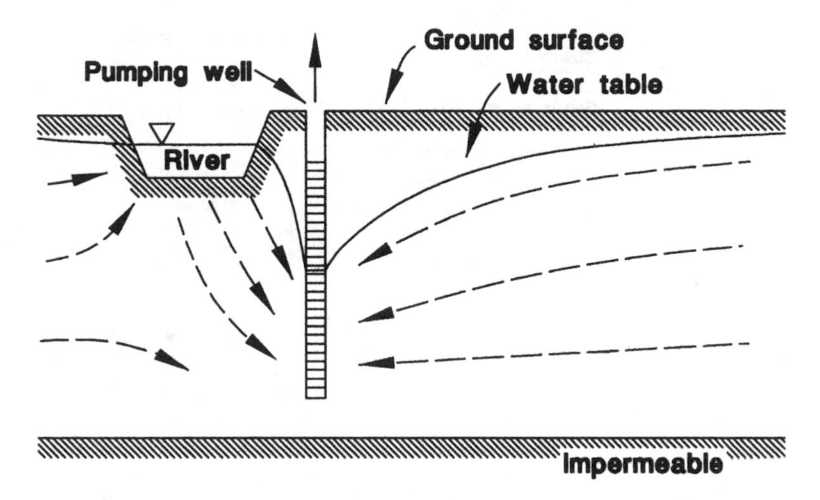


Figure 3.19 Induced recharge resulting from a well extraction.

These include water spreading methods — basins, stream channel, ditch furrow, flooding, and irrigation (see Figure 3.21); the pit method (Figure 3.22); and the recharge well method (Figure 3.23).

Recent interest has been focused on the reuse of municipal wastewater to recharge groundwater aquifers. Almost all uses of this are nonpotable, e.g., irrigation or industrial purposes, because of questionable health effects.

Recharge of wastewater (usually after secondary treatment) improves its quality by removal of physical, biological, and some chemical constituents.

Storage is provided until subsequent reuse reduces seasonal temperature variations and dilutes the recharged water with native groundwater. Land application practices involve irrigation, spreading overland flow, and recharge

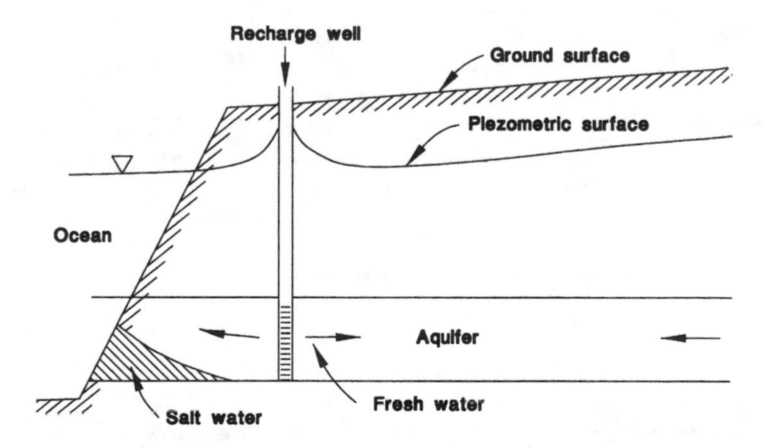


Figure 3.20 Control of seawater intrusion by recharge well.

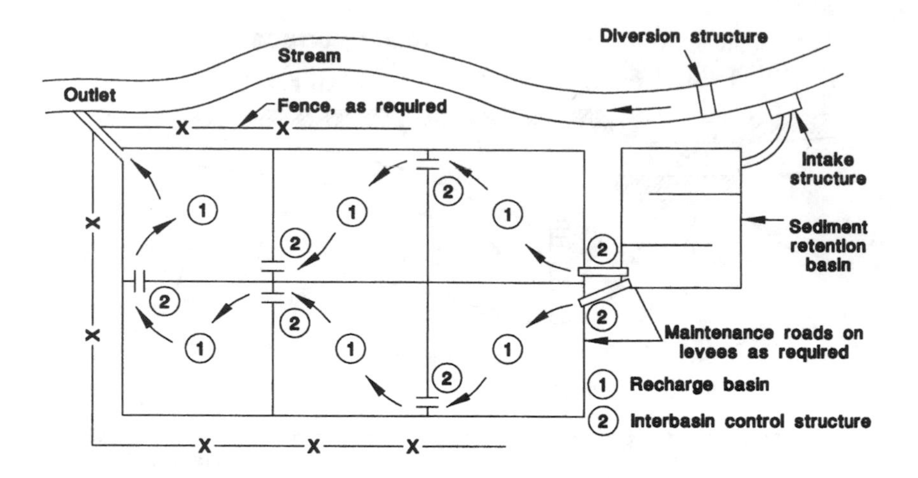


Figure 3.21 Multi-basin recharge method.

wells. Selection of a given system is governed by soil and subsurface conditions, climate, availability of land, and intended reuse of the wastewater.

Groundwater recharge by infiltration from ponds depends on rate of infiltration which subsequently depends on the soil characteristics. The infiltration rate of any soil can be measured by a double ring infiltrometer. The total volume of water infiltrating the soil per unit of surface area can be determined by integrating Horton's equation[51,52]

$$f = f_c + (f_o - f_c)e^{-Kt} \tag{3.60}$$

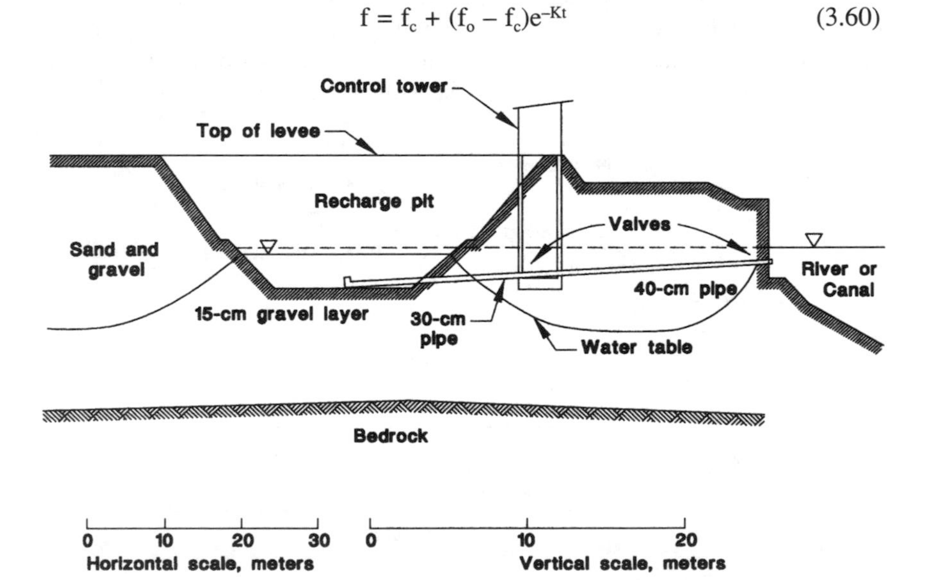


Figure 3.22 Cross section through a recharge pit.

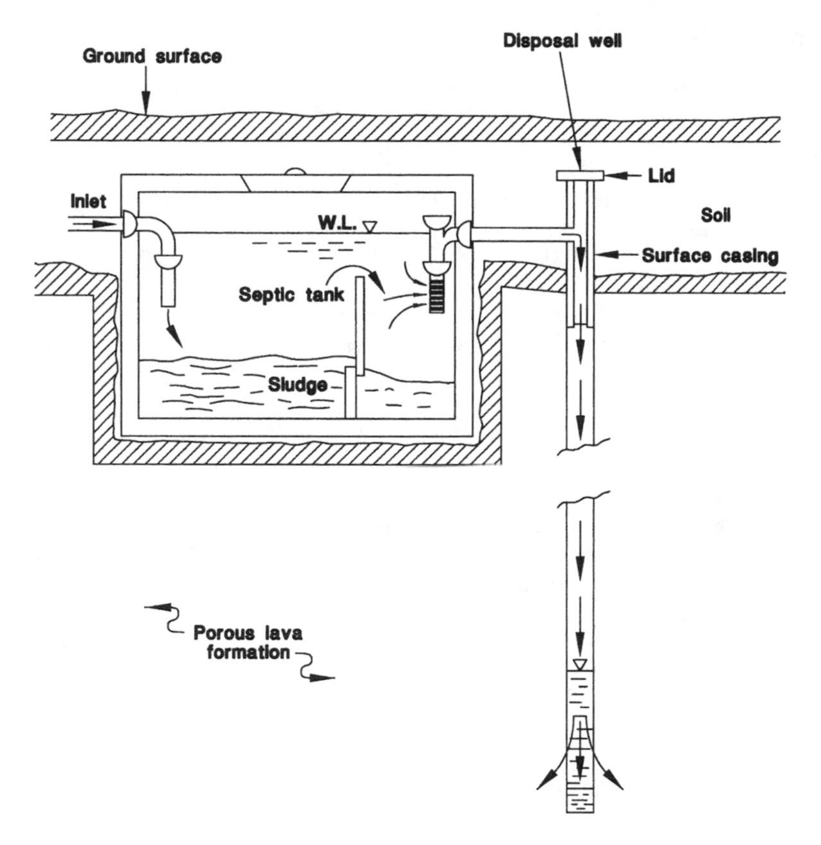


**Figure 3.23 Recharge well for disposal of septic tank effluent into a lava formation.**

and by integrating Eq. 3.60 the total volume of water (F) infiltrating the soil until time t

$$F = f_c t + \frac{1}{K}(f_o - f_c)(1 - e^{-Kt}) \tag{3.61}$$

where f is the rate of infiltration of water into soil at time t; $f_o$ is the initial infiltration rate; $f_c$ is the final infiltration rate; and K is a rate constant. $f_o$, $f_c$, and K can be given for any soil from the infiltration tests.

On the other hand, groundwater recharge by injection or by induction depends on the hydraulic conductivity of the aquifer to be recharged. Hydraulic conductivity of aquifers can be determined by pumping tests as mentioned earlier or by using slug tests.

## REFERENCES


1. Meinzer, O.E., The occurrence of ground water in the United States, with a discussion of principles. United States Geology Survey, Water-Supply Paper 489, Washington Government Printing Office, 321, 1923.
2. Meinzer, O.E., Hydrology — The history and development of ground-water hydrology, *Journal of the Washington Academy of Sciences,* Vol. 24, No. 1, p. 6–32, 1934.
3. Meinzer, O.E., Physics of the earth; Part 9, *Hydrology,* New York, McGraw-Hill Book Co., 712, 1942.
4. Veatch, A.C. and Slichter, C.S., et al., Underground water resources of Long Island, NY, U.S. Geological Survey, Washington, D.C., 394, 1906.
5. Norton, W.H., et al., Underground water resources of Iowa, USGS Water Supply Paper 293, 1912, pp. 68–69.
6. Tolman, C.F., *Ground Water,* McGraw-Hill, Inc., New York, NY, 593, 1937.
7. Givan, C.V., Flow of water through granular materials, *Am. Geophys. Union Trans.,* Washington, D.C., 1934, pp. 572–579.
8. White, W.N., et al., Geology and groundwater resources of the Lufkin area, Texas, USGS Water Supply Paper 849-A, Washington, D.C., 1941, pp. 1–64.
9. Thompson, D.G., Groundwater supplies of the Atlantic City regions, *Bull. 30,* New Jersey Dept. Cons. and Devel., Trenton, NJ, 138, 1928.
10. Muskat, M., The flow of compressible fluids through porous media and some problems in heat conduction, *Physics,* Vol. 5, No. 3, Menasha, WI, 1934a, pp. 71–94.
11. Muskat, M., Two fluid systems in porous media: The encroachment of water into an oil sand, *Physics,* Vol. 5, No. 9, Menasha, WI, 1934b, pp. 250–264.
12. Muskat, M., The seepage of water through porous media under the action of gravity, *Am. Geophys. Union Trans.,* Washington, D.C., 1936, pp. 391–395.
13. Muskat, M., Use of data on the build-up of bottom-hole pressures, *Am. Inst. Min. Met. Eng., Petroleum Division, Trans.,* Vol. 123, New York, NY, 1937a, pp. 44–48.
14. Muskat, M., Flow of homogenous fluids through porous media, McGraw-Hill Book Co., Inc., 1937b.
15. Meyer, W., Deussow, J.P., and Gillies, D.C., Availability of groundwater in Marion County, Indiana, U.S. Geological Survery Open-File Report 75-312, 87, 1975.
16. Morel-Seytoux, H.J. and Daly, D.J., A discrete kernel generator for stream aquifer studies, *Water Resources Research,* Vol. II, Washington, D.C., 1975.
17. Biswas, A.K., *History of Hydrology,* Amsterdam, North-Holland Publishing Co., New York, Elsevier Publishing Co., 336, 1970.
18. Meyer, G., Davis, G., and LaMoreaux, P.E., Historical perspective, in *Hydrogeology, The Geology of North America*, Vol. 0–2, Back, Rosenshein, and Seaber, Eds., The Geological Society of America, 1988.
19. Piper, A.M., Robinson, T.W., and Park, C.F., Jr., Geology and groundwater resources of the Harney Basin, Oregon, USGS Water Supply Paper 841, Washington, D.C., 189, 1940.
20. Wenzel, L.K., Local overdevelopment of groundwater supplies, with special reference to conditions at Grand Island, Nebraska, USGS Water Supply Paper 836, Washington, D.C., 223, 1940.

21. De Wiest, R.J., *Geohydrology,* John Wiley & Sons, New York, 1965.
22. Soliman, M.M., *Environmental Effects on the Arid Coastal Water Sheds in Egypt,* International Symposium of Arid Region Hydrology, San Diego, CA, 1990.
23. Walton, W.C., *Groundwater Resource Evaluation,* McGraw-Hill, New York, 1970.
24. Dupuit, A.J.E.J., *Etudes Theoriques et Pratiques sur le Mouvement des Eaux dans les Canaux Decouverts et a Travers les Terrains Permeables*, Paris, 275, 1863.
25. Daubree, A., *Textbook on Hydrology,* Vol. 3, 1887.
26. Fuller, M.L., Water Supply Paper 160, 1906.
27. Jacob, C.E., *On the Flow Water in an Elastic Artesian Aquifer,* American Geophysical Union *Trans*., 1954.
28. Theis, C.V., *The Relation Between the Lowering of Piezometric Surface and the Duration of Discharge of a Well Using Groundwater Storage,* Am. Geophysical Union Trans., 1935.
29. Freeze, R.A. and Cherry, J.A., *Groundwater,* Prentice Hall, Englewood Cliffs, NJ, 1979, pp. 45–50, 58–61, 152–163.
30. Dupuit, A.J.E.J., *Etudes Theoriques et Pratiques sur le Mouvement des Eaux dans les Canaux Decouverts et a Travers les Terrains Permeables*, Paris, 275, 1863.
31. Soliman, M.M., *Groundwater Management in Arid Regions,* Vol. I, Ain Shams University, 1984.
32. Todd, D.K., *Groundwater Hydrology,* John Wiley & Sons, New York, 1959.
33. Boulton, C.A., The drawdown of the water table under non-steady conditions near a pumped well in an unconfined formation. Proceedings, Institute of Civil Engineers, 3, 3, 1954, pp. 564–579.
34. Boulton, N.S., Analysis of data from nonequilibrium pumping tests allowing for delayed yield from storage. Proceedings, Institute of Civil Engineers, 26 (6693), 1963, pp. 469–482.
35. Boulton, N.S. and Streltsova, T.D., New equations for determining the formation constant of an aquifer from pumping test data. *Water Resources Research,* 11, 1, 148–153, 1975.
36. Boulton, N.S. and Streltosova, T.D., The drawdown near an abstraction well of large diameter under non-steady conditions in an unconfined aquifer. *Journal of Hydrology,* 30, 1976, pp. 29–265.
37. Neuman, S.P., Theory of flow in unconfined aquifers considering delayed response of the water table. *Water Resources Research,* 8, 4, 1031–1045, 1972.
38. Neuman, S.P., Supplementary comments on theory of flow in unconfined aquifers considering delayed response of the water table. *Water Resources Research,* 9, 4, 1102–1103, 1973.
39. Neuman, S.P., Calibration of distributed parameter groundwater flow models viewed as a multiple-objective decision process under uncertainty. *Water Resources Research,* 9, 4, 1006–1021, 1973.
40. Neuman, S.P., Analysis of pumping test data from anisotropic unconfined aquifers considering delayed gravity response. *Water Resources Research,* 11, 2, pp. 329–342, 1975.
41. Neuman, S.P., Perspective on "delayed yield." *Water Resources Research,* 15, 4, 899–908, 1979.

42. Lai, R.Y.S. and Su, C.W., Nonsteady flow to a large well in a leaky aquifer, *Journal of Hydrology,* 22, 333–345, 1974.
43. Muskat, M., *The Flow of Hogeneous Fluids through Porous Media.* McGraw-Hill, New York, 1937.
44. Huntush, M., Nonsteady Flow to Well Partially Penetrating an Infinite Leaky Aquifer, Proceedings, Iraqi Science Society, Baghdad, 1956.
45. Huntush, M.S., *Hydraulics of Wells,* Advanced Hydroscience, Vol. I, 1964.
46. Jacob, C.E., Notes on determining permeability by pumping tests under water table conditions, U.S. Geological Survey, Open File Report, Washington, D.C., 1944.
47. Luthin, I., *Drainage of Agricultural Lands,* American Society of Agriculture, Madison, WI, 1957.
48. Soliman, M.M., Boundary flow consideration in the design of wells, *ASCE Journal of Irrigation and Drainage,* March 1965.
49. Bouwer, H. and Rice, R.C., A slug test for determining hydraulic conductivity of unconfined aquifers with completely or partially penetrating wells, *Water Resources Research,* 12, 3, 423–428, 1976.
50. Lerner, D.N., Arie, S.I., and Ian, S., *Groundwater Recharge,* International Association of Hydrogeologists, Lingen (EMS), West Germany, Vol. 8, 1990.
51. Horton, R.E., Analysis of runoff plate experiments with varying infiltration capacity, *Trans. Am. Geophys. Union*, 1939, pp. 693–711.
52. Green, I.R.A., An explicit solution of the modified Horton equation, *Journal of Hydrology,* 83, 23–27, 1987.

# 4 ENVIRONMENTAL IMPACTS RELATED TO HYDROGEOLOGICAL SYSTEMS

## 4.1 NATURAL AND MANMADE DISASTERS

During the past decade geologists have been requested to participate in an ever expanding role of responsibility regarding the evaluation and protection of the environment. The earliest geologists described and catalogued rocks, fossils, and minerals. Subsequently they used this knowledge to develop mineral, water, and energy resources. During the past twenty years, there has been an increasing demand to use this same expertise to bring about remedial actions on "endangered environments" and to aid environmental planning and development. Geologists are now requested to provide solutions to properly manage hazardous, toxic, and radioactive wastes, as well as to cope with problems of catastrophes, both natural and man-made. With this responsibility, there has evolved a need for risk assessment and guidance for the development of insurance programs that will protect individuals against disaster. What are the risks, what are the chances for a reasonable risk assessment, and what are the limits of liability?

Ancient man must have looked upon a volcano such as Santorini (an island volcano in the Mediterranean), or upon the recent hurricanes such as Hugo or Camille in the USA as calamities totally unfathomable. In prehistoric times, disasters of this magnitude were the basis for a belief in gods' venting their wrath upon mankind. These were similar to the Biblical flood of Noah, caused by a storm of such large magnitude that the entire Tigris and Euphrates valleys were inundated, and life was destroyed over a very extensive area. These were the kinds of natural disasters that were the basis for legends handed down by word of mouth over hundreds of years.

Natural catastrophes take place today; however, civilization has become more knowledgeable about their causes and impacts. Information is collected relative to the causes and effects of these natural phenomena. For example, during the last few years, the scientific community has predicted, with considerable accuracy, hurricanes, earthquakes, volcanic eruptions, and tsunamis. Effective warning systems have been developed that save thousands of lives through emergency planning for the evacuation from danger zones. Special construction of buildings has been mandated in the event that one of these natural catastrophes should take place in a populated area. Modern civilization, however, still has not been able to cope with certain types of catastrophic phenomena. Worldwide, nearly 3 million people have died; some 820 million more have been injured, displaced, or otherwise affected by natural disasters during the past 10 years; and property damage from individual catastrophes has been in the billions of dollars. Hugo's damage was in excess of 2 billion dollars; the October 19, 1989, San Francisco earthquake damage exceeded 10 billion dollars; and the more recent hurricane Camille caused damage to Florida and Louisiana in excess of 13 billion dollars.

In 1980 the eruption of Mt. Saint Helens in the state of Washington awakened many to the impact of volcanic hazards, and in December 1987, the United Nations General Assembly designated the 1990s as the International Decade for Natural Disaster Reduction. One of the first reports from this effort is "Reducing Disasters' Toll," by the United States National Research Council.[1] This is one of a series of publications that have been and will continue to be published in the future as a result of cooperative international programs to reduce the impact of natural hazards. Man is now evaluating more carefully these natural phenomena and measuring their worldwide impact.

From 1650 to 1450 BC, major eruptions of Santorini caused such damage to the natural environment that climate around the world was affected; billions of tons of fine ash were thrown into the air; and day turned into night over much of the Eastern Mediterranean. Ocean tidal waves lashed against the shores of all the Greek Islands, Asia Minor, and North Africa. Therefore, even though this catastrophe was not studied and recorded as are those of today, a record remains of the event over 3,500 years after it occurred. From a hydrogeologists' perspective, the environmental impact was so great, even in a relatively unpopulated world, that it left indelible historical evidence. The eruption was recorded by geologic, chemical, and biological time clocks. There is not a written record, at least none discovered to date; however, all available data from a great variety of sources can be pieced together to interpret the size and impact from this historical event. If this event were to happen in a relatively populated area today, such as the Mediterranean, the Gulf Coast of the USA, Japan, or Indonesia, the loss of life and damage to property would be a tragedy unknown to modern civilization.

The Santorini event was so violent that it has never been equaled in the memory of man. The eruption of Krakatau, the only natural event with gaugeable force, could not equal its violence. It was one of the world's all-time most spectacular natural environmental events.

Another Santorini could happen and the hydrogeologist, geophysicist, and seismologist could predict, with some accuracy, where it would happen and sometimes, with adequate data, approximately when. But how can such a catastrophe be insured? What type of environmental planning must be carried out in these critically sensitive spots of the world?

The benefits from investments in geoscience pay enormous dividends. The U.S. Geological Survey reports' predicting the eruption of Mt. Pinatubo in the Philippines resulted in the safe evacuation of more than 100,000 people and billions of dollars in United States military equipment. But imagine the impact of a volcano in today's world that would disrupt an area the size of the Mediterranean with tsunamis a 100 feet high and with a poisonous ash fallout over 1,000 square miles. Similar catastrophes would be another major earthquake along the San Andreas Fault through San Francisco, California, or a repeat of the recent Hurricane Andrew (1992) in the USA that killed 17 and caused over $17 billion in damage.

There are many natural catastrophes for which hydrogeologists or other earth scientists have been able to identify and provide risk factors, locate potential areas of occurrence, and determine frequency, severity, and potential damage to property and life. These include landslides, mudslides, tsunamis, earthquakes, hurricanes, tornadoes, and catastrophic subsidence or sinkhole collapse.

Many of these catastrophes can be predicted with substantial accuracy using present day scientific knowledge, methodology, techniques, and instrumentation. Funding for research, and therefore accuracy of results, depends to a great extent on the perception of the public and its willingness to support financing of necessary research that will allow the most accurate modeling and predictions of these natural events.

In Japan and the United States, extensive research in the area of volcanology has allowed rather accurate predictions of volcanic events. Mt. Saint Helens, predictions of impending activity allowed the removal of populations and minimized damage of many types and the loss of life. The U.S. Geological Survey of the Department of the Interior has an excellent research effort. In addition to its responsibility for the assessment of energy, minerals, water, and topographic mapping, it carries on a detailed study program as well as monitoring and prediction programs on volcanoes.*

Earthquake phenomena has also been the subject of extensive research in the Soviet Union, China, Japan, the United States, and certain other areas of the world that are impacted by frequency of earthquakes. In the United States, the U.S. Geological Survey and National Oceanic and Atmospheric Administration (NOAA) carries on extensive programs of research including a computerized maintenance of records of epicenters over the world and periodic earthquake probability maps rating the areas by frequency of earthquakes. This subject has received so much attention in the past that there are now textbooks and regular scientific journal publications; for example: NOAA earthquake frequency map; NOAA epicenter determination reports; Episodes, regular

* Information is available from The Branch of Distribution, USGS, 604 South Pickett Street, Alexandria, VA 22304, or USGS Center, Menlo Park, California.

articles in an international geoscience journal; Geotimes, annual summary of the American Geological Institute; Geotimes, monthly summary on "Geologic Phenomena."

One of the best summaries on earthquakes is available from the U.S. Geological Survey Information in a 1991 pamphlet titled "Earthquakes."[2]

Predicting earthquakes, though improved in recent years, has not attained quite the accuracy as predicting volcanic eruptions. However, in the case of both volcanoes and earthquakes, the detailed knowledge of the geology of the earth and its geologic structure and plate tectonics plus recent information from satellite research on natural phenomena have provided tools that have made predictions far more accurate in recent years. Information on earthquake research in the United States can be obtained from The Branch of Distribution, U.S. Geological Survey.

Earthquakes and volcanoes are related to major tectonic features of the earth's crust and can be of such a minor impact as an unobservable deep-seated intrusion of magma on the ocean floor or a quake on the Richter scale of a fraction of 1 that could occur unknown to the population. These incidents, however, are in sharp contrast to a large eruption such as a Santorini or Krakatau or an earthquake such as the 1909 San Francisco catastrophe. These are monitored in much detail by today's scientific community and are being used to gain more accurate knowledge of natural phenomena, some of which will effect the insurance industry much more in the future. There will come a time when all of these events will be the subject of insurance programs that will require risk assessments.

## 4.2 LAND SUBSIDENCE

There are a number of other natural phenomena that scientists can predict with a substantial degree of accuracy as to where, why, and when hazards may occur and the frequency and size of the hazard. This would include landslides, mudslides, and sinkhole collapse or catastrophic subsidence. Let us analyze problems associated with land subsidence.

More than 44,000 $km^2$ of land in 45 states in the United States has been lowered by the types of subsidence considered in this report. Underground mining of coal, groundwater withdrawal, and drainage of organic soils are the principal causes of subsidence with approximately 8,000, 26,000, and 9,400 $km^2$ of land having subsided from each of these causes, respectively. In addition, about 18% of the conterminous United States is underlain by cavernous limestone, gypsum, salt, or marble and is locally susceptible to catastrophic collapse into sinkholes.

Annual costs resulting from flooding and structural damage are in the billions of dollars.[3] Although these costs are small relative to those of many other earth-science hazards, their geographic distribution is not uniform. Thus, localized areas bear disproportionate shares of these costs. In addition to this uneven cost distribution, parties damaged by subsidence associated with resource exploitation commonly are stymied from reimbursement by legal recovery systems that are in conflict with doctrines that establish rights to resource exploitation.

Many examples are available of successful efforts at federal, state, and local levels to mitigate specific subsidence problems. The efforts include public information programs, mapping programs, regulation of resource development, land-use management and building codes, market-based methods, and insurance programs. Despite these successes, continued mitigation of subsidence requires action in three additional areas.

First, basic earth-science data and information on the magnitude and distribution of subsidence are needed to recognize and to assess future problems. Such data include geodetic, geologic, hydrogeologic, hydrologic, soils, and land-use information. These data, in both map and tabular formats, help not only to address local subsidence problems but to identify national problems. Collection of these data in general should be overseen by earth-science agencies, particularly state geological surveys and the U.S. Geological Survey. Channels of communication should be developed to designate levels of government and interest groups advising them of the availability of this information.

Second, research on subsidence processes and engineering methods for dealing with subsidence is needed for cost-effective damage prevention and control. Although general understanding of subsidence processes is well developed, prediction of subsidence magnitudes, rates, and location is commonly impeded by incomplete understanding of specific details of the relevant processes and the inability to determine adequately subsurface conditions and physical properties of the deforming earth materials. Even when a specific subsidence occurrence is well understood, however, U.S. experience with engineering designs to accommodate ground deformation, and with methods to control it, is modest. New funding is needed to support research on subsidence processes by the U.S. Geological Survey, Bureau of Mines, Bureau of Reclamation, and Agricultural Research Service, and on engineering methods by the Federal Highway Administration, Corps of Engineers, Bureau of Reclamation, Federal Housing Administration, and Soil Conservation Service.

And third, although many types of mitigation methods are in use in the United States, studies of their cost-effectiveness would facilitate choices by decision makers. Such studies should be funded by the Federal Emergency Management Agency, National Science Foundation, and industrial and professional organizations.

Catastrophic subsidence takes place in areas underlain by limestone that is sufficiently massive, is pure, and has been subject to certain erosional conditions that resulted in dissolution of large segments of the rock (limestone carried away by solution) as to leave a Swiss cheese appearance in the rock itself. For example, beneath Orlando, Florida, the Floridan Aquifer is made up of three major formations:

1. The Suwannee — limestone
2. The Ocala — limestone
3. The Avon Park — limestone/dolomite

These formations include preferential flow zones that have been extensively dissolved. Large solution openings, cavities, and caves have developed in the rocks over millions of years. Subsequently, during the Pleistocene or Glacial periods, these rocks in the peninsula of Florida have periodically been invaded and covered by the sea and covered by sediments of clay, sand, silt, gravel, and some limestones. This is a general geologic setting for catastrophic subsidence. Subsequently, in time to present, the solution cavities have become filled with water and now this groundwater functions as a buoyant effect on the overlying sediments. Mother nature places things in balance. Evidence of the many solution cavity features occurs in the form of thousands of sinkholes that are filled with water and can be seen when flying over this region.

Natural phenomena, however, such as extensive periods of drought followed by heavy rains from tropical storms, can trigger catastrophic subsidence. The downward movement of the water table from shortage of rainfall over a long period and the loss of buoyant support of the water to the sediments, plus a subsequent heavy torrential rain, may act as the lubricant to the unconsolidated material causing a collapse of this material into the solution system. Where this happens, in an area of farms, commercial buildings, highways, and airports, big holes in the land surface can develop with substantial damage to property, animals, and humans. This situation is not unique to Florida but exists in many other parts of the United States, Europe, Southeast Asia, and the Middle East, in fact in approximately 25% of the area of the earth's landmass where limestones occur.

Extensive studies of this type of catastrophic event by many different scientific groups has provided substantial literature on the subject. Scientific groups involved in such studies include: A Panel on Subsidence of the Commission on Engineering and Technical Systems, Committee on Ground Failure Hazards of the National Academy of Sciences, and the Karst Commission of the International Association of Hydrogeologists. Results have been published in textbooks and volumes of symposia papers. Extensive work has been done by the U.S. Geological Survey, and the phenomena is so well known in Florida that catastrophic subsidence insurance is available. This is possible because of predictability on the basis of research regarding this phenomena and the perception of the public and the government that it comprises enough of a source of damage to warrant insurability. Some states, through cooperative programs with the USGS, have mapped in detail the areas where catastrophic subsidence can take place. Triggering effects that cause catastrophic collapse include:

1. Heavy withdrawals of groundwater by pumpage for industrial, agricultural, and municipal use.

2. Diversion of drainage in a karst area.
3. Excavation or use of heavy equipment.
4. Mine dewatering (for example, in South Africa where sinkholes caused the collapse of a three-story building where 29 men lost their lives).
5. Earthquakes.
6. Use of explosives.

A substantial amount of work is now being done on natural risk phenomena using past modeling and predicting. Attention is called to an interesting article appearing in the *Electric Power Research Institute Journal* entitled "Measuring and Managing Environmental Risk."[4]

## 4.3 CAUSES OF SUBSIDENCE

Subsidence is caused by a diverse set of human activities and natural processes, including mining of coal, metallic ores, limestone, salt, and sulfur; withdrawal of groundwater, petroleum, and geothermal fluids; dewatering of organic soils; pumping of groundwater from limestone; wetting of dry, low-density deposits, which is known as hydrocompaction; natural sediment compaction; melting of permafrost; liquefaction; and crustal deformation. This diversity and the broad range of impacts from subsidence are probably the major causes of a lack of national focus on subsidence. Instead, many industries, professions, and federal, state, and local agencies are independently involved with aspects of subsidence. Most occurrences of subsidence in the United States, however, are induced by human activity. Resource development and land-use practices, particularly underground mining of coal, groundwater and petroleum withdrawal, and drainage of organic soils, are the primary causes.

Land subsidence, the loss of surface elevation due to removal of subsurface support at rates that are of practical significance to man-made structures, affects most of the United States. Subsidence is one of the most varied forms of ground failure affecting the country, ranging from broad regional lowering of the land surface to local collapse. Its practical impact depends on the specific form of the surface deformation. Regional lowering may either aggravate the flood potential or permanently inundate an area, particularly in coastal or river settings. Local collapse may damage buildings, roads, and utilities and either impair or totally destroy them. Fortunately, subsidence is more hazardous to property than to life, because of the typically slow rates of lowering. It has caused few casualties. Subsidence, however, increases the potential for loss of life in flood-prone areas by increasing the magnitude and size of areas susceptible to flooding.

Table 4.1 shows the types of land subsidence that affect parts of at least 45 states (Figure 4.1).[4] More than 44,000 km$^2$ of land, an area equal to half the state of Maine, has been lowered.

**Table 4.1 Types of Land Subsidence**

| Types of Land Subsidence |
|---|
| Collapse into voids |
| Mining |
| Sinkhole |
| Compaction |
| Underground fluid withdrawal |
| Natural compaction |
| Hydrocompaction |
| Liquefaction |
| Drainage of organic soils |
| Melting of permafrost |
| Crustal deformation |
| Volcanism |
| Seismic |
| Aseismic |
| Postglacial deformation |

From Panel on Land Subsidence, National Research Council, National Academy Press, Washington, D.C., 1991.

The common association of land subsidence with either the exploitation of natural resources or land development practices is an important aspect of subsidence. These activities have economic benefits. Problems often arise because those who benefit from the activity that causes subsidence may not bear the full cost. In fact, some parties who incur damage may not profit at all from the activity causing the subsidence. In addition to the equity issue, specific subsidence problems may be aggravated by legal and institutional barriers that prevent legal recourse to injured parties. Legal recovery theories conflict in some states with other doctrines that establish rights to resource recovery.

### 4.3.1 Collapse into Voids — Mines and Underground Cavities[5]

Collapse of surficial materials into underground voids is the most dramatic kind of subsidence. Buildings and other engineered structures may be damaged or destroyed, and land may be removed from productive use by such ground failure.

Underground excavations have been constructed in the United States since the early 1700s. Most of the voids with which subsidence has been associated in the United States were created by coal mining. Abandoned tunnels and underground mining of metallic ores, limestone, and salt contribute to a much smaller extent, although associated problems may be severe in some regions.

In general, coal-mine subsidence is caused by collapse of the mined-out or tunneled void. It occurs as both steep-sided pits (Figure 4.2) and broad,

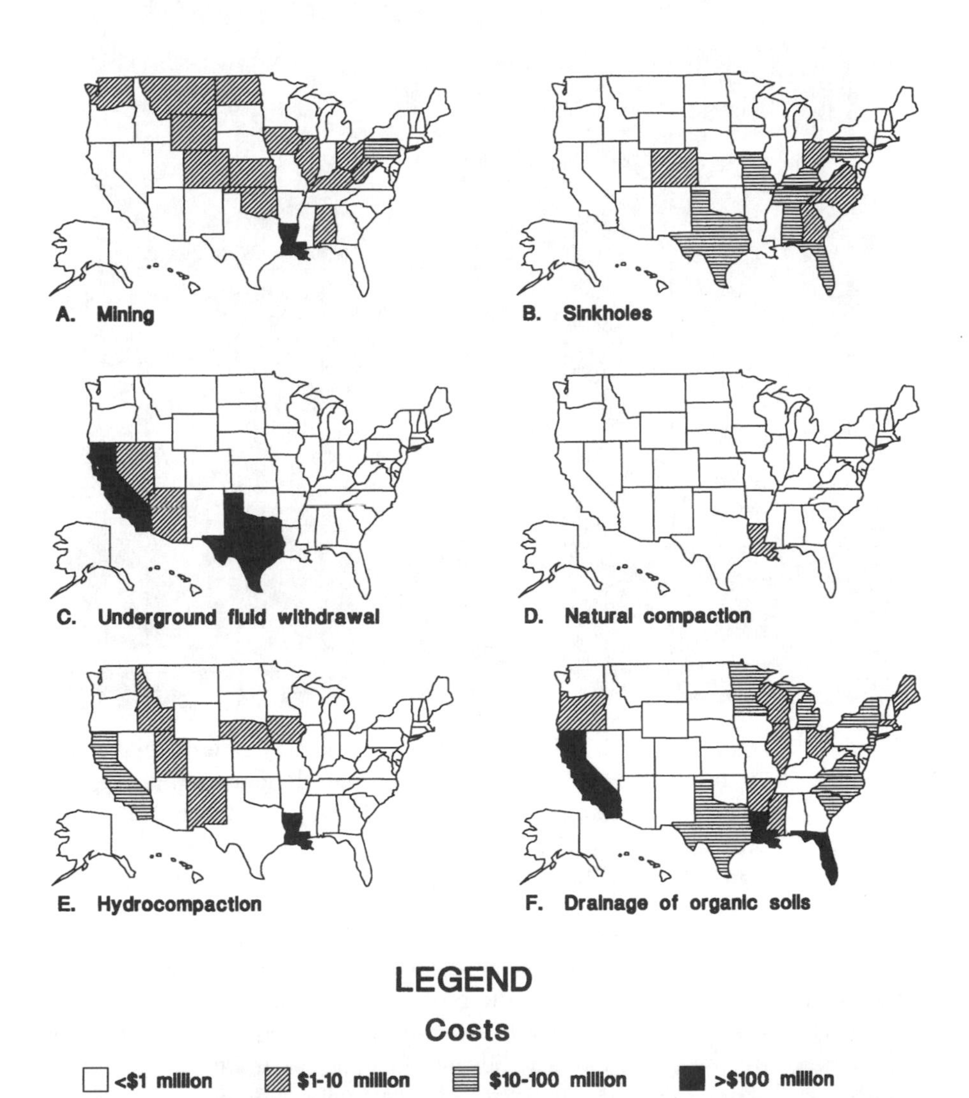


Figure 4.1 National distribution of subsidence problems by state. Costs were compiled from published and unpublished sources for the purpose of providing an order-of-magnitude, state-by-state comparison. Only relative importance is suggested by maps, because the time periods on which estimates are based vary by state, and costs were not converted to constant dollars. In general, costs are conservative estimates. (From Panel on Land, Subsidence, National Research Council, National Academy Press, Washington, D.C., 1991.)[5]

**Figure 4.2 Oblique aerial view of subsidence pits above abandoned coal mines in the Sheridan, Wyoming, area. Diameter of pits ranges from 5 to 50 m. (From Panel on Land Subsidence, National Research Council, National Academy Press, Washington, D.C., 1991.) (Photograph courtesy of C. Richard Dunrud.)[5]**

gentle depressions.[5] Subsidence depends on the number, type, and extent of the voids. For example, it is a planned consequence of the longwall mining method for coal, in which most of the coal seam is removed along a single face, the longwall. By this method, the roof above the mined-out seam is allowed to collapse as the longwall advances laterally as mining progresses. Subsidence above longwall mines is rapid, generally ending within a few months after the removal of subsurface support. Subsidence above mines with partial extraction is usually unplanned. By this method, only parts of the coal, the rooms, are removed. The unmined portions, the pillars, are left to provide support. Collapse into the rooms occurs when the pillars, floors, or ceilings deteriorate. Subsidence resulting from collapse into rooms may take years to decades to manifest itself. Examples of collapse occurring 100 years after mines were abandoned have been documented.

Coal is found in 37 states and mined underground in 22 states of approximately 32,000 km$^2$. Approximately 8,000 km$^2$ of the undermined area, most of which is in the eastern United States, already has experienced subsidence.[5] The U.S. Bureau of Mines estimates that 1,600 km$^2$ of land in urban areas is threatened.[6] Seventy-one percent of this area is in Pennsylvania, Illinois, and West Virginia.[5]

### 4.3.2 Sinkholes[5]

The sudden formation of sinkholes — catastrophic subsidence — is usually caused by movement of overburden into underlying cavities in soluble bedrock (Figure 4.3). Failure of the bedrock is rarely believed to be a major factor in catastrophic subsidence. Most catastrophic subsidence in the United States is associated with carbonates such as limestone, but occasionally it is associated with evaporites such as gypsum and halite (Figure 4.4). Although most historical collapses are man-induced, the cavities in the bedrock usually antedate human activities. This is particularly true of carbonates, because rates of solution are so low. Cavities in halite can be an exception because of its high solubility. For example, several dozen sinkholes have formed in the last 30 years in Kansas as a result of solution of salt beds by leaks through casings of brine-disposal wells. A recent example is a 60 m wide and 33 m deep sinkhole that formed in the summer of 1988 near Macksville, Kansas.[7] Catastrophic subsidence is most commonly induced by water-table lowering, rapid water-table fluctuation, diversion of surface water, construction, use of explosives, or impoundment of water.

**Figure 4.3 Catastrophic subsidence caused by collapse of overburden into voids in limestone near Bartow, Florida, that destroyed two houses. (Photograph courtesy of Florida Sinkhole Research Institute.)[5]**

**Figure 4.4 August 11, 1983, catastrophic subsidence over the Boling salt dome, Texas. Diameter of depression is about 75 m. (Photograph courtesy of Boyd V. Dreyer.)[5]**

Davies et al. indicate that more than 1.4 million $km^2$ of land in 39 states is underlain by cavernous limestone and marble.[8] More than 30,000 $km^2$ of this area lies beneath Standard Metropolitan Statistical Areas inhabited by 33 million people. Fortunately, only a small portion is actually underlain by voids and at risk. Newton estimates that more than 6,000 collapses have occurred in the eastern United States since about 1950.[9] The states with the largest number of active sinkholes include Alabama, Florida, Georgia, Indiana, Missouri, Pennsylvania, and Tennessee.

### 4.3.3 Sediment Compaction[5]

Sediment compaction typically causes broad regional subsidence. Exceptions include ground rupture and hydrocompaction. Rates of subsidence usually are low, ranging from a few millimeters to centimeters per year, but total subsidence may reach several meters as it accumulates over decades.

### 4.3.4 Underground Fluid Withdrawal[5]

The weight of the overburden above underground fluid reservoirs is supported by both fluid pressures and stresses transmitted through the solid framework of the reservoir soil or rock. When fluids are withdrawn, fluid pressures decline and support of the overburden is transferred to the solid framework.

If the reservoir soil or rock is compressible, large and permanent loss of pore volume or compaction will occur as it adjusts to the new stresses. In geothermal reservoirs, significant thermal contraction also may occur as the reservoir cools during exploitation.

Most of this type of subsidence in the United States is caused by pumping of groundwater and petroleum. More than 31 areas in 7 states have subsided. The two largest areas are in the San Joaquin Valley, California, and Houston, Texas, areas, where 13,500 and 12,000 km$^2$, respectively, have subsided because of groundwater withdrawal. Maximum elevation loss from this type of subsidence has been 9 m in the San Joaquin Valley.

Two coastal areas in California and Texas where subsidence caused or threatened inundation and increased flooding potential have suffered the most from this type of subsidence. Petroleum withdrawal in Long Beach, California, caused parts of the city's harbor facility to subside almost 9 m from 1937 to 1966. Groundwater withdrawal in Houston, Texas, has caused some coastal areas to subside by more than 2 m. About 80 km$^2$ of land has been inundated, and several hundred square kilometers, including the 500-unit Brownwood subdivision in Baytown, which was abandoned in 1983, have been added to the area susceptible to flooding by storm surges.

Inland areas are not immune to damage from this subsidence. Changes of surface gradients can affect either the design or operation of canals. For example, canals in the State Water and the Central Valley Projects in California and the Central Arizona Project have been affected. Other causes of damage in inland areas include subsurface deformation and ground rupture. Subsurface deformation shears or crushes well casings and diminishes the productivity of water and oil wells. Ground rupture, including faults and earth fissures, is particularly devastating to man-made structures. Faulting has damaged hundreds of houses in Houston, Texas, and in California in 1963 caused the catastrophic failure of the Baldwin Hills reservoir claiming five lives. Earth fissures (Figure 4.5) are an increasingly common occurrence as groundwater withdrawals increase in alluvial basins in the desert parts of the Sun Belt.[5]

### 4.3.5 Natural Compaction[5]

Sediments compact naturally as they are buried by younger sediment. Probably nowhere in North America is natural subsidence occurring more rapidly than in the Mississippi River Delta area of southern Louisiana, where about 3,900 km$^2$ of land is subsiding, at least in part through natural compaction. Estimated average rates of subsidence range from 8 to 11 mm per century.[10] Maximum rates measured by geodetic surveys are about 12 mm per year. The abandoned town of Balize, approximately 130 km southeast of New Orleans on the tip of the Delta, is an example of the long-term implication of this subsidence. Balize, which was abandoned during a yellow fever epidemic

**Figure 4.5 Tension crack in southcentral Arizona that has been enlarged by erosion into 1-m-wide gully. (Photography courtesy of Thomas L. Holzer.)[5]**

in 1888, was more than 1.2 m below marsh level in 1934.[11] Today the abandoned town sits more than 3 m below sea level.

Increased flooding potential is the principal impact of this type of subsidence, because affected areas commonly are low lying and naturally subject to flooding. Thus, subsidence exacerbates a preexisting problem. The flood problem is particularly acute in coastal areas where long recurrence intervals between large storm surges and tidal floods diminish public perception of the problem. Documenting subsidence problems in coastal areas may be difficult if other types of subsidence are occurring and sea level is changing. For example, subsidence caused by drainage of organic soil and withdrawal of underground fluids is common in many deltaic areas. A well-studied example of this complexity is Venice, Italy, where increases in the incidence of tidal flooding prompted an investigation, which discovered that the mean elevation above sea level had decreased from 130 to 110 cm since the turn of the century. About 14, 41, and 45% of the 20 cm average elevation loss was attributed to natural compaction, sea-level rise, and withdrawal of groundwater, respectively.[12]

Another very important impact from this subsidence is the destruction of productive estuarine marsh and coastal wetlands by either inundation or erosion. Thousands of hectares of low-lying coastal land along the Gulf of Mexico area is converted each year to open water by natural subsidence (Figure 4.6). This process, referred to as coastal land loss, results in a significant loss of habitat for birds, fish, crustaceans, and reptiles and has a profound impact on

**Figure 4.6 Regional land subsidence and diversion of sediment carried by Mississippi River directly to the Gulf are contributing to the disappearance of more than 130 km$^2$ of wetlands annually in southern Louisiana.[5]**

the commercial fishing, shrimping, oystering, and fur trapping industries. In addition, salt-water intrusion into these areas destroys agricultural usage.

The most severe land-loss problem in the United States is in southern Louisiana (Figure 4.6).[5] Channelization of the Mississippi River has caused riverborne sediment, which normally offsets land loss by replenishing beaches and wetlands, to discharge directly offshore into the Gulf of Mexico. The rate of land loss in southern Louisiana currently is about 130 km$^2$ per year; approximately 3,200 km$^2$ of land has been lost in the last 80 years. Nationally, more than 150,000 km$^2$ of coastal marsh has been lost since 1954.[13]

### 4.3.6 Hydrocompaction[5]

Dry, low-density, fine-grained sediment may be susceptible when wetted to a loss of volume known as hydrocompaction (Figure 4.7). These sediments, known as collapsible soils, generally are of two types: mudflow deposits in alluvial fans and wind-deposited, moisture-deficient silt called loess. Most collapsible soils have anomalously low densities because they remained moisture deficient throughout their postdepositional history. When water percolates through the root zone into this type of sediment, the soil structure collapses and the soil compacts. Very localized subsidence, typically 1 to 2 m, may result.

**Figure 4.7 Hydrocompaction of test plot in San Joaquin Valley, California. Water has infiltrated from pond and caused collapsible soils to compact as they became wet. Note extensive ground cracking in background around margin of subsidence depression. Poles were used to measure compaction at different depth intervals. (Photograph courtesy of California Department of Water Resources.)[5]**

Damaging hydrocompaction has been reported in 17 states. The three largest affected areas are the alluvial slopes of the western San Joaquin Valley and loess-covered areas in the Missouri River basin and Pacific Northwest. The major impact has been on design and operation of hydraulic structures — canals and dams. Locally significant impact has been incurred by buildings and highways. Irrigation for agriculture also has caused differential subsidence that required re-leveling of fields.

### 4.3.7 Organic Soil[5]

Drainage of organic soil, particularly peat and muck, induces a series of processes, including biological oxidation, compaction, and desiccation, that reduce the volume of the soil. Biological oxidation usually dominates in warm climates (Figure 4.8). The principal areas of organic soil subsidence in the United States are the greater New Orleans, Louisiana, area; the Sacramento-San Joaquin River Delta, California; and parts of the Florida Everglades. Maximum observed subsidence is 6.4 m in the Sacramento-Jan Joaquin River

**Figure 4.8 Concrete monument set on rock beneath organic soil in Belle Glade, Florida, an active subsidence area. Elevation painted on the monument is feet above mean sea level. Dates show former elevation of land surface. Photograph was taken October 1987. (Courtesy of George H. Synder.)[5]**

Delta. About 9,400 km$^2$ of land underlain by organic soil has subsided in the United States because of drainage. An even larger area is susceptible to subsidence. About 101,000 km$^2$ of the conterminous United States is covered by peak and muck soils;[14] more than 26,000 km$^2$ of organic wetlands is in standard metropolitan statistical areas.

## 4.4 DAMAGE COST AND LEGAL ASPECTS OF LAND SUBSIDENCE[5]

The average annual damage cost from all types of subsidence is conservatively estimated to be at least $125 million (Table 4.2). The costs are dominated by subsidence from underground mining of coal, drainage of organic soils, and withdrawal of underground water and petroleum. These costs consist primarily of direct structural and property losses and depreciation of land values, but they also include business and personal losses that result during periods of repair. Although total annual damage cost of subsidence to the nation is small relative to the nation's economy, subsidence imposes substantial costs on individual cities and neighborhoods.

**Table 4.2 Estimated Annual Losses (in Millions) from Land Subsidence**

| | |
|---|---|
| Mines | $30 |
| Sinkholes | $10 |
| Underground fluid withdrawal | $35 |
| Natural compaction | $10 |
| Hydrocompaction | Not available |
| Organic soils | $40 |
| Total | $125 |

Sources: Jones,[15] Newton,[9] Prokopovich and Marriott,[16] and HRB-Singer.[17]

The dollar value of economic losses from subsidence shown in Table 4.2 reveals only part of the nation's subsidence problem. Inequitable aspects of both economic incentives and the American legal structure are major factors in the nation's subsidence problem, particularly where subsidence is caused by resource extraction or land use. Often existing legal or market incentives do not encourage the developer of a resource to consider the costs imposed on others who do not receive the benefit of development. These are termed external costs.

External costs cause misallocation of society's resources. Social costs, the costs to society as a whole, are greater than private costs when there are external costs. Thus, a mine operator who has caused subsidence on land that he does not own receives all of the benefits of ore production, but he pays only the private cost, a portion of the social cost. The surface property owner affected by subsidence, however, receives none of the benefits and bears the external costs from the actions of the mine operator. This misallocation of resources is corrected when the mine operator pays all of the social costs, private and external, of the mine operation.

In evaluating external costs, several dimensions are worth noting. External costs from subsidence can stem from past activities, as in the case of abandoned mines. These activities can be viewed as a one-time imposition of some level of external costs on the future. Alternatively, past and current activities can cause a continuing subsidence problem. These types of external costs are direct. External costs from subsidence also may be indirect. These include situations where subsidence increases the potential for economic damage from other natural events. Increased susceptibility to flooding from storm surges of low-lying coastal areas is an example.

Another category of external costs potentially presents a special dilemma. This involves situations where the external cost is borne in perpetuity by future generations. For example, consider where drainage of organic soil causes land to subside below sea level or alters the freshwater-saltwater balance. In such cases, land may simply disappear because of inundation, or the ecology may change; and thus a permanent loss is passed on to future generations.

At common law the principal means to prevent externalities by transferring external costs to the actor who created them is the tort system. The tort system generally has a twofold purpose: (1) to compensate worthy victims for damages they have suffered by the negligence of others (thus imposing the costs of the damage on the actor who created them), and (2) to act as a deterrent to negligent activity, that is, to encourage the actor to decide not to engage in an activity if it creates a cost that can be transferred back to that actor in a tort or lawsuit.

There are at least three reasons the legal system does not provide an adequate mechanism for transferring subsidence costs onto the actors that create them.[18] First, in many states, such as those applying the English common law doctrine of absolute ownership of groundwater, the legal system does not allow a party injured by subsidence to recover from the person causing the subsidence. Similarly, the legal system may not provide a recovery mechanism for a surface owner injured by subsidence caused by underground mining. Second, even in those jurisdictions that allow recovery of subsidence damage, lack of public understanding about the causes and effects of subsidence may prevent members of the public from recognizing their damages and the identity of those who caused them. Even if some members of the public can successfully recover their losses, any members of the public who do not or cannot recover their losses will leave external costs that are not internalized. Third, because many of the damages caused by subsidence are indirect, the full cost of subsidence may not be recognized. For example, a flood may inundate 1,000 hectare when it would have inundated only 500 ha in the absence of subsidence. The owners of the property whose inundation was caused by the subsidence may not recognize the role played by subsidence in expanding the flooded area. From an economic point of view, these indirect costs are as significant as direct costs, and, if they are not recognized and recovered, they will not be internalized.

When traditional legal mechanisms fail to provide adequate control over external economic costs, governments may create statutory or regulatory controls to compensate for the deficiencies in the common law. These are not always limited to recovery for negligent performance of an activity. Legislation can remove the element of negligence by establishing strict liability for surface damages. Alternatively, administrative regulation may be used to address local subsidence problems by either imposing taxes that internalize the external costs or regulating resource exploitation or usage.[18]

Given that subsidence commonly is a phenomenon characterized by imposed external cost due to legal conflicts or market failure, what are the barriers to mitigating external costs? At least three barriers inhibit the design of appropriate public policy measures. First, a significant barrier is simply the lack of availability and public understanding of the scientific literature; it is difficult to mitigate the unknown. Second, subsidence is not typically viewed as a catastrophic event or even necessarily the primary cause of large economic damages. As such, on the agenda of preparations for natural hazards, subsidence is low. Third, the potential for long delays in the observance of phenomena causes significant problems for public policy response.

## REFERENCES

1. National Research Council, Reducing Disasters' Toll, National Academy Press, Washington, DC, 1989.
2. USGS, Earthquakes, U.S. Geological Survey Information pamphlet, 1991.
3. Sprigg, W.A., Weather and Outbreaks of Disease, in WSTB, A newsletter from the Water Science and Technology Board, National Research Council, 13, 2, April/May 1996.
4. EPRI, Measuring and managing environmental risk, Electric Power Research Institute, July/August 1985.
5. National Academy of Sciences/National Resource Council Committee on Ground Failure Hazards Mitigation Research, Mitigating Losses from Land Subsidence in the United States, National Academy Press, Washington, DC, 1991, 58 p.
6. Johnson, W. and Miller, G.C., Abandoned Coal-Mined Lands: Nature, Extent, and Cost of Reclamation. U.S. Bureau of Mines Special Publication 6-79, 1979.
7. Kansas Sinkholes, *Geotimes*, 33(11), 1988, p. 15.
8. Davies, W.E., Simpson, J.H., Ohlmacher, G.C., Kirk, W.S., and Newton, J.G., Map Showing Engineering Aspects of Karst in the United States, U.S. Geological Survey (USGS) Open-File Report 76-625, 1976, 1:7,500,000 scale.
9. Newton, J.G., Natural and Induced Sinkhole Development in the Eastern United States, International Association of Hydrological Sciences Publication No. 151, 1986.
10. Penland, S., Ramsey, K.E., McBride, R.A., Mestayer, J.T., and Westphal, K.A., Relative Sea Level Rise and Delta-Plain Development in Terrebonne Parish Region, 1988, Louisiana Geological Survey, Coastal Geology Publication Technical Report No. 4, 1988.
11. Russell, R.J., Howe, H.V., McGuirt, J.H., Dohm, C.F., Hadley, W., Kniffen, F.B., and Brown, D.A., Lower Mississippi River Delta, Reports on the Geology of Plaquemines and St. Bernard Parishes, Louisiana Geological Survey Bulletin No. 8, 1936.
12. Gatto, P. and Carbognin, L., The Lagoon of Venice: Natural Environment Trend and Man-Induced Modification, *Hydrological Science Bulletin,* 16, 4, 1981, pp. 379–391.
13. Gosselink, J., Tidal Marshes, The Boundary Between Land and Water, U.S. Fish and Wildlife Service, Office of Biological Services, 1980.
14. Stephens, J.C., Allen, L.H., Jr., and Chen, E., Organic soil subsidence, in *Man-Induced Land Subsidence,* Holzer, T.L., Ed., Geological Society of America, Reviews in Engineering Geology, Vol. VI, 1984, pp. 107–122.
15. Jones, L.L., External costs of surface subsidence: Upper Galveston Bay, Texas, International Association of Hydrological Sciences Publication No. 121, 1977.
16. Prokopovich, N.P. and Marriott, M.J., Cost of Subsidence to the Central Valley Project, California, Association of Engineering Geologists Bulletin, 20(3), 1983, pp. 325–332.

17. HRB-Singer, Inc., Technical and Economic Evaluation of Underground Disposal of Coal Mining Wastes, Report prepared for U.S. Department of the Interior, Bureau of Mines, Contract No. J0285008, 1980.
18. Amandes, C.B., Controlling land surface subsidence — A proposal for a market based regulatory scheme, University of California at Los Angeles Law Review 31, 6, 1984, pp. 1208–1246.

# 5 KINDS OF WASTE AND PHYSIOGRAPHY OF WASTE DISPOSAL SITES

## 5.1 KINDS AND SOURCES OF WASTES

All kinds of wastes, solid, liquid, and gaseous, have affected the safety and use of our water resources. Some of these wastes are of hazardous nature to man and wildlife. The different kinds of wastes and their main sources are given and discussed in brief in this chapter.

The United States currently faces a very large groundwater contamination problem. Although the total number of contaminated sites is unknown, estimates of the total number of waste sites where groundwater and soil may be contaminated range from approximately 300,000 to 400,000. Recent estimates of the total cost of cleaning up these sites over the next 30 years have ranged as high as $1 trillion.[1]

Recent studies have raised troubling questions about whether existing technologies are capable of solving this large and costly problem. As a result of these studies, there is almost universal concern among groups with diverse interests in groundwater contamination — from government agencies overseeing contaminated sites to industries responsible for the cleanups, environmental groups representing affected citizens, and research scientists — that the nation might be wasting large amounts of money through ineffective remediation efforts. At the same time, many of these groups are concerned that the health of current and future generations may be at risk if contaminated groundwater cannot be cleaned up to make it safe for drinking.

Theoretically, restoration of contaminated groundwater to drinking water standards is possible. However, cleanup of contaminated groundwater is inherently complex and will require large expenditures and long time periods, in some cases, centuries. The key technical reasons for the difficulty of cleanup include the following:[1]

- *Physical heterogeneity*: The subsurface environment is highly variable in its composition. Very often, a subsurface formation is composed of layers of materials with vastly different properties, such as sand and gravel, and even within a layer the composition may vary over distances as small as a few centimeters. Because fluids can move only through the pore spaces between the grains of sand and gravel or through fractures in solid rock and because these openings are distributed nonuniformly, underground contaminant migration pathways are often extremely difficult to predict.
- *Presence of nonaqueous-phase liquids* (NAPLs): Many common contaminants are liquids that, like oil, do not dissolve readily in water. Such liquids are known as NAPLs, of which there are two classes: light NAPLs (LNAPLs), such as gasoline, are less dense than water; dense NAPLs (DNAPLs), such as the common solvent trichlorethylene, are more dense than water. As a NAPL moves through the subsurface, a portion of the liquid will become trapped as small immobile globules, which cannot be removed by pumping but can dissolve in and contaminate the passing groundwater. Removing DNAPLs is further complicated by their tendency, due to their high density, to migrate deep underground, where they are difficult to detect and where they remain in pools that slowly dissolve in and contaminate the groundwater.
- *Migration of contaminants to inaccessible regions*: Contaminants may migrate by molecular diffusion to regions inaccessible to the flowing groundwater. Such regions may be microscopic (for example, small pores within aggregated materials) or macroscopic (for example, can serve as long-term sources of pollution as they slowly diffuse back into the cleaner groundwater).
- *Sorption of contaminants to subsurface materials*: Many common contaminants have a tendency to adhere to solid materials in the subsurface. These contaminants can remain underground for long periods of time and then be released when the contaminant concentration in the groundwater decreases.
- *Difficulties in characterizing the subsurface*: The subsurface cannot be viewed in its entirety, but is usually observed only through a finite number of drilled holes. Because of the highly heterogeneous nature of subsurface properties and the spatial variability of contaminant concentrations, observations from sampling points cannot be easily extrapolated, and thus knowledge of subsurface characteristics is inevitably not complete.

Both the complex properties of the subsurface environment and the complex behavior of contaminants in the subsurface interfere with and retard the ability of conventional pump-and-treat systems to achieve drinking water standards for contaminated groundwater.

At hazardous waste sites nationwide, industries and government agencies are spending millions of dollars trying to clean up contaminated groundwater. These cleanups are required by federal and state laws passed in the last two decades — mostly in response to public concern that drinking contaminated groundwater may affect public health and the environment. The laws require that, in most instances, the contaminated groundwater be restored to a condition that meets state and federal drinking water standards.

Recently, some have begun to question current approaches to groundwater cleanup. Evidence suggests that restoring contaminated groundwater to drinking water standards poses considerable technical challenges that may sometimes be insurmountable. For example, at one New Jersey site, a computer manufacturing company spent $10 million removing toxic solvents from groundwater, but not long after the cleanup system was shut down the solvent concentrations in some locations returned to levels higher than before cleanup began. This company's effort and others like it have raised concern about whether the amount spent to clean up groundwater is proportionate to the benefits society receives. Businesses and government agencies paying for the cleanups are calling for reconsideration of whether returning all contaminated groundwater to drinking water standards is a realistic goal.

In 1980, prompted by the Love Canal incident, Congress, for the first time, made groundwater cleanup a high national priority with the passage of the Comprehensive Environmental Response, Compensation, and Liability Act (CERCLA), commonly known as the Superfund Act. CERCLA established a $1.6 billion federal fund (which has since grown to $15 billion), the Superfund, to pay for cleaning up abandoned hazardous waste sites.[2] CERCLA also provided authority for the Environmental Protection Agency (EPA) to sue parties responsible for the contamination to recover cleanup costs; these groups have since become known as "potentially responsible parties."

In 1984, Congress broadened the nation's groundwater cleanup program by amending the Resource Conservation and Recovery Act (RCRA) to require cleanup of contamination at active facilities that treat, store, or dispose of hazardous waste. To continue handling wastes, operators of active RCRA site must agree to clean up existing pollution. RCRA also covers cleanup of contamination from leaking underground storage tanks containing petroleum products and other organic liquids.

Since the passage of CERCLA and the 1984 RCRA amendments, virtually all states have enacted laws granting them authority to require cleanup of sites with contaminated groundwater.[3] CERCLA and RCRA have strongly influenced the state laws, although some state laws are more stringent than the federal versions.

The primary type of groundwater contamination of concern in the United States today is contamination from hazardous chemicals. The use of such chemicals is ubiquitous: substances identified in contaminated groundwater are used in everything from lumber treating to electronics manufacturing, fuels, food production, and agricultural chemical synthesis. When used as storage or disposed of on land, these chemicals may eventually migrate to the groundwater.

Common causes of groundwater contamination are accidental spills; intentional dumping; and leaks in storage tanks, industrial waste pits, and municipal or industrial landfills. In addition, significant quantities of contaminants may be released through routine activities such as washing engines and rinsing tanks. Standard application of agricultural chemicals is also a source of groundwater contamination. The EPA estimates that about 1% of all drinking water wells in the United States exceed a health-based limit for pesticides.[4] Although pesticide application is a potentially important source of contamination, this report focuses on the point sources of contamination found at hazardous waste sites and other sites where hazardous chemicals have leaked or spilled into the environment. Because point sources affect only a limited area, they present a more manageable problem than contamination of large areas of land with agricultural chemicals, which might far exceed the limits of cleanup technologies. Table 5.1 ranks chemicals found at hazardous waste sites in order of prevalence and gives common sources for these chemicals.

Because of the widespread use and disposal of hazardous chemicals on land, the groundwater contamination problem is potentially very large. However, estimates of the total number of contaminated sites have varied.

Table 5.2 shows estimates of the number of sites in each of these categories as compiled from three different sources. As this table shows, the total number of sites where groundwater may be contaminated is likely to be in the range of 300,000 to 400,000. However, it is extremely important to recognize that the magnitude of the contamination problem varies widely at these sites. Groundwater contamination from a single leaking underground storage tank at a gas station may affect a relatively small area. On the other hand, contamination of CERCLA sites and at major DOE installations may be widespread and very difficult to clean up. The differences between these types of sites are illustrated by the costs of cleaning them up. According to recent EPA data, the average cost of cleaning up a leaking underground storage tank is $100,000, while the average cost of cleaning up a Superfund site is $27 million.[5] According to the EPA (1993),[5] the cost of cleaning up underground storage tank leaks varies widely and may be as low as $2,000 for some sites and as high as $1 million for others. By far the bulk of the sites listed in Table 5.2 are contaminated from leaking underground storage tanks. The larger sites posing the greatest hazard to public health and the environment represent a relatively small portion of the total potential number of sites.

### 5.1.1 Solid Wastes

It is expected that by the end of this century, the solid wastes produced by the USA alone will approach 475 million tons.[6] This is equivalent to 17 lbs/capita/day. Table 5.3 shows the sources of this waste in percentages.

**Table 5.1 The 25 Most Frequently Detected Groundwater Contaminants at Hazardous Waste Sites**

| Rank | Compound | Common Sources |
|---|---|---|
| 1 | Trichloroethylene | Dry cleaning; metal degreasing |
| 2 | Lead | Gasoline (prior to 1975); mining; construction material (pipes); manufacturing |
| 3 | Tetrachloroethylene | Dry cleaning; metal degreasing |
| 4 | Benzene | Gasoline; manufacturing |
| 5 | Toluene | Gasoline; manufacturing |
| 6 | Chromium | Metal plating |
| 7 | Methylene chloride | Degreasing; solvents; paint removal |
| 8 | Zinc | Manufacturing; mining |
| 9 | 1,1,1-Trichloroethane | Metal and plastic cleaning |
| 10 | Arsenic | Mining; manufacturing |
| 11 | Chloroform | Solvents |
| 12 | 1,1-Dichloroethane | Degreasing; solvents |
| 13 | 1,2-Dichloroethene | Transformation product of 1,1,1-trichloroethane |
| 14 | Cadmium | Mining; plating |
| 15 | Manganese | Manufacturing; mining; occurs in nature as oxide |
| 16 | Copper | Manufacturing; mining |
| 17 | 1,1-Dichloroethene | Manufacturing |
| 18 | Vinyl chloride | Plastic and record manufacturing |
| 19 | Barium | Manufacturing; energy production |
| 20 | 1,2-Dichloroethane | Metal degreasing; paint removal |
| 21 | Ethylbenzene | Styrene and asphalt manufacturing; gasoline |
| 22 | Nickel | Manufacturing; mining |
| 23 | Di(2-ethylhexyl)phthalate | Plastics manufacturing |
| 24 | Xylenes | Solvents; gasoline |
| 25 | Phenol | Wood treating; medicines |

*Note:* This ranking was generated by the Agency for Toxic Substances and Disease Registry using groundwater data from the National Priorities List of sites to be cleaned up under CERCLA. The ranking is based on the number of sites at which the substance was detected in groundwater.

The solid wastes produced by sewerage systems are not included in this section but are incorporated with the liquid waste in the following section. Table 5.3 shows that the daily waste production from nonindustrial and municipal sites is over 50% of the total waste generated. Urban dwellers produce more waste than their rural counterparts. Table 5.4 shows the material components of this waste stream.

**Table 5.2 Number of Hazardous Waste Sites Where Groundwater May Be Contaminated**

| | Source of Estimate | | |
|---|---|---|---|
| Site category | EPA, 1993 | Russell et al., 1991 | Office of Technology Assessment, 1989 |
| CERCLA National Priorities List | 2,000 | 3,000 | 10,000 |
| RCRA corrective action | 1,500–3,500 | NA | 2,000–5,000 |
| Leaking under-ground storage tanks | 295,000 | 365,000 | 300,000–400,000 |
| Department of Defense | 7,300 (at 1,800 installations) | 7,300 | 8,139 |
| Department of Energy | 4,000 (at 110 installations) | NA | 1,700 |
| Other federal facilities | 350 | NA | 1,000 |
| State sites | 20,000 | 30,000 | 363,000–466,000 |
| Total | 330,150–332,150 | NA | 363,000–466,000 |

*Note:* The numbers presented in this table are estimates, not precise counts. In addition, at some of these sites, groundwater may not be contaminated. For example, the EPA (1993) estimates that groundwater is contaminated at 80% of CERCLA National Priorities List sites. There is also some overlap in site categories. For example, 7% of RCRA sites are federal facilities, and 23 DOE sites are on the CERCLA National Priorities List (EPA, 1993). NA indicates that an estimate comparable to the other estimates is not available from this source.

**Table 5.3 Percentage of Solid Wastes Generated in the USA/Day**

| | |
|---|---|
| Residential | 34 |
| Commercial | 14 |
| Bulky Wastes | 4 |
| Industrial | 48 |
| Total | 100 |

Most of this refuse has been disposed of in landfill sites within or near the community generating the waste, or by incineration. In the past, most landfill sites were little more than open dumps and the incinerators produced ash and hazardous and noxious gases which were introduced into the atmosphere.

Future solid waste generation in industrial and institutional sectors are dependent upon the types of industry located in the urban areas and the number of employees they bring into these areas. Large industrial development could

**Table 5.4 Sources and Percentage of Refuse by Weight of an Average Municipal Refuse from Studies Made by Purdue University[6]**

| Component rubbish | % of all refuse by weight |
|---|---|
| Paper | 42.0 |
| Wood | 2.4 |
| Grass | 4.0 |
| Brush | 1.5 |
| Geens | 1.5 |
| Leaves | 5.0 |
| Leather | 0.3 |
| Rubber | 0.6 |
| Plastics | 0.7 |
| Oil, paints | 0.8 |
| Linoleum | 0.1 |
| Rags | 0.6 |
| Street sweepings | 3.0 |
| Dirt | 1.0 |
| Unclassified | 0.5 |
| Total | 64.0 |
| Food wastes | 12.0 |
| Noncombustibles (metals, glass, and ashes) | 24.0 |
| Total | 100.0 |

greatly increase quantities of waste, and the population associated with these industries will also increase quantities of waste from hospitals, schools, and other institutions.

The main sources of solid wastes in industry are wood factories, paper mills, steel and aluminum factories, all kinds of packing companies, glass factories and industries that deal with metallurgy, food, and chemicals.

Urban areas generate a variety of solid waste materials from households, hospitals, and clinics which may include bottles, syringes, toxic materials, radioactive substances, and dressings.

Agricultural solid wastes may be produced in rural areas as crop wastes, agricultural processing wastes, animal manure, and hazardous pesticide containers. Some crop wastes are disposed of by being plowed back into the soil, burned on site, or used as feed; however, in all instances, the waste moves in a continuous cycle and eventually into the air, soil, or water. Returning the wastes to the soil is a low cost method that returns organic matter and nutrients to the land. Burning crop wastes is an effective method of controlling plant diseases and weeds, and eliminates excessive crop wastes detrimental to the soil if plowed under. However, burning crop wastes creates problems of air pollution.

Abandoned transformers, batteries, vehicles, large appliances, and furniture are considered solid wastes which are produced in both urban and rural areas.[7]

Dumping these solid wastes in disposal sites[8] or landfills may affect the water resource systems. If they are near streams or above groundwater aquifers, the leachate produced from these sites will move into the subsurface[9] and groundwater resources. A more complete analysis of this subject is given in Chapter 7.

### 5.1.2 Liquid Wastes

Liquid wastes are the main sources which contaminate the water resource systems either through the surface water networks such as rivers, canals, and drains, or into groundwater aquifers because they can easily migrate downward into the porous soils and fractured rocks to the groundwater reservoirs.

Some principal waste sources contaminating water resource systems include:

1. industrial waste water that is contained in surface impoundments (lagoons, ponds, pits, and basins);
2. liquids derived from municipal and industrial solid refuse and sludge that are disposed of on land;
3. sewage wastes from urban areas that discharged to septic tanks and cesspools;
4. storm water runoff in rural and urban areas that is collected, treated, and discharged to the land;

5. brine from petroleum exploration and development that is injected into the ground or stored in evaporation pits;
6. solid and liquid wastes from mining operations that are disposed of in tailing piles or lagoons or discharged to land;
7. domestic, industrial, agricultural, and municipal waste water that is disposed of in wells; and
8. animal feed lot waste that is disposed of on land and in lagoons.

The sources of potential contaminants and their various routes to the water resources systems are shown in Figure 5.1. Table 5.5 lists the waste disposal sources and their relative impact on the water resources environment.

Industrial wastewater impoundments[11] are a serious source of groundwater contamination because of their large number and their potential for leaking hazardous substances which are relatively mobile in the groundwater environment. In some heavily industrialized sections, regional problems of groundwater contamination have developed where the areal extent and the toxic nature of the contaminants have prevented the use of groundwater from shallow aquifers. Contaminated groundwater originating from impoundments at industrial establishments can be even more important because of the potential for migrating to local water supply wells.[12]

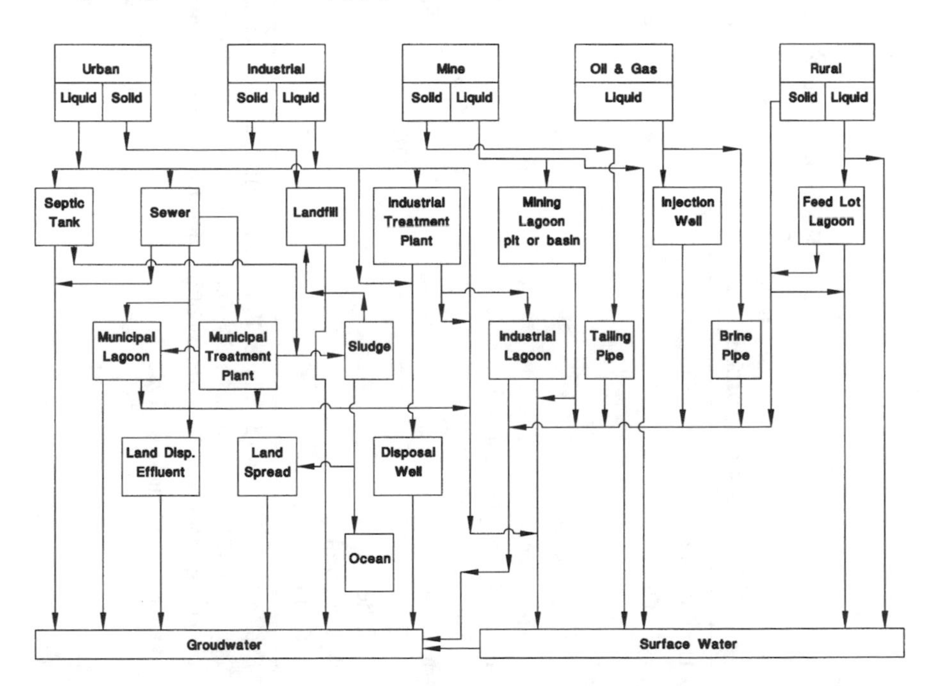


**Figure 5.1 Waste disposal routes of contaminants from solid liquid wastes.**

**Table 5.5 Waste Disposal Sources and Their Impacts**

| Waste disposal practice | Frequency reported | Principal contaminants | Typical hazard to health | Typical size of affected area |
|---|---|---|---|---|
| Land disposal of solid waste | | C, I, $H_m$, P | | |
| Municipal | I | | I | b |
| Industrial | III | | I | b |
| Municipal waste water | | N, C, $H_m$ | | |
| Sewer System | III | | II | a |
| Treatment Lagoons | III | | II | b |
| Petroleum exploration and development | | S, C, O | | |
| Wells and pits | II | | III | b |
| Mine waste | | Sa, I, $H_m$ | | |
| Coal | III | | III | b |
| Other | III | | II | b |
| Agricultural | | N | | |
| Cattle and other | III | | III | c |

Explanation of Table 5.5: I, high; II, moderate; III, low; A = Acid; B = Bacteria; C = Chloride; $H_m$ = Heavy metals; I = Iron; M = MBAS; N = Nitrate; O = Oil; P = Phenols; S = Sodium; Sa = Sulfuric acid; T = Temperature; a, small area but can be regional due to high density of individual sources; b, can affect adjacent properties; c, contained one property.

Modified from Shuckrow, 1982.[10]

Solid waste disposal sites can be sources of groundwater contamination because of the generation of leachate caused by water percolating through the bodies of refuse and waste materials. Precipitation falling on a site either becomes runoff, returns to the atmosphere via evaporation and transpiration, or infiltrates the landfill which produces leachate. This leachate is a highly mineralized fluid containing such constituents as chloride, iron, lead, copper, sodium, nitrate, and a variety of organic chemicals. Where manufacturing wastes are included, hazardous constituents are often present in the leachate (e.g., cyanide, cadmium, chromium, chlorinated hydrocarbons, and PCB). The composition of the leachate is dependent upon the industry using the landfill or dump.

## 5.2 TYPES OF WASTE

### 5.2.1 Urban Wastes

Septic tanks and cesspools[13] rank highest in total volume of waste water discharged directly to groundwater and are the most frequently reported sources of contamination. However, most problems are related to individual homesites or subdivisions where recycling of septic fluids through aquifers has affected private wells used for drinking water. Except in locations where such recycling is so quick that pathogenic organisms can survive, the overall health hazard from on-site domestic waste disposal is only moderate, with relatively high concentrations of nitrate representing the principal concern.

### 5.2.2 Municipal Wastes

Municipal waste water follows one of three routes to reach groundwater: leakage from collecting sewers, leakage from the treatment plant during processing, and land disposal of the treatment-plant effluent. In addition, there are two indirect routes: effluent disposal to surface water bodies which recharge the aquifers and land disposal sludge which is subject to leaching. The volume of waste water entering the water resources systems from these various sources is substantial. There have been many documented cases of hazardous levels of constituents of sewage or storm water affecting well-water supplies.[14,15]

Municipal and industrial sludge is the residue remaining after treatment of waste water. Sludge may be a product of physical, biological, or chemical treatment or a combination thereof. Groundwater quality degradation can be caused by land spreading of sludge because organisms (such as viruses) and chemical ions and compounds can be leached by precipitation and percolate to groundwater.

### 5.2.3 Petroleum Wastes

Disposal of brine from oil and gas[17] production activities has been a major source of groundwater contamination in areas of intense petroleum exploration and development. Most oil-field brines today are disposed through injection into oil producing zones or deep saline aquifers by utilizing old production wells or brine injection wells. Many of these wells are poorly designed for injection, however, and offer the opportunity for the salt water to enter the freshwater formations through ruptured or corroded casings.

As the American economy expanded during the 1950s and 1960s, industries sought new methods for managing wastes. The toxic effects of many chemicals were recognized and industries wanted an inexpensive method of management that would isolate the chemicals from the environment. Deep-well injection was a disposal method that had developed in the oil industry during the 1930s. This technology was soon adopted by petrochemical industries.

During the 1960s and 1970s, the deep-well injection industry grew until there were at least 209 wells injecting industrial waste in 1974. Even though the industry experienced no major operations problems or contamination incidents during this time, the American public had become quite concerned with both air and water pollution by the late 1960s. Consequently, environmental laws to control air and water were passed in the early 1970s.[18]

In 1984 and 1985, two publications (cited in LaMoreaux[19]) were released which brought deep-well injection to the forefront of the hazardous waste debate. In the first publication, waste generators and managers were surveyed and the quantities of hazardous waste produced and managed were estimated. The survey revealed that deep-injection wells received almost 60% of the hazardous waste disposed. The second publication contained a negative assessment of the technology and history of deep-well injection.

When the Safe Drinking Water Act (SDWA) was enacted in 1974, the law required the EPA to set minimum standards which the states were to adopt in their Underground Injection Control (UIC) Programs. These new regulations, adopted in 1980, contained a classification system for injection wells. This classification system (Table 5.6) is based on three factors: the industries involved, the fluid being injected, and the location of the injection zone in relation to potable water.[18,19]

### 5.2.4 Mining Wastes

All forms of mining can result in products and conditions that may contribute to groundwater contamination. The patterns of groundwater recharge and movement responsible for the distribution of contaminants are highly variable and inherently dependent upon the mining practice itself and local conditions of geology, drainage, and hydrology.

Associated with both surface and underground mining, refuse piles[20] and slurry lagoons are probably the major potential sources of groundwater contamination. Where aquifers underlie these sources, water with a low pH and an elevated level of total dissolved solids can percolate to groundwater.

The importance of resource development to the world economy demands a full understanding of the impact from extraction of natural resources to the

**Table 5.6 Classification of Injection Wells**

| | |
|---|---|
| *Class I* | Wells used by generators of hazardous waste, or owners or operators of hazardous waste management facilities to inject hazardous wastes beneath the lowermost formations containing, within one-quarter (1/4) mile of the well bore, an underground source of drinking water. Class I includes other industrial and municipal disposal wells which inject fluids beneath an underground source of drinking water. |
| *Class II* | Wells that inject fluids: (1) which are brought to the surface in connection with conventional oil or natural gas production; (2) for enhanced recovery of oil or natural gas; and (3) for storage of hydrocarbons which are liquid at standard temperature and pressure. |
| *Class III* | Wells that inject for extraction of minerals including: (1) mining of sulfur by the Frasch process; (2) *in-situ* production of uranium or other metals (but only from ore bodies which have not been mined conventionally); and (3) solution mining of salts or potash. |
| *Class IV* | Well used to dispose of hazardous or radioactive wastes into or above formations which, within one-quarter (1/4) mile, contain an underground source of drinking water. |
| *Class V* | Wells not covered under the other classes. These include air conditioning return flow wells; certain multiple use cesspools; cooling water return flow wells; drainage wells; dry wells; recharge wells; salt-water intrusion barrier wells; sand or other backfill wells; multiple use septic systems; subsidence control wells; radioactive waste disposal wells (other than Class IV); wells associated with recovery of geothermal energy; wells for solution mining of conventional mines (such as stopes leaching); wells for injection of spent brine after extraction of halogens or their salts; injection wells used in experimental technologies; wells used for *in-situ* recovery of lignite, coal, tar sands, and oil shale. |

From 40 CFR 146.5.[18]

human environment as a basis for decision making on the part of government as well as the extraction industry. Complete protection of the environment is sometimes impossible; thus, the implementation of decisions to mine balanced against benefit to be gained is important.

Minerals are the basis for materials upon which civilization depends, The environmental impact of mining is complex and involves geographic, geologic, technical, and socio-economic parameters.

A recent text, *Environmental Effects of Mining*, by Ripley, Redmann, and Crowder[21] provides an excellent background on environmental hydrogeology as a further reference. The book characterizes the sequential stages of mining activities:

1. *Exploration*, which may involve geochemical or geophysical techniques, followed by the drilling of promising targets and the delineation of orebodies.
2. *Development*, i.e., preparing the minesite for production by shaft sinking or pit excavation, building of access roads, and constructing of surface facilities.
3. *Extraction*, i.e., ore-removal activities that take place at the minesite itself, namely extraction and primary comminution (or crushing).
4. *Beneficiation* (or concentration), which takes place at a mill usually not far from the minesite; at this point (except in the case of coal), a large fraction of the waste material, or gangue, is removed from the ore.
5. Further processing, which includes *metallurgical processing* and one or more phases of *refining*; it may be carried out at different locations.
6. Since every orebody is finite, a final *decommissioning* stage is required through which the disturbed area is returned to its original state or to a useful alternative.

As the authors state, these stages are generally not discrete but may overlap or take place simultaneously. In subsequent chapters of the book, however, each is discussed in its context of environmental effects.

### 5.2.5 Industrial Wastes

Industrial waste, sewage effluent, spent cooling water, and storm water are discharged through wells into fresh- and saline-water aquifers in many parts of the world. The greatest attention in existing literature has been given to deep disposal of industrial and municipal wastes through wells normally drilled a thousand feet or more into saline aquifers.

Some environmental impacts associated with subsurface injection through wells are:[22]

1. groundwater contamination
2. surface water pollution
3. alternation of hydraulic conductivity
4. land subsidence and/or earthquakes
5. contamination of mineral resources

The generation and disposal of large quantities of animal waste at locations of concentrated feeding operations is another environmental problem.

There are three primary mechanisms of groundwater contamination from animal feed lots[23] and their associated treatment and disposal facilities: (1) runoff and infiltration from feed lots, (2) runoff and infiltration from waste products collected from the feed lots and disposed of on land, and (3) seepage or infiltration through the bottom of a waste lagoon. The principal contaminants are phosphate, chloride, nitrate, and, in some cases, heavy metals.

Figure 5.1 illustrates the sources of liquid contamination and the different activities involved in disposing of them into different environmental sites.

## 5.3 GASEOUS WASTES

### 5.3.1 Industrial Wastes

Industrial activities produce two classes of contaminants or pollutants into the atmosphere. The first include solid and liquid particles which may be designated as a particulate matter, e.g., dusts found in smoke of combustion and as droplets of cloud and fog. The term *aerosol* describes the entire system of particles of colloidal size (of a maximum of 1 micron) together with the gas in which they are suspended. The second class includes the compounds in the gaseous state which are known as *chemical pollutants*. One group of chemical pollutants of industrial and urban areas, produced directly from a source on the ground (of primary origin), includes the following gases: nitrogen oxides (NO, $NO_2$, $NO_3$), Co, $SO_2$, and hydrocarbon compounds. In polluted air, certain chemical reactions take place among the components injected into the atmosphere, generating a secondary group of pollutants; e.g., $SO_2$ may combine with oxygen to produce $SO_3$ which in turn reacts with suspended droplets of water to yield sulphuric acid which is both irritating to organic tissues and corrosive to many inorganic materials.

Also, isolated industrial activities can produce pollutants far from urban areas; e.g., processing of sulphide ores, by heating in smelters close to the mine, leads to the release of enormous concentrations of sulfur compounds that can be destructive to vegetation. Another example of industrial wastes that leads to "Bergkrankheit" (a mountain disease that causes mortality from lung cancer among pitchblende miners in Austria) is the radioactive gas radon, which emits alpha particles.[24]

The ozone layer (11–15 mi) is a region of concentration of ozone molecules ($0_3$) which is produced by the action of ultraviolet rays upon ordinary oxygen atoms. The ozone layer serves as a shield, protecting the earth's surface from most of the ultraviolet radiation found in the sun's radiation spectrum. If excessive ultraviolet radiation reaches the earth's surface, it will destroy all exposed bacteria and severely burn animal tissues and plants. Therefore, the presence of the ozone layer is essential in the environment of the biosphere. Conversely, the ozone, which produces a toxic and destructive gas by the action of sunlight upon nitrogen oxides and organic compounds (e.g., photochemical reactions brought about by the presence of sunlight on hydrocarbon compounds), may lead to a toxic product known as ethylene.[25]

Nuclear test explosions inject into the atmosphere a wide range of particulates, including many radioactive substances capable of traveling thousands of miles in the atmospheric circulation. Some of these pollutants belong to a very special and dangerous class as they are the sources of ionizing radiation, which is capable of tearing off electrons from atoms that intercept radiation.[25b]

### 5.3.2 Radon Risk

Radon is a cancer-causing radioactive gas. It comes from the natural (radioactive) breakdown of uranium in soil, rock, and water and gets into the air which we breathe. High levels of indoor radon are found in every state of the USA and are estimated to cause many thousands of deaths each year.

Radon gas is found in nearly all soils, and it typically moves up through the ground to the air above and into our homes through cracks and other holes in the foundation. Although radon from soil gas is the main cause of radon problems, it sometimes enters the house through the well water, as it can be released into the air when water is used for showering and other household uses.

The geology of the site and its radon potential may give an idea about the radon in a house and if it has an indoor radon problem. Scientists evaluate the radon potential of an area and create a radon potential map by using a variety of data which includes the uranium or radium content of the soils and underlying rocks and the permeability and moisture content of the soils.

The amount of radon in the air is measured in "picocuries per liter of air," or pCi/L. Sometimes it is expressed in working levels (WL) rather than picocuries per liter (pCi/L).

Because the level of radioactivity is directly related to the number and type of radioactive atoms present, a house having four picocuries of radon per liter of air (4 pCi/L) has about eight or nine atoms of radon decaying every minute in every liter of air inside the house. A 1000 square foot house with 4 pCi/L of radon has nearly two million radon atoms decaying in it every minute.

Radon levels in outdoor air, indoor air, and groundwater can be very different. Outdoor air averages about 0.2 pCi/L. Radon in indoor air averages about 2 pCi/L. Radon in soil air ranges from 20 to more than 100,000 pCi/L. The amount of radon dissolved in groundwater ranges from about 100 to nearly three million pCi/L.

Testing for radon in homes is easy and can be carried out by radon test kits which are available at hardware stores and other retail outlets.

There are many publications available to provide information on radon gas. Several most recent on this subject include: Measurement and Determination of Radon Source Potential by Allan B. Tanner,[26] Development of EPA's Map of Radon Zones by White, Gundersen, and Schumann in the 1992 International Symposium on Radon and Radon Reduction Technology,[27] and Radon in the Geological Environment by Reimer and Tanner.[28]

### 5.3.3 Forest Growth Reduction by Air Pollution[29]

Little is known about the influence of ozone gas on forest growth although it is the most widespread gaseous air contaminant that influences U.S. forests today. Photosynthetic response to ozone shows a certain reduction in carbon dioxide assimilation following exposure to ozones for a certain time.

Elevated atmospheric fluoride can reduce the growth of trees contaminated from industrial sources as indicated by research conducted in the western United States. Also, the influence that $SO_2$ exerts on plant photosynthesis has received man's research attention more than any other air pollutant, because it may cause, together with other factors, large reductions in the photosynthetic rate of the alfalfa plant.

### 5.3.4 Acid Rain[30,31]

The world's total fossil fuel reserve is estimated to contain about 10 trillion tons of carbon. The temperature of the lower atmosphere is estimated to increase from 2.5 to 3°C for every doubling of $CO_2$ in the atmosphere.

The acidified wet (acid rain) and non-acidic precipitation leads to acid disposition. The former is enriched in sulfuric, nitric, carbonic, chloride, fluoride, and organic acids and may react with soils and surface waters in such a way as to increase leaching of metals and nutrients.

The quantification of corrosion of limestones carried by acid precipitation depends on many factors, namely the amount of precipitation and its pH value, temperature, global radiation, effective catalysts, buffer capacity of the soil, soil moisture, and geological features of the saturated zones. The hydrogen ion concentration is greatly affected by climatic and meteoric influences in space and time. Figure 5.2 illustrates the impacts of acid rain.

### 5.3.5 Mines

There are several environmental impacts of mineral and fuel extraction processing. A few of these are as follows:

#### *Acid Mine Drainage*[33]

Acid mine drainage (AMD) is one of the earliest recognized water pollution sources and has been the focus of research activity for decades. Acid mine drainage causes pollution problems for the coal industry as well as for other mining industries.

Several methods of treatment and ways of controlling surface water drainage were considered to mitigate or minimize acid drainage.

Neutralization of acid mine drainage water with lime and/or limestone results in generation of a waste slurry (sludge) containing the precipitates from

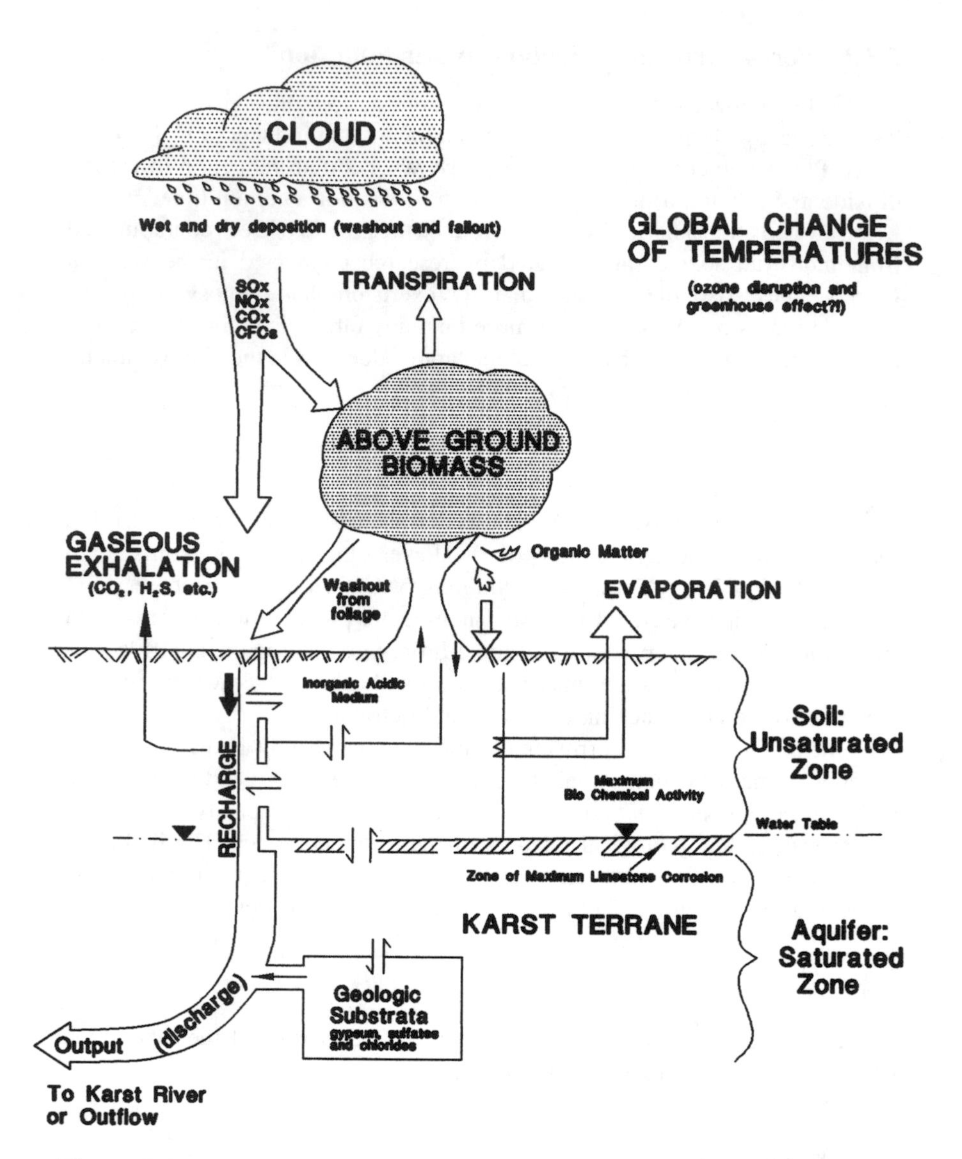


**Figure 5.2 A flow diagram of acid rain (after Kramer et al., 1986).[25b]**

the neutralization reactions as well as unreacted reagents. The main constituents of the solid sludge are iron, aluminum, sulfates and hydroxides, and unused calcium carbonate or hydroxide.

Several factors are to be considered prior to the applicability of spray irrigation of neutralization sludge:

1. Erosion of the sludge during high and medium intensity rainfall is readily apparent and undesirable.

2. No gross pollution is observed in the runoff from sludge irrigated areas during mild precipitation events.
3. Flat topography on sites proposed for spray irrigation of neutralization sludge is essential to prevent erosive loss of sludge during precipitation events.
4. Sludge application appears to have a slight beneficial effect upon establishment and maintenance of a vegetative cover.

Studies on the use of drying beds of acid mine drainage sludge indicated that the drainage rate through the sludge and sand averaged 26 liters/day/sq. m (0.6 gal/day/sq.ft.), whereas the sludge solids appeared to stabilize near 20% within 20 days drying time when using lime-neutralized coagulant-treated sludge in summer operations.

Acid mine drainage may be controlled by: (a) air sealing of the underground mine to prevent oxygen from entering the mine, which would stop the oxidation of pyrite in the mine and reduce the production of iron and acidity; (b) reducing the amount of pollution source, rechannelling streams to establish drainage away from the mines, and constructing solid "dry" seals in portals through which water could not pass; and (c) revegetating disturbed areas to prevent erosion and stabilize the backfills.

### *Coal Mining*

The mining of coal has a strong environmental impact on the land and water of the coal fields and upon the miners. Coal mining has caused a terrifying record of death and injury from explosion, fires, and cave-ins. Another hazard lies in the inhalation of coal dust produced in mines causing the prevalent black lung disease pneumoconiosis.

An explosion that killed 78 employees in the Farmington, West Virginia, mine in 1968 triggered an angry reaction from coal miners across the nation, who marched to Charleston, West Virginia, and Washington, DC, demanding protection at both the state and federal levels. Within months, the U.S. Congress passed the Coal Mine and Health Safety Act, which took effect on March, 30, 1970.

America's mining community observed the 25th anniversary of the Coal Mine Health and Safety Act on March 30, 1996, with special ceremonies and remembrances. "Coal miners were almost five times as likely to be killed in the mines in 1969 as they are today," J. Davitt McAteer, assistant labor secretary for mine safety and health, told the audience at Department of Labor headquarters in Washington, DC.

Before the law, about 250 workers died each year in coal-mining accidents. Between 1992 and 1994, the average number of coal-mining deaths dropped to fewer than 50 a year. Incidences of black lung disease, caused by exposure to respirable dust in coal mines, has been reduced during the past quarter century by an average of 75%, and the prevalence of the disease among miners has declined by more than two thirds.[34]

### *Mining of Radioactive Ores*

A comparatively new environmental hazard is ionizing radiation which is derived from the mining and milling of uranium ores. Very high radiation doses (of 5600 mrem/yr) measured in the uranium mines may be considered about 50 times as much of the average radiation dose from all sources. Besides the direct gamma radiation from the uranium ore, there is a radiation hazard in the inhalation of radon gas which is a radioactive product of uranium decay. The gas produces other radioactive derivative products which lodge in the lungs and cause cancer, e.g., some 50% of the miners of pitchblende (an ore of uranium and radium) died of lung cancer due to radiation exposure.

### 5.3.6 Hydrocarbons

Marine oil pollution events have frequently occurred in the recent years with detrimental effects upon the ecosystem of the shore zone, polluting beaches and damaging marine life of the coast. Evidence of widespread marine oil pollution comes from the collection of floating oil-tar lumps over wide reaches of the oceans. These lumps represent the nonvolatile residue of crude oil spilled in oil transportation.

Together with the major marine oil pollution to the seas and oceans, mention should be made of the tremendous number of oil wells (~800 wells) in the state of Kuwait which were set on fire by the Iraqis at the end of the Gulf Crisis in January 1991. A firm decision should be taken by the Security Council of the United Nations to avoid such dangerous and irresponsible action towards humanity, animal, and plants, as well as to the living organisms in the sea.

## 5.4 HAZARDOUS WASTES

### 5.4.1 Definition

Hazardous wastes are defined as: "Any waste material or a mixture of wastes which is toxic, corrosive, flammable and irritant; a strong sensitizer, which generates pressure through decomposition, heat or other means; such a waste or mixture of wastes may cause substantial personal injury, serious illness or harm to wildlife during or as a proximate result of any disposal of such waste or mixture of wastes."[35] Hazardous wastes which are the residues of normal industrial processes cannot be prevented but must be controlled by industry. The Environmental Protection Agency (EPA) estimates 30 million tons/year are being generated in this manner.

One of the most interesting recent summaries on hazardous and radioactive waste management appeared in the August 1996 issue of *Geotimes*.[36] These three articles: "Why Have Earth Scientists Failed to Find Suitable Nuclear Waste Disposal Sites," by John B. Robertson, "Unsaturated-Zone Characterization for Low-Level Radioactive Waste Disposal in Texas," by Bridget R. Scanlon, and "Beneath the Surface: Geophysical Aspects of Radioactive Waste Disposal," by Don W. Steeples highlight some of the serious problems faced by environmental geoscientists in managing wastes of all types but particularly those of hazardous or radioactive nature.

Management of hazardous waste requires an understanding from many scientific disciplines: geography, geology, hydrology, hydrogeology, and biology. In addition to the scientific knowledge required, there must be an understanding of technical regulations, economics, permitting, and institutional and public policy issues, for example, the Resource Conservation and Recovery Act (RCRA) of 1976 and the Comprehensive Environmental Response, Compensation, and Liabilities Act (CERCLA) of 1980. These statutes, plus others passed at federal level by Congress, must also be implemented with appropriate regulations and legislation by state statutes.

In addition, new techniques for waste management are evolving and must be understood and implemented as they become acceptable practices, for example, such state-of-the-art treatments in chemical, physical, and biological treatments. Incineration has become an effective technique for destroying certain types of hazardous waste.

Another problem relates to the storage and transport of waste. One of the most difficult problems in waste management relates to temporary storage and the generation of waste products and leachates that become difficult in the environment.

Finally, a major problem with regard to waste management practices of all types relates to the contamination of surface water, soil, and groundwater. Groundwater waste problems can be of the most insidious types.

In a recent text, Hazardous Waste Management Engineering by Martin and Johnson (1987),[37] a table is presented that provides some sample alternatives for waste management at an individual site (see Table 5.7).

### 5.4.2 Toxic Materials

Thompson et al.,[38] studied the sources of hazardous industrial wastes and stated that many industrial processes involve the use of materials which are potentially toxic if released into the environment. These materials are usually concentrated by waste treatment systems into a semi-solid or sludge that should be disposed without degrading the surrounding environment. Many of the most dangerous waste products are organic chemicals. Incineration or other techniques are generally applied for oxidizing organic materials in order to produce nontoxic or less toxic compounds. Inorganic contaminants which cannot be destroyed

Table 5.7 The Black-Box Approach

| Management option | Effluent $X_1$ | Air emission $X_2$ | Residue $X_3$ |
|---|---|---|---|
| Incinerator | 0 | Fugitive stack, PIC[a] | Ash and scrubber water |
| Secure L.F. | Leaks (leachate) | Air emission | Remainder after lifetime |
| Land treatment | Leachate and runoff | Air emission | Remainder on land |
| Storage tanks | Leakage | Air emission | Pump out |
| Piles | Leaching and runoff | Air emission | Pile residue |
| S.I. | Leaching | Air emission | Residue in impoundment |
| Containers | Leakage and spillage | 0 | Container contents |
| Treatment removal | | | |
| Conservative pollutants | Effluent + PIT[b] | Air emission | Residue |
| Nonconservative pollutants | Effluent + PIT[b] | Air emission | Residue |
| Destruction | Effluent + PIT[b] | Air emission | Residue not destroyed |
| Fixation/stabilization | Leachate | Air emission | Fixed solids |
| Encapsulation | Leachate | 0 | Fixed solids |

[a] PIC = Products of incomplete combustion.

[b] PIT = Products of incomplete treatment.

From Martin and Johnson.[37]

must also be disposed in a way that limits their migration to the environment. Table 5.8 lists examples of industries that produce large amounts of hazardous inorganic wastes containing high concentrations of toxic metals, e.g., cadmium, chromium, mercury, and lead. Although these materials are present in most industrial sludges as insoluble hydroxides or sulfides, changes in the pH or oxidation conditions of the environment can lead to their mobilization.

### 5.4.3 Soil Hazardous Wastes

Vegetables grown in areas with contaminated soils have proven to have high concentrations of lead (10–100 ppm in the ash). The development of highly sophisticated analytical techniques, such as the atomic absorption spectrophotometer which permits rapid microchemical studies of cells, soil, and water, facilitates investigation into the sources of trace elements and their

**Table 5.8 Examples of Major Industries which Produce Major Amounts of Hazardous Industrial Wastes and their Toxic Constituents[38]**

| Industry | Type of waste | Typical hazardous components |
|---|---|---|
| Electroplating and metal finishing | Plating wastes and spent liquors | Chromates, copper, nickel, phosphate |
| Metal refining and smelting | Slags | Aluminum fluoride, zinc, copper, lead |
| NiCad battery industry | Production sludges | Nickel, cadmium, lead |
| Leather production | Tanning liquors | Chromium, sulfide |
| Electronics industry | Spent etching solutions and solder | Copper, nickel, cadmium, lead |
| Dyestuff, petrochemical | Spent catalysts | Cobalt |
| Metallurgical industry | Mine tailings, sludges | Most heavy metals, selenium, antimony, arsenic |
| Dye manufacturer | Dye liquors | Sulfide, polysulfides |
| Chemical industry | Effluent treatment sludges | Copper, iron, nickel, fluoride, chromium, lead |
| Gas purification | Spent oxide and sulfur residues | Arsenic, antimony, sulfides, polysulfides |
| Power generation | Flue gas cleaning sludges, fly and bottom ash | Heavy metals, calcium sulfite and sulfate |

actual role in physiological function. Trace elements, which are usually, but not always, present in the body in parts per million, may cause pathological conditions when found in a slight deficiency or in a slight excess of a particular element, as in the case of selenium and fluorine.

However, the fact that an element is present in soil does not mean that it is necessarily available to plants or that it will dissolve in water as its physical state and its mineralogical criteria may determine its availability. Various soil waste sources include animal wastes, fertilizers, pesticides, plant residues, and saline waste waters. LeGrand[39] discussed the movement of agricultural pollutants in groundwater and concluded that there were sufficient safeguards to minimize groundwater pollution. The unsaturated zone above the water table reduces almost all of the foreign bodies that are potential pollutants of the underlying groundwater. Fertilizers used to improve crop productivity may result in leaching of nitrogen and phosphorous to the groundwater.

The groundwater contamination of silica and nitrate may have resulted from the percolation of nitrate fertilizer and the leaching of silica through water-logged soil. Factors which tend to reduce the pollution of groundwater from wells and springs may be summarized as follows:[40]

1. A deep water table that allowed adsorption slowed subsurface movement of pollutants and facilitated oxidation,
2. Sufficient clay in the path of pollutants to favor retention or sorption of pollutants,
3. A gradient beneath a waste site away from nearby wells, and
4. A great distance between wells and wastes.

Pesticides contribute significantly to improve crop productivity, but they pose risks to humans and the environment. Two pesticides discovered in 1979 and 1982 in groundwater in certain areas of the United States were dibromochloropropane (DBCP) and ethylene dibromide (EDB), respectively. By 1986, a total of 19 different pesticides had been detected in groundwater in 24 states. The health effects of chemicals may be classified as follows:

1. Acute effects resulting from contact with high levels of a chemical over a short period of time are usually easier to identify because they occur soon after exposure, e.g., nausea, skin irritation, etc.[41]
2. Chronic effects, which generally occur as a result of long-term exposure to low levels of a chemical, are harder to document, particularly because of the long-term interval between exposure and outcome.

Concentrations of pesticides in groundwater have been found at low levels; therefore, most of the concern has been focused on the potential for chronic effects such as cancer, mutations, birth defects, and immunological changes. However, there are great variations in the chemical properties of pesticides which control their tendency to leach to groundwater.

Local agricultural practices which may affect the potential for pesticides to contaminate groundwater vary greatly across the USA depending on the crops grown, local pest control needs, and the preferences of the grower.[41] Also, animal wastes in soil which lead to groundwater contamination with nitrates are a result of leaching from livestock feeding operations.

### 5.4.4 Radioactive Wastes

Halcomb[42] states that the nuclear power and radioisotope industry in the twentieth century have developed greatly and presented a continuous problem in dealing with the management of the radioactive waste materials. Natural decay is the only means of destroying radioactivity and as the various waste radionuclides have decay rates ranging from days to thousands of years, treatment and processing (such as solidification techniques) become an important factor in radioactive waste management. The main two classes of radioactive wastes (radwastes) for which solidification techniques have been developed are high level and low level wastes. The high level wastes are the wastes of fission products with a high level of penetrating radiation and high rates of heat generation, and they include the liquids of high activity radioactive wastes or their solidified products. Because of their high hazard and long decay times, considerable money and effort have been devoted to the removal of high-level wastes from solution and its solidification and shipping to centralized repositories for permanent emplacement. The danger and expense of handling these wastes can be reduced by allowing them to decay before final disposal, thereby reducing the radioactivity and heat production in the waste. Present policy allows ten years of interim storage before final disposition.

The low level wastes include a variety of materials with low levels of radiation or those radioactively contaminated wastes that are generated from the nuclear fuel cycle operations and facilities not specifically designated high level. The majority of these wastes are generated by nuclear power plants and are usually in the form (prior to solidification) of processed wastewater, evaporator concentrates, sludges, and filter aids. These wastes include a broad spectrum of materials varying widely in chemical and radioactive content.

## 5.5 PHYSIOGRAPHY OF WASTE SITES[43,44]

In the early 1970s, a period which has been termed "a decade of environmental action," the subject of radioactive waste management continued to be of increasing public concern. Extensive research and development have been supported by the Atomic Energy Committee (AEC) to safely store certain types and quantities of solid and liquid radioactive waste materials in the deep underground formations (1,000–3,000 feet for the deep hydrologic environment and about 3,000–12,000 feet for the petroleum industry).

There are many types of injection wells that are used to dispose of different forms of liquid wastes. The geological and hydrological criteria must be considered in siting wells used for the disposal of hazardous wastes. For siting a hazardous waste injection well, attention should be given to the nature of the host interval into which wastes will be injected and the adequacy of the confining units that separate the wastes in the host interval from the environment and drinking water supplies. Site-specific geologic, hydrologic, and structural conditions are to be considered for the selection of a site for a hazardous waste disposal injection well. *Stratigraphically*, the injection zone should be bounded at the top and below by a low permeability confining zone that prevents the upward and downward migration towards fresh water aquifers. Both the injection and confining zones should have enough thickness and extend laterally far enough so that wastes remain in the injection zone and do not reach the discharge areas. *Lithologically*, the permeability and porosity parameters should be determined for both the injection and confining intervals to decide the best suited rock type for each interval. The injection intervals which receive and house the wastes (without artificial fracturing by high pressure) should have high porosity and a permeability on the order of ≈0.1 – 1.0 cm/sec. For acidic wastes, which must be neutralized after injection in order to render them nonhazardous, a carbonate-rich injection zone is desirable, because the reaction generates $CO_2$ gas which is generally dissolved in the groundwater of the injection zone. In contrast, the confining zone must be of very low permeability, in the order of $10^{-6} – 10^{-9}$ cm/sec, for example, clay, shale, anhydrides, and possibly dense and unfractured carbonates (except for acidic wastes). *Structurally*, the hazardous waste injection sites are considered to be safe if faults and fractures represent pathways for waste migration through the injection zone but not into and through the confining units. Faults can also represent barriers to migration of groundwater (and wastes) sealed with secondary materials, but generally they are considered as potential pathways.

Pre-waste injection hydrofracturing should be avoided as it may extend into the confining units and create pathways out of the injection zone. Also, faults and other fractures should be avoided because when a faulted injection horizon is under stress the fluids enter the fault system, thus facilitating movement and seismic activity. A full investigation of structural and tectonic complexities is recommended during the siting of disposal wells, especially in areas of tectonic activity.

*The hydrochemical criteria* deals with the compositions of the groundwater in the injection interval, as well as in the confining units and is important because: (1) reactions may occur between the wastes and the formation water in the injection zone, and (2) the composition of groundwater may reflect the history, age, and mixing of groundwater. Age dating of groundwater could be determined by using radioactive isotopes, such as $C^{14}$ or $Cl^{36}$.

The *hydrologic aspects* mainly include the hydrophysical characteristics of the injection zone and the hydrologic characteristics of the confining units. The former deals with the aquifer geometry parameters (e.g., thickness, areal extent, and form), the hydrophysical parameters of the host formation(s) (e.g., permeability "k," porosity "Ø," pressure "P," temperature "T," saturation "S"), and its physiochemical parameters: chemistry (including total dissolved solids, TDS), pressure, temperature, density, and compressibility.

Structural, physical, and hydraulic parameters of the confining and semiconfining formations adjacent to the host formation determine the boundary conditions of the host hydraulic system; i.e., the degree of waste confinement. The host hydraulic system which includes the aquifer and confining units may be an open or a closed system. Larger amounts of waste can generally be injected into open hydraulic systems because the liquid within the aquifer can be displaced; but contamination can occur at the open boundaries with adjacent hydraulic systems and should be avoided for injection of highly persistent and refractory wastes. Injection into more closed systems requires use of relatively high injection pressures to compress subsurface materials and to create storage space. Excessive injection pressures can fracture confining units and can open the system boundaries, allowing uncontrolled contaminant migration.

### 5.5.1 Permeable Formations (3,000–12,000 feet) Containing Connate Brine

Disposal of large volumes of waste in porous beds, which are interstratified with impermeable beds in a synclinal structure, is of a particular interest. Deep disposal wells have been utilized by the chemical, petrochemical, and steel industries. The application of deep-well injection for certain types of radioactive wastes needs more caution and precise studies from the reservoir engineering point of view, in order to determine if there should be sufficient porosity or permeability in the receiving strata to receive "x" number of gallons of waste per day, if the well and the pumping equivalent should be properly protected against corrosion, or if the disposal well is properly cased and cemented from the surface to the disposal zone.[43]

Hydrogeologic information of the receiving formation is also required to define the rate of movement and direction of flow and, in short, to understand the effects of disturbing the hydrologic system by injecting the wastes, e.g., the potential effects of increasing the formation pressure, such as fracturing confining beds and changing the hydraulic gradients. Continuing surveillance is quite necessary for all toxic waste injection wells to avoid any accidental contamination of overlying freshwater aquifers.

### 5.5.2 Impermeable Formations

#### *Hydraulic Fracturing in Shale Formations*

In the petroleum industry the technique of hydraulic fracturing is widely used to increase oil recovery in reservoir rocks of low permeability. For waste disposal purposes, a well is drilled into a shale formation rather than into an oil reservoir rock, and the aqueous wastes are mixed with preblended dry solids (principally containing cement). Then, the resulting slurry is pumped down the well and out into a conformable, horizontal fracture in the thick shale formation at depths usually on the order of 1,000–1,500 feet. The well, which is cased by a strong steel casing, is prepared for the injection by perforating the casing at the desired depth and pressurizing the well with water. This pressure induces a fracture in the rocks into which the waste slurry is pumped, causing the fracture to extend. After completion of the pumping phase, the cement slurry is allowed to harden under pressure to form a thin horizontal grout sheet. This procedure is then repeated successively up the well, creating a stack of horizontal grout sheets.[44]

### *Storage in Crystalline Bedrock*[44]

Storage in deep impermeable basement rocks is sometimes found necessary for the disposal of high level wastes below a site. For example, high level radioactive liquid wastes at a chemical reprocessing plant in South Carolina are stored in the crystalline rock where it is separated from the overlying unconsolidated sediments (at a depth of approximately 1,000 feet) by a clay bed known as saprolite, which is a residual weathered product of the crystalline rock. A long-term disposal facility that consisted of a concrete-lined vertical shaft (≈15 feet in diameter) would pass through the unconsolidated sediment and about 500 feet into the crystalline rock where six tunnels (each approximately 3500 feet long with a cross section of 26 feet wide by 28 feet high) were constructed. Waste feed and instrument lines from the surface would go into each of the tunnels through separately drilled service shafts.

Prior to repository in crystalline rocks, an exploratory drilling program should be conducted to determine the hydraulic and physical characteristics of the basement rock and the overlying sedimentary aquifer. These hydraulic and physical characteristics can be determined by obtaining continuous rock-core samples from the overlying sediments. Hydrologic studies, including drill stem testing (DST) to calculate the bottomhole pressures (BHP), pumping tests of packed-off sections of the hole, and piezometer measurements of the several main aquifers that overlie the bed rock, can be conducted. From these tests, the permeabilities of the rock and the regional permeability of the fractured-rock zones can be estimated together with the hydraulic gradient and extensive tritium tests can also be made.

### *Storage in Basalt Flows*[44]

At the AEC Hanford Plant in Richland, Washington, approximately 25% of the tank-stored high-level wastes were in the form of salt cakes and sludges. This form was the result of Hanford's in-tank solidification that reduced the high-salt-content waste to this form by successive passes from the tank through an evaporation cycle.

The bulk of the high level wastes could be immobilized as salt cakes in the existing underground tanks and could be safely isolated in the underground concrete-tank structures for hundreds of years. The principal protection needed would be from erosion, requiring minimal surveillance. However, the salt cakes are retrievable and their storage in underground tanks is considered an interim measure until the long-term safety of this method is determined to be acceptable.

The AEC has started investigating the structure, stratigraphy, and hydrology of the basalt flows underlying the Hanford site for possible consideration as a long-term storage environment for the Hanford wastes. The information collected shows that more than 100 basalt flows, some as much as 200 feet in thickness separated by weathered basalt zones, are present to a depth greater than 10,000 feet in the center of the site.

### *Storage in Salt Mines*

The AEC's ground disposal research and development program was directed in the past 30 years to use the underground salt formations for the isolation of high-level solidified radioactive wastes. Salt beds are generally located in areas of low seismic activity and widely separated from flowing aquifers. They are normally dry and impervious to water and have considerable compressive strength, being similar to concrete in this respect.

It may be interesting to mention that salt movements and diapirism of salt bodies penetrating younger beds are characterized by two phenomena: halokinesis and halotectonics.[45] Halokinesis leads to the formation of salt domes and/or to piercing diapirs when there is a continuous supply of mother salt or where there are pressure contrasts. Halokinematics was proposed to define salt movements more precisely according to their cause.[46] Because of the plasticity of salt, fractures seal or close rapidly.

In locations where salt dome structures exist, large spaces can be mined out and, even at depths of 1,000 feet, two thirds of the salt can be removed with only slight deformation of the support pillars.

## 5.6 ENVIRONMENTAL CONCERNS ON HYDROGEOLOGICAL SYSTEMS

### 5.6.1 Man-Made Earthquakes[47,48]

There is a close relationship between man and the surrounding physical environment which is actually a dynamic system accommodated by man's activities. The living or biological systems of the earth are progressively changing as well as the nonliving physical aspects of the earth which undergo change at an equal or greater rate. As a matter of fact, a dynamic system is more difficult to adopt man's activities to a constantly changing situation than to an unchanging or static system. Man's activities have made great physical and chemical changes to our planet in the past 50 years.

Earthquake activity can be explained by the theory of plate tectonics which states that the earth's crust is made up of rigid plates floating on an underlying layer and moving relative to each other in response to convection currents and resulting in the formation of subduction and rift zones. Man-made earthquakes may also result from pumping of waste fluids into deep disposal wells as it is believed that this injection raises the fluid level pressure in the reservoir (or the injection interval) for some distance from the well borehole and during the shut down. This elevated pressure equalizes throughout the reservoir and at increasing distance from the well where the fluid pressure reduces the frictional resistance in fractures, and eventually movement takes place and small earthquakes resulted.

### 5.6.2 Transport of Polluted Waters by Subterranean Karst Flow Systems

In karst regions, the water flow may be partially or totally turbulent which leads to the contamination process with little or no filtration and/or precipitation to take place and also leads to the ionic retardation processes of adsorption or desorption of pollutants. As hydraulic conductivity and water-flow velocity are high in karst regions, the flow velocity of pollutants is also high and therefore karst water can be polluted very quickly in different pathways and directions in different velocities. The fluid movement and the pollutant migration depend greatly on the spatial distribution of faults and fissures. There is little contact between rock matrix and water in karstic aquifers where the water flow takes place through the solution channels (of irregular dimensions and directions).

In karst terrains, the passage of rainwater into the subterranean cavities normally carry polluted materials without being filtered as in other rocks and the velocity of flow of the groundwater is high in the karst in the order of several hundreds or thousands of meters/day, whereas outside the karst regions, they are usually of several meters/day.

Emili[49] discussed the transport of intestinal pathogenic germs by subterranean karst systems and stated that the subterranean water courses are one of the characteristics of the karst which are of interest to geologists, hydrogeneticists, and also to epidemiologists, concerning the question of whether the intestinal pathogenic bacteria, which are the main causative agents of typhoid, can be transported by karst rivers over large karst areas. Subterranean connections between sinkholes and springs have been proven in several occasions by the use of dyes. Emili also established the earlier known rule that the waters of karst at points of observations (sources) are less polluted with bacteria of the coliform group than the waters at the sites of the sinkholes. This is a clear proof that in the subterranean course of karst rivers, there occurs a retention of bacteria, which are much more resistant in an open environment than the intestinal pathogenic germs. His opinion was that transport of intestinal germs over greater distances by subterranean rivers in the karst area did not come into consideration practically. If, however, hydric epidemics of typhoid fever in the karst region are so frequent, such pollution of drinking water with pathogenic germs occurs as a rule in the vicinity of the water supply plants.

## REFERENCES

1. Committee on Groundwater Cleanup Alternatives, *Alternatives for Groundwater Cleanup*, National Research Council, National Academy Press, Washington, DC, ISBN No. 0-309-04994-6, 1994, 315 pp.

2. EPA, Superfund Fact Sheet: Who Pays for Superfund? EPA Washington, DC, 1990.
3. EPA, An Analysis of State Superfund Programs: 50-State Study — 1993 Update, EPA, Office of Emergency and Remedial Response, Washington, DC, 1994.
4. EPA, National Pesticide Survey, Update and Summary of Phase II Results. EPA 570/9-91-020, EPA, Office of Pesticide Programs, Washington, DC, 1992.
5. EPA, Cleaning Up the Nation's Waste Sites: Markets and Technology Trends, EPA 542-R-92-012, EPA, Office of Solid Waste and Emergency Response, Washington, DC, 1993.
6. REA, Modern Pollution Control Technology, Vol. II, Staff of Research and Education Association (REA), 1980.
7. Klickat Regional Council, Klickat County Solid Waste Management Plan, unpublished report, 1977.
8. Blebs, R.T. and Scaro, Ed. C., Cost Accounting for Landfill Design and Construction Past and Present, Proceedings of Waste Tech 85, Boston, October 28, 1985.
9. Prothro, M.G., An Overview Status of the Surface Water Toxic Control Program, Proceedings of the Industrial Waste Symposium, San Francisco, 1989.
10. Shuckrow, A.J., Pajack, A.P., and Touhill, C.J., Hazardous Waste Leachate Management Manual, Noyes Data Corporation, New Jersey, 1982.
11. Miller, D.W., *Waste Disposal Effects on Groundwater*, Premier Press, New Jersey, 1980.
12. U.S. Environmental Protection Agency, Handbook for Monitoring Industrial Waste Water, EPA, Washington, DC, 1973.
13. Patterson, J.W., et al., Septic Tanks and the Environment, Illinois Institute for Environmental Quality, Chicago, 1971.
14. Soliman, M.M., Groundwater Quality Model of Greater Cairo, Research Report presented to the Egyptian Academy of Science, 1982.
15. Soliman, M.M., Environmental Impacts of Irrigation by Sewage Water, P.P. International Conference on Advances in Groundwater Hydrology, Tampa, FL, Nov. 1988.
16. Page, A.L., Fate and effects of trace elements in sewage sludge when applied to agricultural lands, U.S. EPA 670/2-74-005, 97.
17. American Petroleum Institute, Annual statistical review, American Petroleum Institute, Washington, DC, 1975, p. 79.
18. Moffett, T.B., LaMoreaux, P.E., Smith, J.Y., and Dismukes, M.B., *Management of Hazardous Wastes by Deep-Well Disposal*, The University of Alabama Environmental Institute for Waste Management Studies, Tuscaloosa, Ch. 3, 1987.
19. LaMoreaux, P.E. and Vrba, J., Eds., Hydrogeology and Management of Hazardous Waste by Deep-Well Disposal, International Association of Hydrogeologists, a report of the Commission on Hydrogeology of Hazardous Wastes, Vol. 12, University of Alabama Press, 1990.
20. Martin, J.F., Quality of Effluents from Coal Refuse Piles, First Symposium on Mine and Preparation Plant Refuse Disposal, Washington, DC, 1974, 26–37.
21. Ripley, E.A., Redmann, R.E., and Crowder, A.A., *Environmental Effects of Mining*, St. Lucie Press, Delray Beach, FL, 1996.
22. Stallman, R.W., Subsurface waste storage — the earth scientist's dilemma, in American Association of Petroleum Geologists Memoir 18, Underground Waste Management Implications, Tulsa, OK, 1972, pp. 6–10.

23. Anderson, R.K., Feedlot wastes, System Management Division, Office of Solid Waste Management Programs, U.S. EPA, unpublished report, 1975.
24. Strahler, A.N. and Strahler, A.H., Environmental Geoscience — Interaction between Natural Systems and Man, Hamilton Publishing, Santa Barbara, CA, 1973, pp. 46–54, 151–154, 255–256.
25. Spencer, J.M., Geological influence on regional health problems, in *The Environmental Geology — Text and Readings*, Tank, R.W., Oxford University Press, New York, 459–460, 1983.
25b. Kramer, J.R., Andren, A.W., Smith, R.A., Johnson, A.H., Alexander, R.B., and Ochlent, G., Streams and Lakes, in Committee on Monitoring Assessments of Trends in Acid Deposition, Long Term Trends, National Academy Press, Washington, DC, 1986, pp. 231–249.
26. Tanner, A.B., Measurement and Determination of Radon Source Potential. A Literature Review, U.S. Natl. Inst. Standards and Technology Report, NISTIR-5399, 1994, 190 pp.
27. White, S.J., Gundersen, L.C.S., and Schumann, R.R., Development of EPA's map of radon zones, in *The 1992 International Symposium on Radon and Radon Reduction Technology*, U.S. Environmental Protection Agency, Research Triangle Park, NC, paper VIII-4, 1992, 14 pp.
28. Reimer, G.M. and Tanner, A.B., Radon in the geological environment, in *Encyclopedia of Earth System Science*, Nierenberg, W.A., Ed., Academic Press, San Diego, California, 1991, pp. 705–712.
29. Smith, W.H., *Air Pollution and Forests — Interactions between Air Contaminants and Forest Ecosystems*, Springer-Verlag, New York, 1981, 44–50, 204–210.
30. Weinberg, A.M., Is nuclear energy necessary?, Outlook of the future, in *The Environment, Geology — Text and Readings*, Tank, R.W., Oxford University Press, New York, 1983, pp. 345–353, 532–540.
31. Flohn, H., Possible Climatic Consequences of a Man-Made Global Warning, Presented at the Commission of the European Community Conference, Dublin, Ireland, Oct. 1979.
32. Grube, E., Jr. et al., Disposal of Sludge from Acid Mine Drainage Neutralization, EPA, Paper present at the NCA/BCR Coal Conference & Expo III, Louisville, KY, Oct. 1976, pp. 1–18.
33. Hill, R.D., Elkins Mine Drainage Pollution Control Demonstration Project, Paper presented at the 3rd Symposium on Coal Mine Drainage Research, Pittsburgh, PA, May 20, 1970.
34. LaMoreaux, P.E., Twenty-fifth anniversary of the Coal Mine Health and Safety Act, *Environmental Geology*, 27, 1996, pp. 71–72.
35. Doyle, R.D., Use of an extruder/evaporator to stabilize and solidify hazardous wastes, in *Toxic and Hazardous Waste Disposal*, Pojasek, R.B., Ed., Ann Arbor Science Publishers, Ann Arbor, MI, Vol. 1, 1979, pp. 65–66.
36. News & Features, *Geotimes*, 41, 8, August 1996, pp. 16, 22 and 26.
37. Martin, E.J. and Johnson, J.H. Jr., Eds., *Hazardous Waste Management Engineering*, Van Nostrand Reinhold Company, New York, 1987, p. 3.
38. Thompson, D.W. et al., Survey of available stabilization technology, in *Toxic and Hazardous Waste Disposal*, Pojasek, R.B., Eds., Ann Arbor Science Publishers, Ann Arbor, MI, 1, 1979, pp. 9–11.

39. LeGrand, H.E., *Movement of Agricultural Pollutants with Groundwater, Agricultural Practices and Water Quality*, Willrich and Smith, Eds., Iowa State University Press, Ames, 1970, pp. 303–313.
40. Todd, D.K. and McNulty, D.E.O., Agricultural pollution, in *Polluted Groundwater — A Review of the Significant Literature*, A Water Information Center, Huntington, NY, 1976, pp. 48–51.
41. EPA, Pesticides in groundwater: the concern, in *Agricultural Chemicals in Groundwater: Proposed Pesticide Strategy*, U.S. Environmental Protection Agency, Office of Pesticides and Toxic Substances, 1987, pp. 21–29.
42. Holcomb, W.F., An overview of the available methods of solidification for radioactive wastes, in *Toxic and Hazardous Waste Disposal,* Pojasek, R.B., Ed., Ann Arbor Science Publishers, Ann Arbor, MI, 1, 1979, pp. 23–25.
43. Belter, W.G., Deep disposal systems for radioactive wastes, in *Underground Waste Management and Environmental Implications*, Cook, J.D., Ed., Proceedings of the Symposium, Houston, TX, 1971, pp. 341–347.
44. Stow, S.H. and Koleju, V., Geologic and hydrologic criteria for design of a deep disposal system, in *Hydrogeology and Management of Hazardous Waste by Deep Well Disposal*, a report of the Hydrogeology of Hazardous Waste Commission of the International Association of Hydrogeologists, Vol. 12, 1990, pp. 21–38.
45. Assaad, F.A., A further geologic study on the Triassic formations of North Central Algeria with special emphasis on halokinesis, *J. Petro. Geol. London,* 4, 2, 1981, pp. 163–176.
46. Assaad, F.A., An approach to halokinematics and interplate tectonics (North Central Algeria), *J. Petro. Geol. London,* 6, 1, 1983, pp. 83–88.
47. U.S. Department of Energy, Geothermal energy and our environment, in *Environmental Geology, Text and Readings,* Tank, R.W., Ed., Oxford University Press, New York, 1983, pp. 35–82, 532–540.
48. Evans, D.M., Manmade earthquakes in Denver, in *Environmental Geology — Text and Readings*, Tank, R.W., Ed., Oxford University Press, New York, 1983, pp. 108–118.
49. Emili, H., Possibility of transport of intestinal pathogenic germs by subterranean karst rivers, in KRS Jugoslavic Carsus Iugoslaviae, Zagreb, Petrik, M., Zagreb, 1969, pp. 465–467.

# 6 ENVIRONMENTAL IMPACTS ON WATER RESOURCE SYSTEMS

## 6.1 INTRODUCTION

As the world's population continues to grow, it is evident that we must think very carefully in terms of development that meets the needs of the present without compromising the ability of future generations to meet their own needs and aspirations, i.e., truly sustainable development. It is clear, however, that some quite unsustainable development policies and practices, particularly concerning water management, have been followed. Economic and social change necessitate development of water resources based on sound environmental principles. A sound scientific understanding should form the foundation upon which rational decisions regarding water resources management should be made.

The essential role of science in continued socio-economic development, an area in which water resources are essential, is not a simple one. In managing our resources, it is evident that nature not only affects man, but man's activities can sometimes have devastating results on nature. It has become evident now, for example, that some of man's activities appear to be leading to possible major climate changes. The probable consequences are not yet known with certainty, but it is clear that a climate change would result in a redistribution, in time and space, of our water resources. In order to understand this change and be able to cope with it, we must have a much stronger scientific understanding of the processes involved. The approaching problems, coupled with the many existing environmental stresses (including, for example, land and water pollution, erosion and sedimentation, natural and man-caused hazards), emphasize the need for continued development of human potentials, education, training, and public understanding, as essential elements in a major international effort.

The responsibility of water scientists and engineers, then, must be, with full consideration of the changing environment, (1) to develop and maintain information on the availability of water resources; (2) to assess, monitor, and predict the resulting quality of water bodies and the water-related environment; (3) to develop a better scientific understanding of the effects of man's activities which influence hydrological regimes (especially those resulting from climate change); and (4) to provide decision makers with the necessary information in properly constructed formats such that they will understand the problems and the importance of the hydrological sciences as a basis for proper environmental management, especially the water resources, and to react appropriately.

## 6.2 CLIMATIC CHANGES AND THEIR EFFECT ON THE WATER RESOURCES

The climatic changes[1] foreseen and the probable consequent changes in the physical environment mean that the present understanding of water resources and of the hydrological cycle is not sufficient. Prediction of the new hydrological regimes will require a better understanding of the systems and their capability for quantitative analysis than is now available. More than ever, an awareness of water problems must be emphasized with decision makers. They must be given recommendations for action based upon sound scientific rationale.

Hydrology and water resources for sustainable development in a changing environment open a new era in the development of water sciences and management and have been designed to provide an international focal point for a broad coordinated effort in hydrology and the scientific bases for water management. They represent the combined efforts of national, regional, and international governmental organizations.

## 6.3 SURFACE WATER POLLUTION

The water quality aspects of water resources are becoming increasingly important. During the last decade attention has been focused on many new problems and approaches for their prediction. Besides the pollution from point and nonpoint sources such as agriculture and deforestation, we must also consider the problem of pollutant precipitation from the atmosphere. Simply adding water quality parameters to classical hydrological formulas is no longer sufficient. Detailed small-scale hydrological investigations of a different nature are needed to predict pollution pathways. It is necessary to research the way in which polluting substances, under certain conditions, are transformed, combined with other substances, temporarily retarded somewhere, and then suddenly released. Nonpoint pollutant transport and transformation processes are greatly influenced by vegetation, land use, and soil processes including snowmelt aspects, and this continues in the unsaturated and saturated zones of soils, in rivers and lakes, and finally in the estuaries in the brackish interface between the river and the sea.

Flow patterns on hillslopes and within the soil profile depend upon the interaction between the modified rainfall inputs described above and drainage basin geology, pedology, and topography. The nature of these interactions determines the division of hillslope outputs into stream quickflow and baseflow and, thus, has important implications for the form of the storm hydrograph and for sediment and solute losses from arid and forest catchments. A fundamental distinction occurs between infiltration-excess overland flow and the subsoil and subsoil-related processes which are highly dependent on site-specific soil moisture and moisture retention characteristics.

The next important hydrological division is at the soil surface. If rainfall intensities exceed the infiltration capacity of the soil, the unabsorbed water (minus any losses from evaporation of surface water stores and once surface detention capacity has been exceeded) runs off downslope via the hydrological process known as infiltration-excess overland flow (Figure 6.1). In forest areas, however, infiltration rates are usually high as a consequence of the high hydraulic conductivity of the forest litter layer,[3] good soil aggregate structure, and the presence of macropore channels formed by roots and soil fauna activities.[4] Thus infiltration-excess overland flow is a relatively limited phenomenon, occurring only where litter dynamics produce ephemeral patches of bare ground,[5] where soil becomes exposed through treefall, and where landsliding exposes the regolith,[6] such as on steep slopes under high rainfalls or seismic triggering. Elsewhere, surface wash and associated soil losses are usually insignificant. However, in some localities, the interactions between rainstorm inputs and soil properties may induce natural surface flow, as in a location[7] where highly transmissive surface soils (saturated hydraulic conduc-

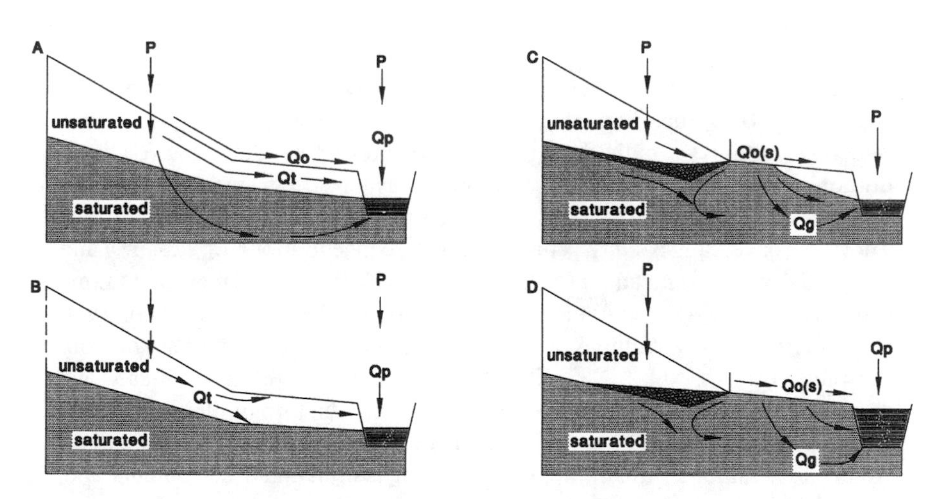


**Figure 6.1 Flowpaths[2] of streamflow sources and changing contributions to streamflow through the storm hydrograph.**

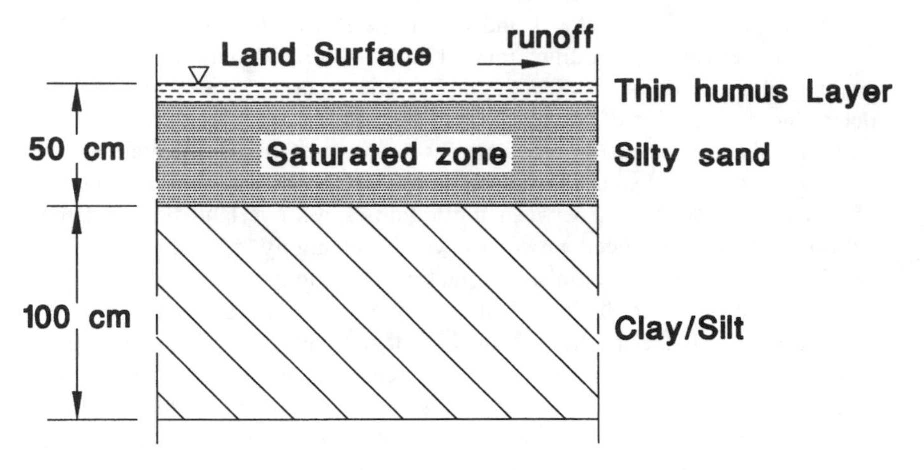


**Figure 6.2 Runoff draining a surface and soil horizons at one of the tropical rain forest sites.**

tivity K = 20 m/d) are underlain at shallow depth by a relatively impermeable subsoil (K = 0.02–0.16 m/d). With prolonged heavy storm rainfall, a perched water table rises to the surface and saturated overland flow results (Figure 6.2).[8] Similar natural occurrences of overland flow have been recorded from the Ivory Coast.[9] Clearly such settings pose a high erosion risk.

Furthermore, hydrological processes interact with sediment transport effects as small-scale landslides in saturated streamhead hollows in humid temperate environments appear to lead to channel network extension.

Different flow routes clearly have different residence times and, therefore, might be expected to show different solute concentrations (Figure 6.3). In particular, macropores create preferred paths along which water moves as channelized flow, thus bypassing the soil matrix. This biphasic flow regime (where water flows rapidly through the large pores while remaining relatively immobile in the fine pore spaces) is important as it restricts leaching of solutes from the soil matrix.[11] At Reserva Ducke, Amazonas, Lal[12] and Nortcliff and Thornes[13] have suggested that the low solute concentrations of stream water (wet and dry season mean: ($\Sigma$Ca + Mg + Na + K = 0.03 meg/l) represent rapid filling and emptying of the macropore system, masking the smaller contribution in terms of flow volume of the longer residence time/higher solute concentration ($\Sigma$Ca + Mg + Na + K = 0.3 meg/l) water within the meso- and micropores. It follows from this study that variations in throughflow chemistry, and thus understanding of hydrological pathways, are unlikely to be detected simply by sampling streamflow outputs from instrumented catchments and, furthermore, that such outputs are unlikely to give a correct picture of nutrient availability to plant rooting systems. Finally, as organic matter decomposition

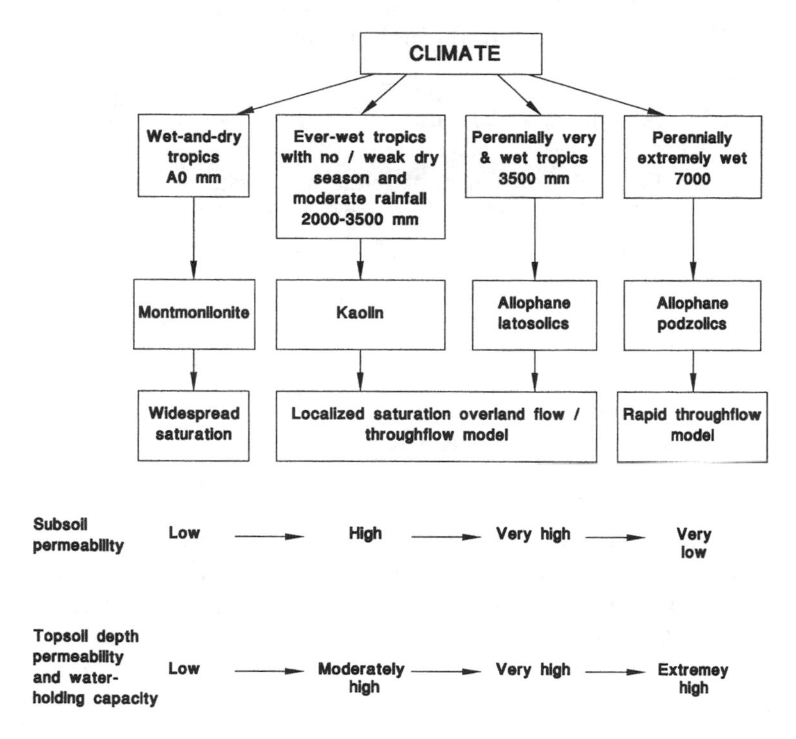


**Figure 6.3 Relationships between climate, soil type, and hydrological model. (From Burt, T.P., in *Solute Processes,* Trudgill, S.T., Ed., Wiley, Chichester, 1986, pp. 193–249. With permission.)**

rates are controlled, at least in part, by degree of saturation, there are likely to be linkages between soil organic matter dynamics and fluctuations in the extent of streamside and floodplain saturated areas.[13]

Surface water pollution by man's activities has been covered by many books, articles, and research papers. Since the main objective of this book is to cover the geohydrologic part related to the environmental impacts, the surface water pollution issue is confined to this limit.

## 6.4 GROUNDWATER POLLUTION

The movement of contaminants in groundwater is a particularly active area of research. Models have been developed to study saltwater intrusion as well as leachate migration from waste disposal sites. Groundwater pollutants can be categorized as bacteria, viruses, nitrogen, phosphorus, metals, organics, pesticides, and radioactive materials. This section covers information on subsurface transport in a general way.

### 6.4.1 Migration of Pollutants in Aquifers

Movement of contaminants in groundwater occurs not only by advection but also by dispersion. *Advection* (also referred to as convection) refers to the transport of a solute at a velocity equivalent to that of groundwater movement.[14] It is considered to be the movement of solute with a rate equal to the average pore water velocity due to the hydraulic gradient which has the form:

$$F_c = q \cdot c \tag{6.1}$$

where $F_c$ = mass flux ($M/T/L^2$)
$q$ = average pore water velocity ($L/T$)
$c$ = concentration ($M/L^3$)

*Dispersion* refers to mixing and spreading caused in part by molecular diffusion and in part by variations in velocity within the porous medium. For many field problems, dispersion caused by molecular diffusion and by flow around grains in the porous medium is negligible in comparison with dispersion caused by large scale heterogeneities within the aquifer. It can simply be defined as the movement of solute due to varying velocity from pore to pore at high velocities and then the dispersive mass flux in X direction:

$$F_x = -D_L \frac{\partial c}{\partial x} \tag{6.2}$$

where $F_x$ = dispersive flux in x direction ($M/T/L^2$)
$D_L$ = dispersion coefficient in the longitudinal direction and has the dimension ($L^2/T$).

Also the dispersive mass flux in y direction becomes,

$$F_y = -D_T \frac{\partial c}{\partial y} \tag{6.3}$$

where $F_y$ = dispersive flux in y direction ($M/T/L^2$)
$D_T$ = dispersion coefficient in the transverse direction ($L^2/T$)

Experiments have demonstrated that, in an isotropic medium, the longitudinal and transverse components of dispersion in Eq. 6.2 and 6.3 are linearly dependent on the average speed of groundwater flow. For a uniform flow field with an average linear velocity equals $V_x$.

$$D_L = a_L V_x \tag{6.4}$$

and

$$D_T = a_T V_x \tag{6.5}$$

where $a_L$ = dispersivity in the longitudinal direction (L)
$a_T$ = dispersivity in the transverse direction (L)

Eq. 6.2 and 6.3 are equivalent to Fick's law. The solute transport governing equations can be obtained in the same way as obtaining the governing equation of groundwater flow. The equation governing the solute transport can be developed by utilizing a conservation of mass approach and employing Ficks law of dispersion. The equation in statement form is:

| net rate of change of mass of solute within | = | flux of solute out of the element | – | flux of solute into the element | ± | loss or gain of solute mass due to reactions |
|---|---|---|---|---|---|---|

The one-dimensional form of the equation for a nonreactive, dissolved constituent in a homogeneous, isotropic aquifer under steady state, uniform flow is:

$$D\frac{\partial^2 c}{\partial x^2} - u\frac{\partial c}{\partial x} = \frac{\partial c}{\partial t} \tag{6.6}$$

where D = coefficient of dispersion in x direction ($L^2/T$)
u = average linear groundwater velocity (L/T)
c = concentration ($M/L^3$)

The two-dimensional equation for a nonreactive, dissolved chemical species in groundwater flow becomes:

$$\frac{\partial}{\partial x}\left(D_L\frac{\partial c}{\partial x}\right) + \frac{\partial}{\partial y}\left(D_T\frac{\partial c}{\partial y}\right) - q_x\frac{\partial c}{\partial x} - q_y\frac{\partial c}{\partial_y} \pm Qc' = \frac{\partial c}{\partial t} \tag{6.7}$$

where $D_L$ and $D_T$ = the hydrodynamic dispersion coefficients in x and y directions respectively.
$c'$ = concentration of the solute of a source/sink of a strength, Q, (assumed to be known)
$q_x$, $q_y$ = effective pore water velocities in x and y directions respectively $\left(\text{generally } q = \frac{Vi}{n_e}\right)$ where $V_i$ is the flow per unit area and $n_e$ is the effective porosity

The analytical solution of Eq. 6.6 where the solute transport problem is one-dimensional flow, was solved[15] for a soil column of length $\ell$ (See Figure 6.4). The boundary conditions represented by the step function input are described mathematically as:

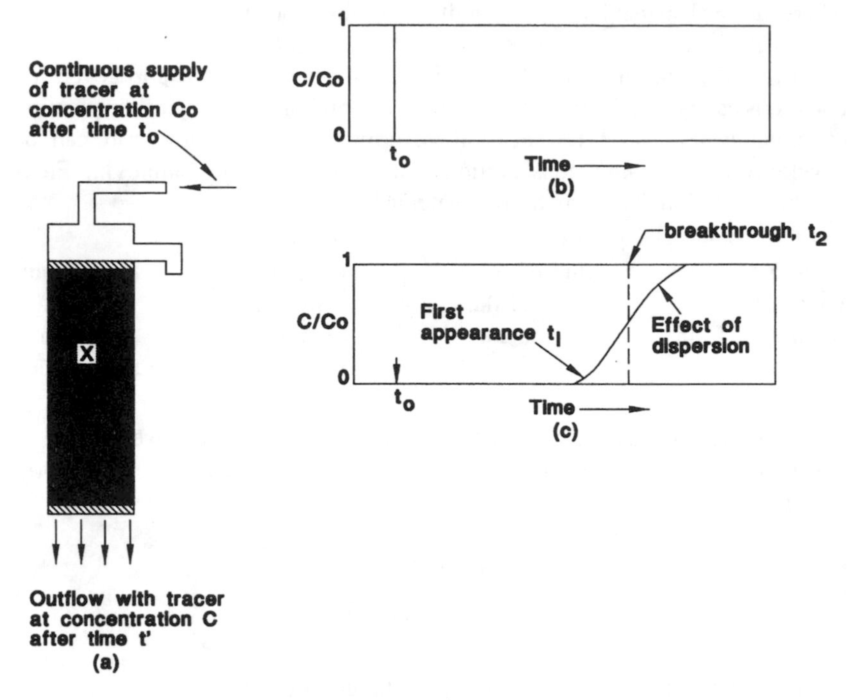


**Figure 6.4 Longitudinal dispersion of a tracer passing through a column of porous medium; (a) column with steadyflow and continuous supply of tracer after time $t_o$; (b) input of tracer; (c) relative tracer concentration in outflow from column (dashed line indicates plug flow condition and solid line illustrates effect of mechnical dispersion and molecular diffusion). (From Shamir, V. and Harleman, D.R.F., *Water Resource Res.*, 3, 2, 1967, 557–581. With permission.)**

$$C(\ell, o) = o \quad \ell \geq o$$
$$C(o, t) = Co \quad t \geq o$$
$$C(\infty, t) = o \quad t \geq o$$

The analytical solution derived[16] is:

$$C = \frac{Co}{2}\left[\exp\left(\frac{ux}{D_L}\right)\operatorname{erfc}\left(\frac{x + ut}{2\sqrt{D_L t}}\right) + \operatorname{erfc}\left(\frac{x - ut}{2\sqrt{D_L t}}\right)\right] \tag{6.8}$$

where u =average linear velocity
$D_L$ =longitudinal dispersion and
$D_T$ = o because the problem is one-dimensional flow, and

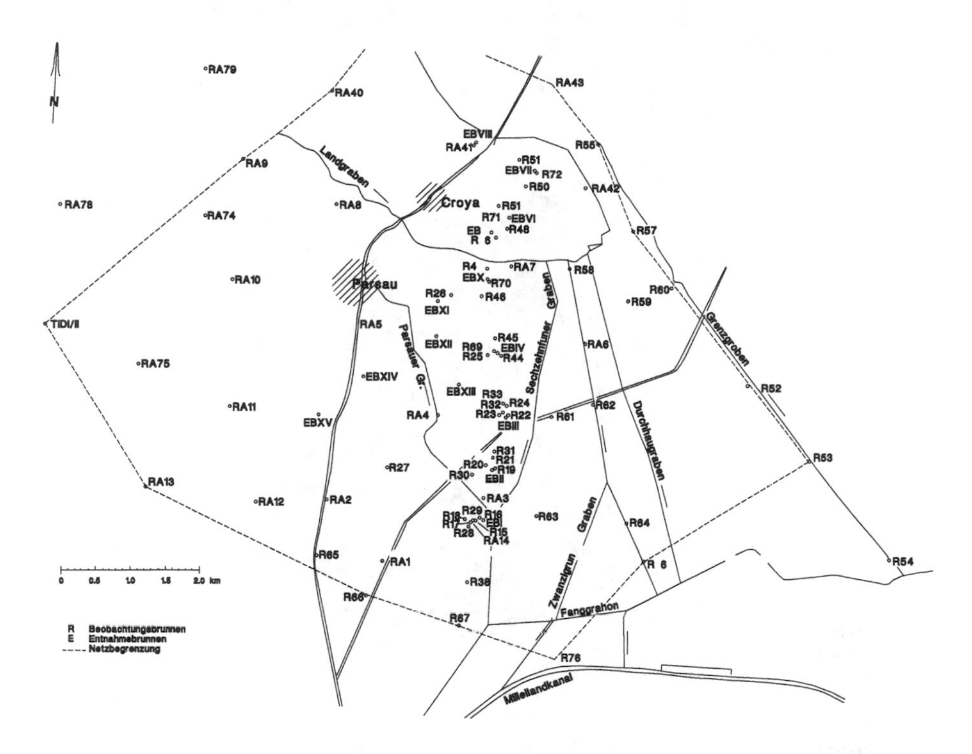


**Figure 6.5 General layout of Ruehn region, Germany.**

$$\mathrm{erfc}(z) = \frac{2}{\sqrt{\pi}} \int_{2}^{\infty} e^{-u_2} du \tag{6.9}$$

Soliman and Hassan[17] applied the finite element methods[18] to solve both the groundwater equation and the solute transport equation for the two-dimensional flow. The finite element model was applied to two aquifers. The first aquifer is located in the Ruehn Valley in Germany (Figure 6.5). The model area was discretized into 1450 elements and 780 nodes (Figure 6.6). The model was constructed to simulate the propagation of a contaminant plume created by injection at a point in the middle of the aquifer thickness and within the zone of influence of a well group. The rate of injection was 10 kg/h for a total time of 2 hours. Figure 6.7 shows the computed concentration isolines three weeks after the injection time.

The second aquifer is located east of the Nile Delta in Egypt (Figure 6.8). The model area is about 4750 sq. km. The Delta aquifer is composed generally of unconsolidated sand and gravel with occasional clay lenses. The aquifer thickness varies between 200 and 700 m. The upper boundary of this formation is a clay cap aquitard with a variable thickness ranging from 0.0 m up to 15 m. The model area was discretized into 543 elements and 310 nodes (Figure

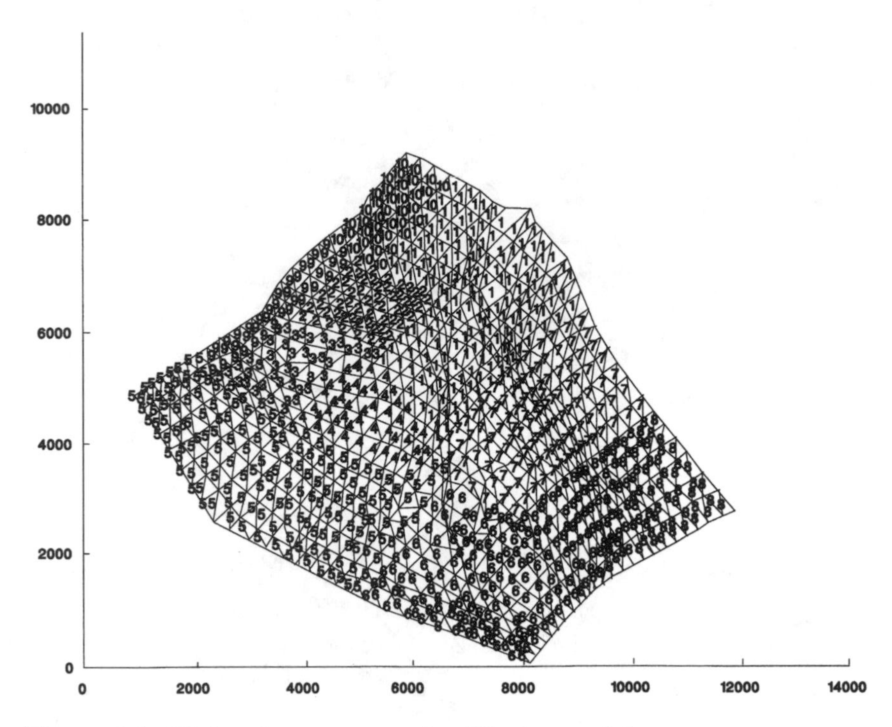


**Figure 6.6 Finite element network of Ruehn model.**

6.9). The same grid was used for both groundwater flow model and solute transport model. Figure 6.10 illustrates the simulated field situation in the form of concentration isolines. For each run the computed concentration values were compared with the observed values at some points where data is available. The calibrated values of the model — longitudinal dispersivity (northwards) and lateral dispersivity (eastwards) — amount to 100 and 10 km, respectively. Figure 6.11 shows the result of computation in isolines form.

## 6.4.2 Saltwater Intrusion

Saltwater intrusion along sea coasts creates a region separating the fresh water from the salt water. For simplicity, this region is sometimes taken as a separating surface and, therefore, may be named as an interface between the fresh water and salt water which is stable only underground. It is being broken up in open basins by diffusion and mixing due to motions resulting from very small potential gradients. The source of the fresh water is either rain water or irrigation water which seeps through the ground surface.

Recently a great deal of emphasis has been placed upon the law of Ghyben-Herzberg. Each of these authors independently made the discovery that in wells near the sea coast the salt water was not encountered at sea level, as they had expected, but at a depth below sea level on the order of forty times

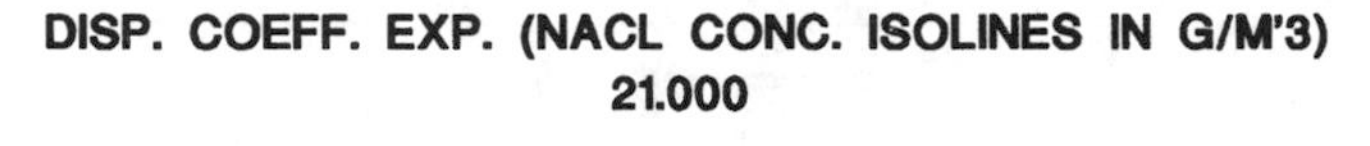


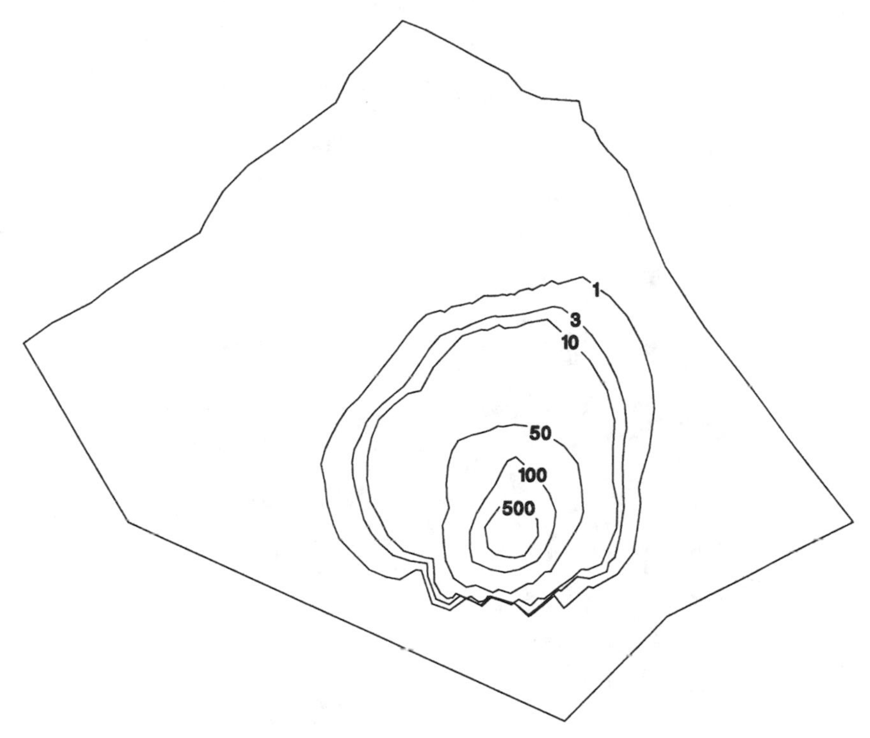


**Figure 6.7 TDS contour map after 21 days (in gm/m³).**

the height of the fresh water above sea level. For this phenomenon each deduced the same explanation, namely, that a static equilibrium existed between the fresh water and the salt water. Following this reasoning, we should have[19] (Figure 6.12).

$$H_s = \frac{e_f}{e_s - e_f} h \tag{6.10}$$

where $H_s$ = depth of freshwater flow below sea level
$e_f$ and $e_s$ = freshwater and saltwater density, respectively
h = freshwater depth above sea level
For, $e_s$ = 1.025 gm/cm³ and $e_f$ = 1 gm/cm³ Eq. 6.10 becomes:

$$H_s = 40 h \tag{6.11}$$

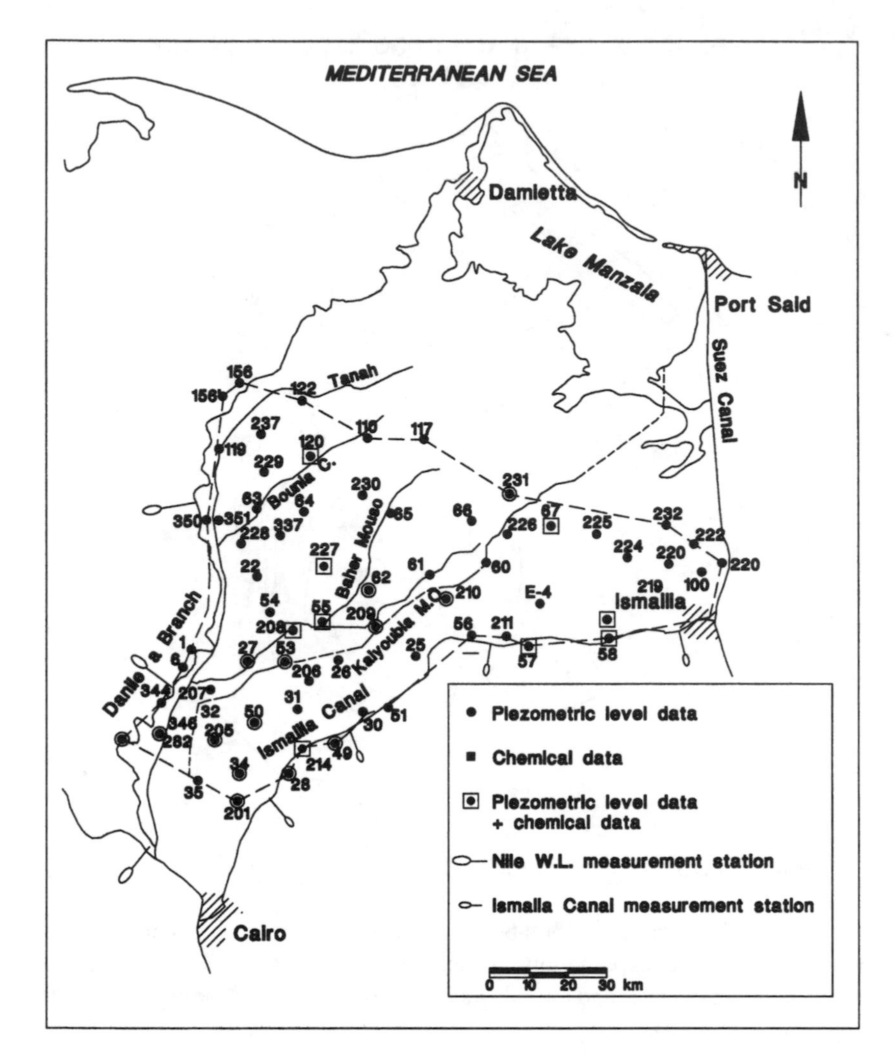


**Figure 6.8 Map of the eastern region of the Nile Delta.**

The assumption upon which Eq. 6.10 is derived is that we are dealing with a case of hydro-static equilibrium, and to the extent that this assumption is valid, the equation is correct.

In groundwater problems, however, the assumption is not valid at all because the fresh water is not at a constant potential but in a state of continuous motion. If no additional water were added by precipitation or irrigation water, the flow of the fresh water would continue until it had all been dissipated, and only salt water with a water table at sea level would remain. Therefore, the potential flow theory should be applied to define the dynamic depth of fresh

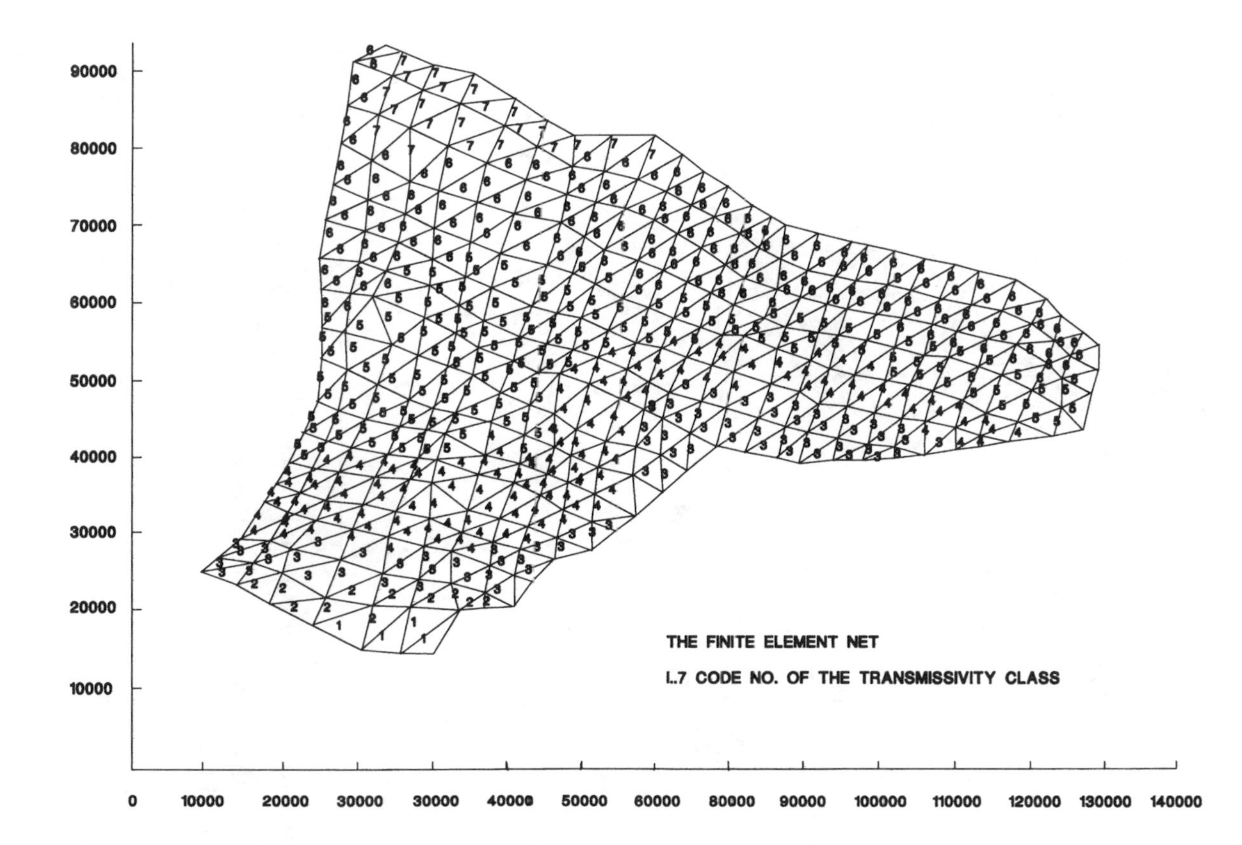


Figure 6.9 Finite element network of Nile Delta model.

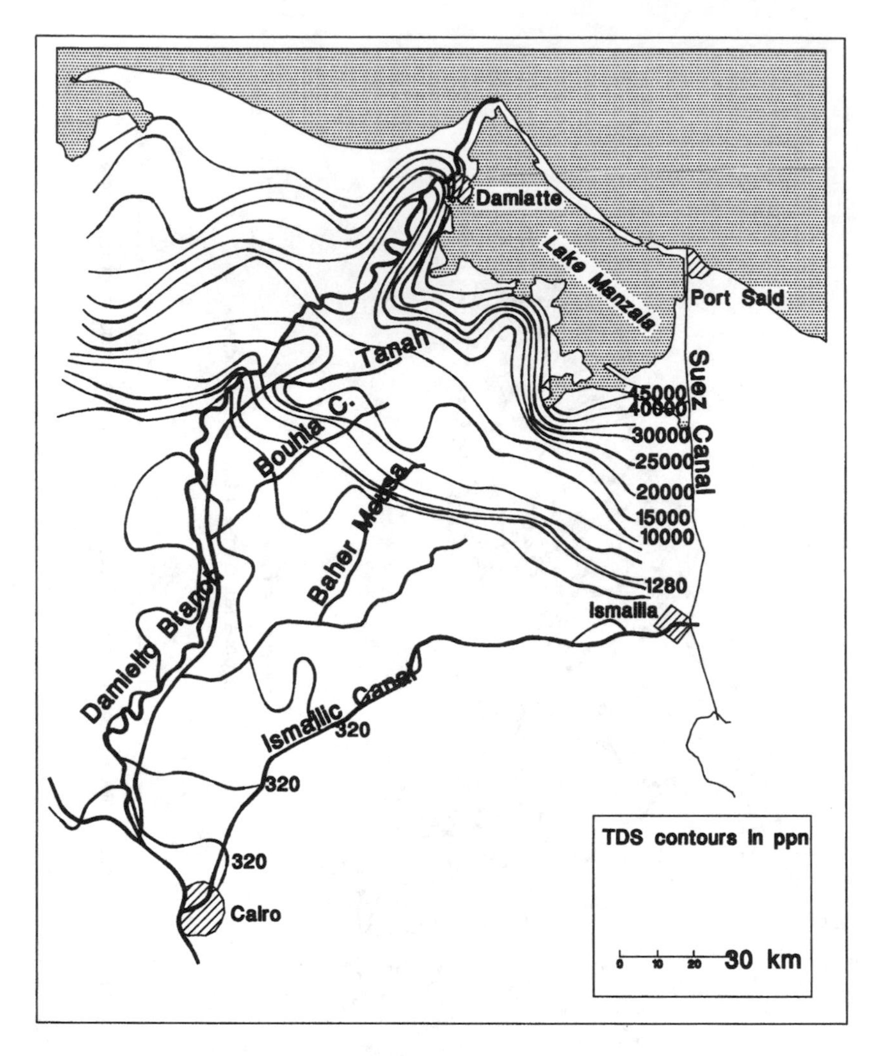


**Figure 6.10 TDS field data in isolines form (in ppm).**

water. Following this procedure,[19] the actual depth (H) accordingly becomes (Figure 6.13):

$$H = \frac{e_s}{e} H_s + \Delta z \tag{6.12}$$

and

$$\Delta z = -\frac{e}{e_s - e} \Delta h \tag{6.13}$$

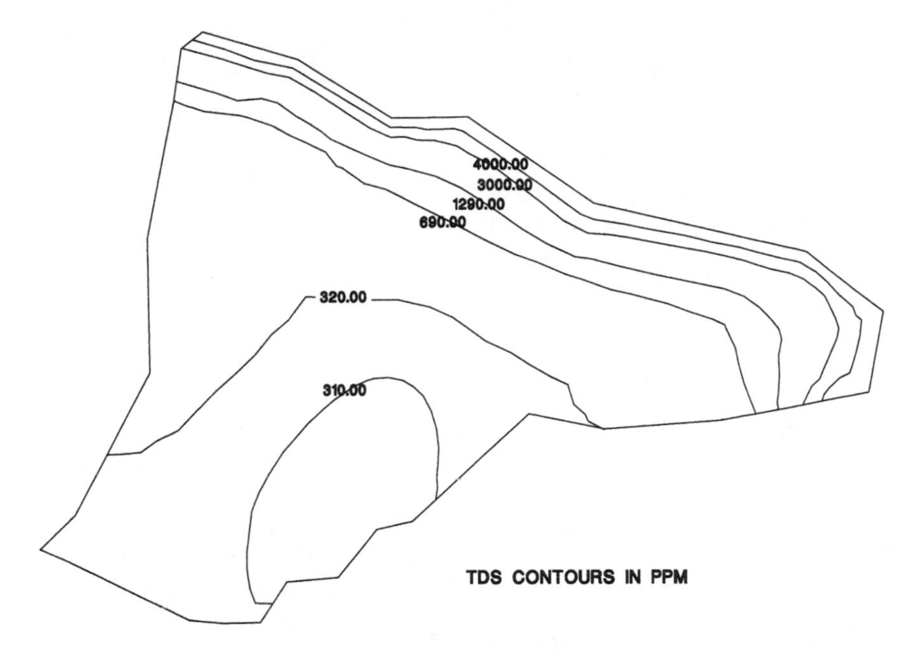


**Figure 6.11 TDS from numerical solution (in ppm).**

where $\Delta z$ = difference between potentials at the interface
$\Delta h$ = difference between water levels

Eq. 6.12 can also have the form,

$$H = H_F + \Delta Z \tag{6.14}$$

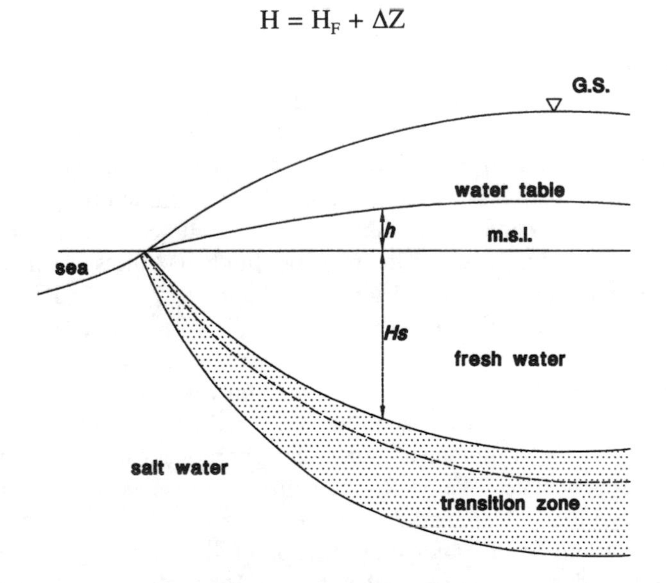


**Figure 6.12 Ghyben-Herzberg freshwater body.**

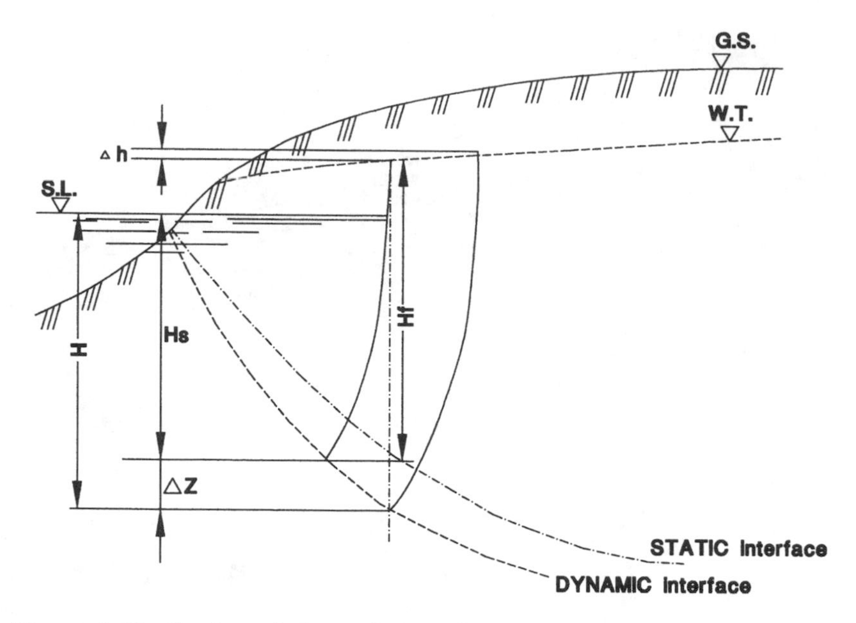


**Figure 6.13 Static and dynamic interfaces.**

This means that the difference between the dynamic depth, H, and the static depth, $H_s$, is $\Delta Z$.

A number of investigations have been carried out over the past few years, most of them based on the Ghyben-Herzberg relationship, in which the fresh water and saline water are considered to be immiscible fluids. In fact, they are miscible, and the sharp interface is not realistic especially when the width of the dispersion zone is considerable. Henry[20] was one of the first investigators who solved the coupled flow and mass transport equation for steady state idealized confined aquifer. Henry[21] also developed the first analytical solution that included the effect of dispersion in confined coastal aquifer under steady state conditions. Shamir and Harleman[22] presented a finite difference method for solution of dispersion problems in steady three-dimensional potential flow field in porous media, in which the miscible fluids had the same density and viscosity. Pinder and Cooper [23] determined the movement of salt water front in confined coastal aquifers including the effect of dispersion. The method of characteristics was used to solve the solute transport equation. Pinder and Frind[24] used Galerkin's procedure in conjunction with the finite element technique to simulate seawater intrusion into confined coastal aquifers.

Lee and Cheng[25] formulated a finite element model using stream function to obtain a steady state solution for the convective-dispersive transport equation. Segol and Pinder[27] applied the finite element technique to the solution of transport equation. Frind[26] studied the case of a fully confined aquifer overlain by an aquitard that extends out under the sea. He applied the Galerkin finite element technique with linear rectangular elements. The solution was based on a linear interpolation for the potential heads and the concentration between nodal points. Kawatani[28] presented a two-dimensional model for the behavior intrusion in layered coastal aquifers. He addressed the instability problems which occurred when the convective terms exceeded a certain criterion related to the dispersion coefficients and the size of the finite elements. Kawatani took the longitudinal dispersion coefficient as a constant to avoid the instability and other numerical problems. Pandit and Anand[29] did a parametric study on an idealized confined aquifer with the same boundary conditions as applied by Henry. Their results indicated that the depth of the aquifer influenced the extent of saline water intrusion into the aquifer. It was also concluded that cyclic flow existed when the characteristic velocity at the seaward boundary was bigger than longitudinal dispersivity. Huyakorn et al.[30] developed a three-dimensional finite element model for simulation of saltwater intrusion in single and multiple coastal aquifers with either a confined or phreatic top aquifer. The model employed the Picard sequential solution algorithm with special provisions to enhance convergence of the alternative solution.

Some of the above studies have reported the existence of cyclic flow near the sea boundary (Figure 6.14). Because of its higher density, the seawater migrates to the bottom of the aquifer and mixes with the fresh water. This mixed water is of lower density, and it finds its way back again to the sea from the upper part of this boundary.

The finite element model (2D-FED) was designed by Sherif[31] for the saltwater intrusion problem north of the Nile Delta in Egypt.

The Nile Delta groundwater reservoir extends over six million acres and is naturally bounded northward by the Mediterranean Sea and eastward by the Suez Canal (Figure 6.15). The western boundary extends well into the desert. The Nile Delta aquifer system is a complex groundwater system. It is a leaky one, with an upper semipermeable boundary (clay cap) and lower impermeable boundary.

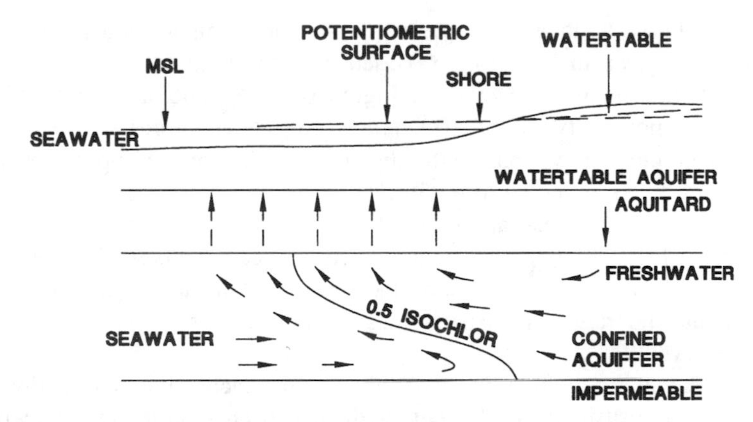


**Figure 6.14 Cyclic flow near the sea side.**

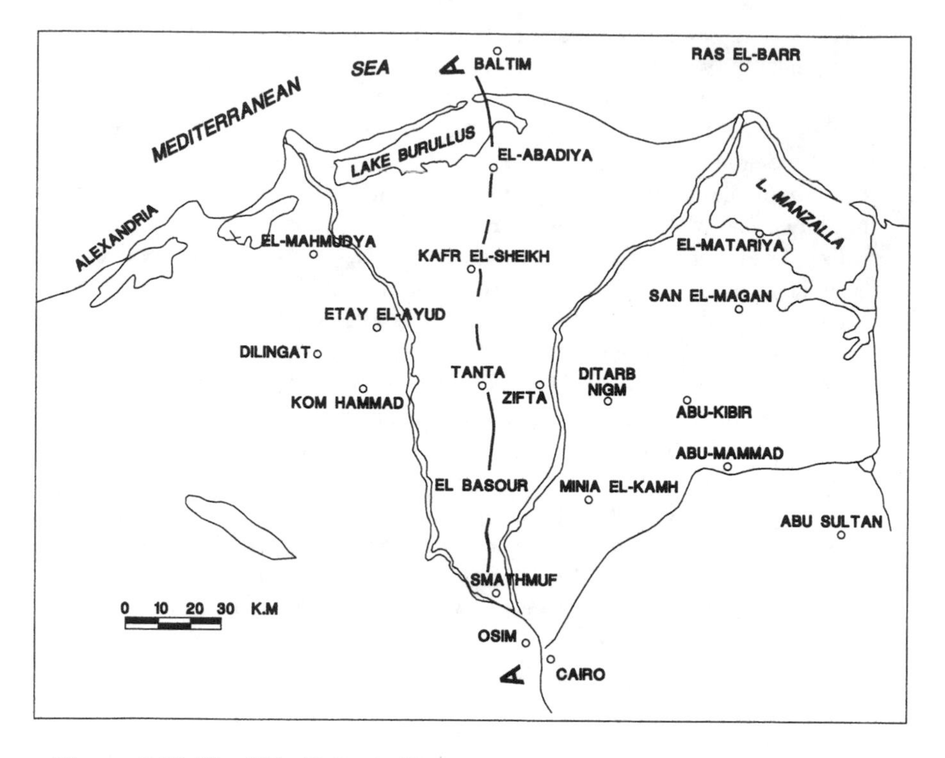


**Figure 6.15 The Nile Delta in Egypt.**

The Nile Delta aquifer is subjected to the problem of saltwater intrusion from both the Mediterranean Sea to the north and the Suez Canal to the east. As with any coastal aquifer, an extensive saltwater wedge has intruded into the coastal part of the Nile Delta aquifer forming the major constraint against aquifer exploitation.

The 2D-FED model[31] is applied to the longitudinal geological cross-section A-A given in Figure 6.16. Based on some field data, the depth and length of the domain are given in Figure 6.17. A calibrated value of the hydraulic conductivity, K, of 100 m/d is considered representative of the aquifer medium. The vertical hydraulic conductivity for the upper semi-pervious layer, $K_z$, is set equal to 0.05 m/d.

The piezometric head at the land boundary is 14.0 m above sea level. A piezometric level of 0.6 m was observed near the sea boundary. The free water table level is given for some stations and is assumed linear between them. The longitudinal and transverse dispersivities $a_L$ and $a_T$ are assumed 100.0 and 10.0 m, respectively.

The domain is divided into five subdomains; each subdomain is divided into a number of triangular elements with smaller areas in the regions where the variation in the concentration gradient is relatively high. An intensive grid is adopted near the sea boundary. The domain is finally represented by a

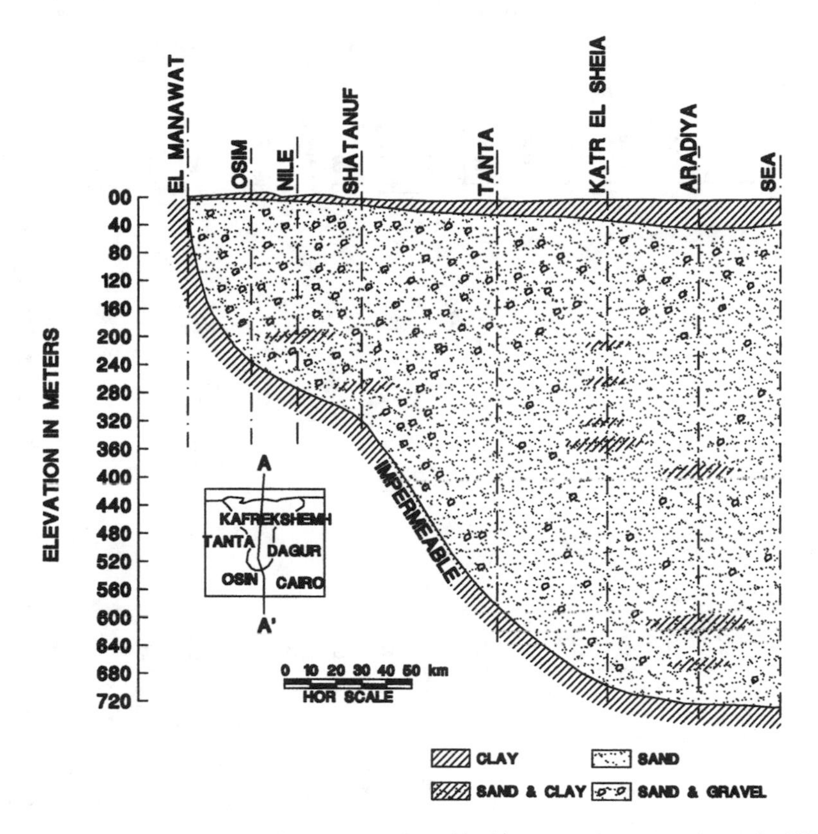


**Figure 6.16 Lithological cross section (A-A) normal to the shore in Nile Delta aquifer.**

nonuniform grid with 4020 nodes and 7600 triangular elements. The solution converges to an accuracy of $10^{-5}$ after fourteen iterations.

It can be concluded from the shape of the equipotential lines that the depth of the window at the seaside is about 350 m (Figure 6.17). There is some upward flux of the mixed water through the upper semipervious layer within a distance of 22.0 km from the sea boundary. Strong cyclic flows in the Nile Delta aquifer occur at the sea boundary.

## 6.4.3 Landfill Leachate

Solid wastes deposited in a landfill degrade chemically and biologically to produce solid, liquid, and gaseous products. Ferrous and other materials are oxidized; organic and inorganic wastes are utilized by microorganisms through aerobic and anaerobic synthesis. Liquid waste products of microbial degradation, such as organic acids, increase chemical activity within the fill. Food wastes degrade quite readily, while other materials, such as plastics, rubber, glass, and some demolition wastes, are highly resistant to decomposition.

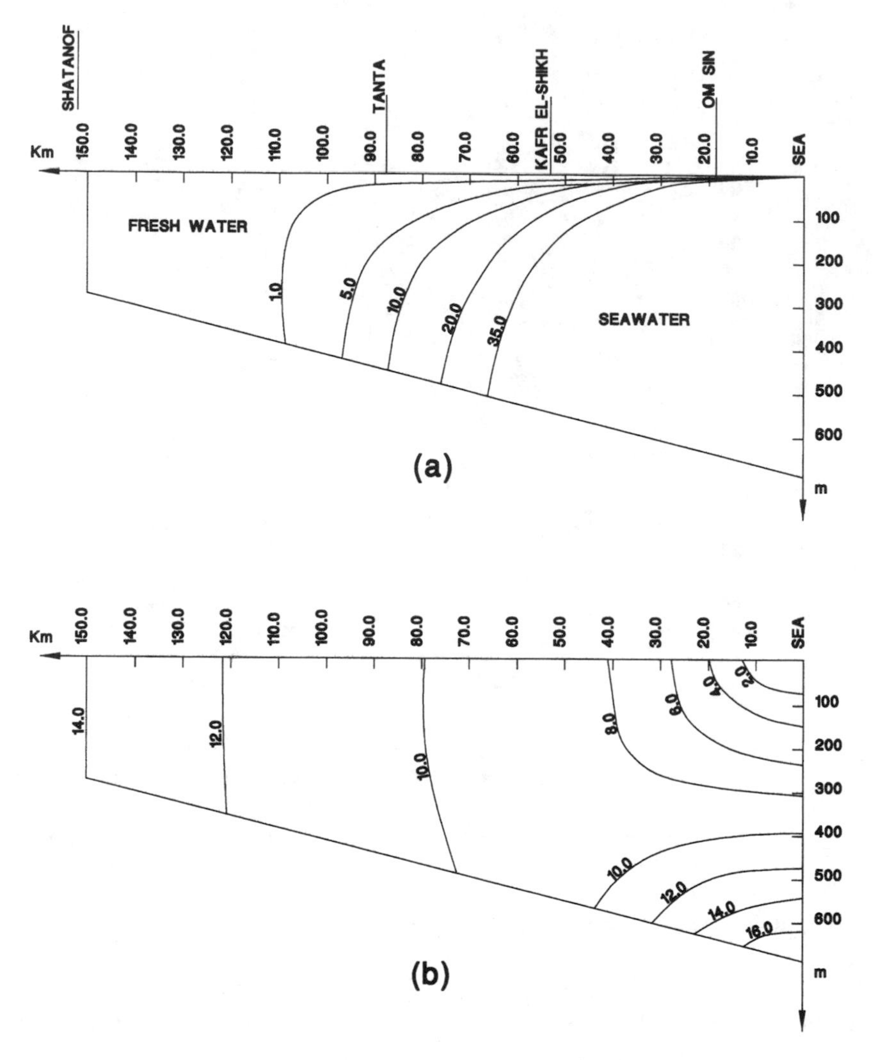


**Figure 6.17 (a) Equi-concentration lines (× 1,000 ppm); (b) equi-potential lines from the model in Nile Delta aquifer.**

Some factors that affect degradation are the heterogeneous nature of the wastes; their physical, chemical, and biological properties; the availability of oxygen and moisture within the fill; temperature; microbial populations; and type of synthesis. Since the solid wastes usually form a heterogeneous mass of nonuniform size and variable composition and other factors are complex, variable, and difficult to control, it is not possible to accurately predict contaminant quantities and production rates.

Biological activity within a landfill generally follows a set pattern. Solid waste initially decomposes aerobically, but as the oxygen supply is exhausted, facultative and anaerobic microorganisms predominate and produce methane gas, which is odorless and colorless. Temperature rises to the high mesophilic-low thermophilic range (15 to 66°C) because of microbial activity.

Characteristic products of aerobic decomposition of waste are carbon dioxide, water, and nitrate. Typical products of anaerobic decomposition of waste are methane, carbon dioxide, water, organic acids, nitrogen, ammonia, and sulfides of iron, manganese, and hydrogen.

Leachate percolating through soils underlying and surrounding the solid waste is subject to purification (attenuation) of the contaminants by ion exchange, filtration, adsorption, complexing, precipitation, and biodegradation. It moves either as unsaturated flow, if the voids in soil are only partially filled with water, or as a saturated flow, if they are completely filled. The type of flow affects the mechanism of attenuation as do soil particle size and shape and soil composition.

Attenuation of contaminants flowing in the unsaturated zone is generally greater than in the saturated zone because there is more potential for aerobic degradation, adsorption, complexing and ion exchange or organics, inorganics, and microbes. Aerobic degradation of organic matter is more rapid and complete than anaerobic degradation. Because the supply of oxygen is extremely limited in saturated flow, anaerobic degradation prevails. Adsorption and ion exchange are highly dependent on the surface area of the liquid and the solid interface.

Leachate travel in the saturated zone is primarily controlled by soil hydraulic conductivity and hydraulic gradient, but a limited amount of capillary diffusion and dispersion do occur. The leachate is diluted very little in groundwater unless a natural geologic mixing basin exists.

Bouwer[32] notes that landfills behave essentially as point sources of pollution. When leachate from a landfill reaches the groundwater, it travels in the general direction of the groundwater movement. The high chloride ion concentration usually provide early indication of the presence of a leachate in groundwater.[33]

The shape and areal extent of a leachate contaminant plume for a thoroughly characterized landfill are shown by contours in Figure 6.18.

Figure 6.19 shows the progress of the movement of leachate in anisotropic aquifer system with significant variation in hydraulic conductivities of the subsurface geologic material. This figure shows the time sequence of the movement of plume and the network of the monitoring system.

The movement of the plume in saturated carbonate rocks is controlled by fracture and solution channels. The movement of the contaminated groundwater (plume) is generally directional along discontinuities and dissolution-widened joints. Figure 6.20 illustrates the movement of groundwater and plume in a fractured and karstified carbonate rock.

Additional examples as related to aquifer degradation, landfill leachate, and industrial contaminants plume, configuration, and movement are covered in Chapter 7, the Appendices, and diskette.

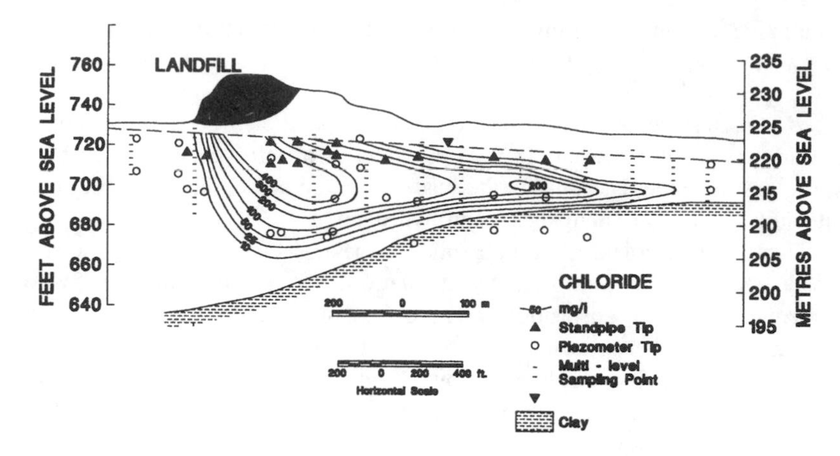


Figure 6.18 Example leachate Water Quality Plume-Field determined. (From Sara, M.N., *Standard Handbook for Solid and Hazardous Waste Facility Assessments,* Lewis Publishers, Boca Raton, FL, 1994, pp. 10–68. With permission.)

## 6.5 GROUNDWATER MONITORING

Water quality characteristics must be established and trends observed. A comprehensive investigational program for evaluation of these factors is essential. Groundwaters should be sampled regularly on a continuing basis. Sam-

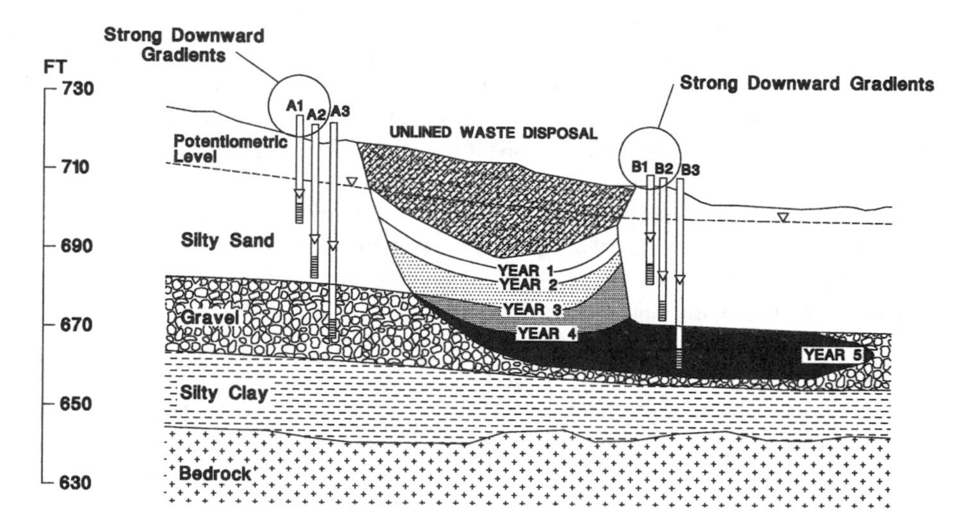


Figure 6.19 Progress of Leachate Plume in Anisotropic aquifer. (From Sara, M.N., *Standard Handbook for Solid and Hazardous Waste Facility Assessments,* Lewis Publishers, Boca Raton, FL, 1994, pp. 10–68. With permission.)

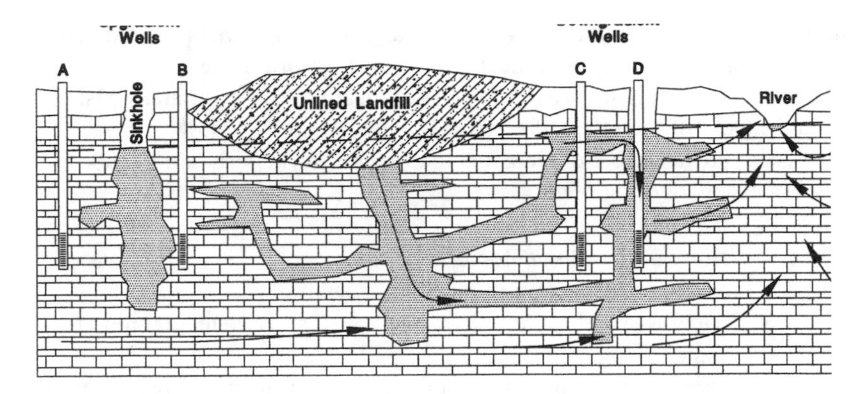


**Figure 6.20 Movement of Leachate Plume in karstified carbonate rock. (From Sara, M.N., *Standard Handbook for Solid and Hazardous Waste Facility Assessments,* Lewis Publishers, Boca Raton, FL, 1994, pp. 10–68. With permission.)**

pling points should include shallow and deep wells, selected with due consideration to areal distribution, geologic and hydrologic influences, pattern of pumpage, and waste disposal. Practice samples should be obtained at the peak of the pumping season, when water levels are at their lowest and also, if possible, during the period of replenishment when levels are high.

In areas vulnerable to sea-water intrusion, particular attention should be given to changes in chloride concentration. Total dissolved solids content is significant where salt balance problems exist. Phenols, boron, and heavy metals are usually critical where large volumes of industrial waste are involved. Domestic sewage disposal particularly influences nitrate, phosphate, and detergent content. These water quality observations, in order to be properly evaluated and interpreted, must be augmented with adequate information on geology, hydrology, rate and pattern of extractions, and sewage disposal practices.

In general, a groundwater management program should be used to indicate potential problems so that timely corrective action can be taken. The collection of basic data should not become an end in itself, but should be an integral part of a comprehensive, balanced program of continuing examination and interpretation of all factors involved in protection of the aquifer and maintenance of groundwater quality.

The adequacy of a groundwater monitoring program[35] hinges, in large part, on the quality and quantity of the hydrogeologic data used in designing and implementing the program. For example, sites with more heterogeneous subsurfaces will require more hydrogeologic information to provide a reasonable assurance that well placements will intercept contaminant migration pathways. Likewise, investigating techniques that may be appropriate in one setting (given certain waste characteristics and geologic features) may be inappropriate in another.

Once contaminant leakage in any site has been detected via detection monitoring efforts, a more aggressive groundwater program called assessment monitoring must be undertaken. Specifically, the researchers must determine the vertical and horizontal concentration profiles of all the hazardous waste constituents in the plumes escaping from waste management areas beside the establishment of the rate and extent of contaminant migration.

There are a number of elements that the researchers should include in the assessment monitoring plan:

1. description of hydrogeologic conditions
2. description of detection monitoring system
3. description of a short term statistical analysis of detection monitoring data
4. description of approach for conducting assessment as direct or indirect methods of investigation or as mathematical modeling of contaminant movement
5. discussion of number, location, and depth of wells
6. information on well design and construction
7. description of sampling and analysis procedures
8. determinations of the rate of migration and the extent and hazardous constituent composition of the contaminant plume
9. specifying a schedule of implementation of the assessment plan

There are several different t-tests which can be used to analyze interim status detection monitoring groundwater data. Because of its simplicity and reliability, a t-test termed the Averaged Replicate (AR) t-test is presented in this section.

In AR t-test the background mean and variance must be calculated for the first year. This is done by first averaging the replicate measurements and then using these replicate averages to calculate the background mean and variance as described below.

Background Mean:

$$X_{iJ} = \sum_{K=1}^{n} (X_{iJ})_K / n \tag{6.15}$$

$$X_b = \sum_{i=1}^{n_1} \sum_{J=1}^{n_2} (X_{iJ} / M) \tag{6.16}$$

where n = The number of replicate measurements; $n_1$ and $n_2$ are the number of wells and the number of sampling periods, respectively. M is equal to $n_1$ times $n_2$

$X_b$ = Background mean

Background variance:

$$S_b^2 = \sum_{i=1}^{n_1} \sum_{j=1}^{n_2} (X_{ij} - X_b)^2 / (M - 1) \tag{6.17}$$

The data for each parameter from each monitoring well, from each sampling event (upgradient and downgradient wells from the site) after the first year, must be compared individually with the background data collected during the first year. At least four replicate measurements should be taken from each well for each indicator parameter (IP) during every quarter to semi-annual sampling event. These monitoring data are used to calculate a mean and variance for every IP at every monitoring well each time the well system is sampled.

The mean ($X_m$) for monitoring well m for the AR t-test is:

$$X_m = \sum_{k=1}^{N_m} X_{km} / N_m \tag{6.18}$$

where $X_{km}$ = the Kth replicate measurement from the mth monitoring well where K = 1 to $N_m$

$N_m$ = number of replicate measurements from the monitoring well m

The AR t-statistic is calculated as follows:

$$t^* = \frac{X_m - X_b}{S_b^2 \; 1 - 1/M} \tag{6.19}$$

The critical t-statistic ($t_c$) is obtained directly from Table 6.1. The $t_c$ is the value from Table 6.1 which corresponds to M – 1 degrees of freedom (Note: If pH is being tested, use the two-tailed critical values; otherwise use the one-tailed critical values).

The t* is then compared with $t_c$ using the following decision rules:

- If total organic carbon (TOC) and total organic halogens (TOX) are being evaluated and if t* is less than $t_c$, then there is no statistical indication that the IP concentrations are larger in the well under comparison than in the background data. If t* is larger than $t_c$, there is a statistical indication that IP concentrations are larger in the well under comparison.
- If pH is being evaluated and if |t*| (which is the absolute value of t*) is less than $t_c$, then there is no statistical indication that pH has

changed. If $|t^*|$ is larger than $t_c$, there is an indication that pH has changed statistically. If $t^*$ is negative, then pH increased; if $t^*$ is positive, pH decreased.

**Table 6.1 One- and Two-Tailed Critical t Values at the 0.1 Level of Significance**

| Degrees of freedom | One-tailed | Two-tailed |
|---|---|---|
| 1 | 31.821 | 62.657 |
| 2 | 6.965 | 9.925 |
| 3 | 4.541 | 5.841 |
| 4 | 3.747 | 4.604 |
| 5 | 3.365 | 4.032 |
| 6 | 3.143 | 3.707 |
| 7 | 2.998 | 3.499 |
| 8 | 2.896 | 3.355 |
| 9 | 2.821 | 3.250 |
| 10 | 2.764 | 3.169 |
| 11 | 2.718 | 3.106 |
| 12 | 2.618 | 3.055 |
| 13 | 2.650 | 3.012 |
| 14 | 2.642 | 2.977 |
| 15 | 2.602 | 2.947 |
| 16 | 2.583 | 2.921 |
| 17 | 2.567 | 2.898 |
| 18 | 2.552 | 2.878 |
| 19 | 2.539 | 2.861 |
| 20 | 2.528 | 2.845 |
| 21 | 2.518 | 2.831 |
| 22 | 2.508 | 2.819 |
| 23 | 2.500 | 2.807 |
| 24 | 2.492 | 2.797 |
| 25 | 2.485 | 2.787 |
| 26 | 2.479 | 2.779 |
| 27 | 2.473 | 2.771 |
| 28 | 2.467 | 2.763 |
| 29 | 2.462 | 2.756 |
| 30 | 2.457 | 2.750 |
| 40 | 2.423 | 2.704 |
| 60 | 2.390 | 2.660 |
| 120 | 2.358 | 2.617 |
| ∞ | 2.326 | 2.576 |

Adapted from Table III, Statistical Tables for Biological, Agricultural, and Medical Research.[35]

## REFERENCES

1. UNESCO, Hydrology and Water Resources for Sustainable Development in a Changing Environment, 1990.
2. Ward, R.C., On the response to precipitation of headwater streams in humid areas, *Journal of Hydrology,* 79, 1984.
3. Walsh, R.P.D. and Voigt, P., Vegetation litter, an underestimated variable in hydrology and geomorphology, *Journal of Biogeography,* 4, 1977, pp. 253–254.
4. Lal, R., *Tropical Ecology and Physical Edaphology,* Wiley, London, 1987.
5. Spencer, T., Douglas, I., Greer, T., and Sinun, W., Vegetation and fluvial geomorphic process in Southeast Asian tropical rain forests, in *Vegetation and Erosion,* Thornes, J.B., Ed., Wiley, Chichester, 1990, pp. 451–469.
6. Garwood, N.C., Janos, D.P., and Brokaw, N., Earthquake-caused landslides: A major disturbance to tropical forest, *Science,* 205, 1979, pp. 997–999.
7. Bonell, M. and Gilmore, D.A., The development of overland flow in a tropical rain forest catchment in Northeast Queensland, *Earth Surface Processes and Landforms,* 8, 1983, pp. 253–272.
8. Douglas, I. and Spencer, T., *Environmental Change and Tropical Geomorphology,* Allen and Unwin, London, 1985, pp. 39–73.
9. Wierda, A., Veen, A.W., and Hughes, R.W., Infiltration at the Tai rain forest (Côte d'Ivoire): measurements and modelling, *Hydrological Processes,* 3, 1989, pp. 371–382.
10. Burt, T.P., Runoff and denundation rates on temperate hillslopes, in *Solute Processes,* Trudgill, S.T., Ed., Wiley, Chichester, 1986, pp. 193–249.
11. Solins, P. and Radulovich, R., Effects of physical structure on solute transport in a weathered tropical soil, *Journal of the Soil Science Society of America,* 52, 1988, pp. 1162–1173.
12. Lal, R. and Russell, E.W., *Tropical Agricultural Hydrology,* John Wiley & Sons, New York, 1981, pp. 37–57.
13. Nortclif, S. and Thorns, J.B., The dynamics of a tropical flood plain environment with reference to forest ecology, *Journal of Biogeography,* 15, 1988, pp. 49–59.
14. Roberts, P.V., Reinhard, M., and Valocchi, A.J., Movement of organic contaminants in groundwater: implications for water supply, *Journal of the American Water Works Association,* August 1982, pp. 408–413.
15. Bear, J., *Hydraulics of Groundwater,* McGraw-Hill, New York, 1979, chap. 8.
16. Canter, L.W., Knox, R.C., and Fairchild, D.M., *Groundwater Quality Protection,* Lewis Publishers, Boca Raton, FL, 1987, chap. 6.
17. Soliman, M.M. et al., Groundwater quality model with applications to various aquifers, *Environmental Geology Water Science,* 17, 3, Springer-Verlag, New York, 1991, pp. 201–208.
18. Huyakorn, P.S. and Pinder, G.F., *Computational Methods in Subsurface Flow,* Academic Press, New York, 1983.
19. Soliman, M.M., *Groundwater Management in Arid Regions,* Ain Shams University Press, Vol. 1, 1984, pp. 179–187.
20. Henry, H.R., Saltwater intrusion in coastal aquifers, International Association of Scientists, Hydrol. Publ., 52, 1960, pp. 478–487.

21. Henry, H.R., Effect of Dispersion on Salt Encroachment in Coastal Aquifers, U.S. Geol. Surv., Water Supply Paper, 1613-C, 1964.
22. Shamir, V. and Harleman, D.R.F., Numerical solution for dispersion in porous media, *Water Resource Research,* 3, 2, 1967, pp. 557–581.
23. Pinder, G.F. and Cooper, H.H., A numerical technique for calculating the transient position of the saltwater front, *Water Resources Research,* 6, 3, 1970, pp. 875–882.
24. Pinder, G.F. and Frind, E.O., Application of Galerkin's procedure to aquifer analysis, *Water Resources Research,* 8, 1, 1972, pp. 108–120.
25. Lee, C.H. and Cheng, R.T., On seawater encroachment in coastal aquifers, *Water Resources Research,* 10, 5, 1974, pp. 1039–1043.
26. Segol, G. and Pinder, G.F., Transient simulation of saltwater intrusion in southeastern Florida, *Water Resources Res.,* 12, 1976, 65–70.
27. Frind, E.O., Seawater Intrusion in Continuous Coastal Aquifer-Aquired System, Proceeding, 3rd Int. Conf. Finite Elements Water Resources, Univ. of Mississippi, Oxford, 1980.
28. Kawatani, T., Behavior of Seawater Intrusion in Layered Coastal Aquifers, Proc. 3rd Int. Conf., Finite Element Water Resources, Univ. of Mississippi, Oxford, 1980.
29. Pandit, A. and Anand, S.C., Groundwater Flow and Mass Transport by Finite Elements — A Parametric Study, Proc. 5th Int. Conf., Finite Elements Water Resources, Univ. of Vermont, 1984.
30. Huyakorn, P.S., Anderson, P.F., Mercer, J.W., and White, H.O., Saltwater intrusion in aquifers: Development and testing of a three-dimensional finite element model, *Water Resources Research,* 23, 2, 1987, pp. 293–312.
31. Sherif, M.M., Singh, V.P., and Amer, A.M., A two-dimensional finite element model for dispersion (2D-FED) in coastal aquifers, *Journal of Hydrology,* 103, 1988, pp. 11–36.
32. Boumer, H., *Groundwater Hydrology,* McGraw Hill, New York, 1978.
33. Freeze, R.A., and Cherry, J.A., *Groundwater,* Prentic-Hall, Englewood Cliffs, NJ, 1979.
34. Sara, M.N., *Standard Handbook for Solid and Hazardous Waste Facility Assessments,* Lewis Publishers, Boca Raton, FL, 1994, pp. 10–68.
35. RCRA, *Ground-Water Monitoring Technical Enforcement Guidance Document,* EPA, Washington, DC, August 1985.

# 7 WASTE MANAGEMENT FOR GROUNDWATER PROTECTION

## 7.1 PRIMARY CONCEPT

Environmental effects are considered the major problem with groundwater management[1] and protection. The major factors in considering the suitability of a water supply are water quality requirements and limitations associated with its uses. Various criteria have been developed covering all categories of water quality, including bacterial characteristics and chemical constituents.

The removal or neutralization of undesirable chemical characteristics is often both difficult and expensive. Criteria of general application for use in evaluating the chemical aspects of water quality should be generally considered as guides and indicators of desirable water quality and not as absolute standards for all applications.

The United States Public Health Service has developed standards for physical, chemical, and bacterial quality of drinking water as shown in Table 7.1. These standards have been widely adopted.

Bacterial standards are expressed in a complex relationship between the number of samples to be analyzed and the allowable number of coliform organisms in these samples. In effect, average monthly coliform is limited to a most probable number (MPN) of one per 100 ml of sample.

General specifications further provide that there shall be no objectionable odors or tastes; turbidity shall not exceed 10 ppm (silica scale); and color shall not exceed 20 (platinum-cobalt scale). In addition these standards, concentration of nitrate shall not be in excess of 10 ppm, as nitrogen (44 ppm as $NO_3$) in domestic water supplies has been determined to be harmful to infants.

Although hardness does not ordinarily affect the provision of drinking water, it is important in general for industrial water usage. Excessive hardness causes increased consumption of soap and induces the formation of scale in pipes and fixtures. The following standards have been formulated by the U.S. Geological Survey:

| Class | Range of hardness in ppm |
|---|---|
| Soft | 0–55 |
| Slightly hard | 56–100 |
| Moderately hard | 101–200 |
| Very hard | 201–500 |

The suitability of water for irrigation use depends upon such factors as soil texture and composition, crop types, irrigation practices, and chemical characteristics of the water supply. Sodium can be a significant factor in evaluating irrigation water quality because of its potential effect on soil structure. A standard of classification based upon the total salinity and the relative proportion of sodium in irrigation water has been developed by the Salinity Laboratory of the USDA. The classification makes use of the sodium adsorption ratio (SAR) of soil solution which is defined as:

$$SAR = \frac{Na^{+}}{\sqrt{(Ca^{2+} + Mg^{2+})/2}} \tag{7.1}$$

in which Na, Ca, and Mg are expressed in meg/l.

**Table 7.1 Limiting Concentrations of Mineral Constituents for Drinking Water[15]**

| Constituent | Limits in ppm |
|---|---|
| **Mandatory Limits** | |
| Fluoride (F) | 1.0 |
| Lead (Pb) | 0.1 |
| Selenium (Se) | 0.05 |
| Hexavalent chromium ($^{+6}Cr$) | 0.05 |
| Arsenic (As) | 0.05 |
| **Nonmandatory Limits (But Recommended)** | |
| Iron (Fe) and Manganese (Mn) together | 0.3 |
| Magnesium (Mg) | 125 |
| Chloride (Cl) | 250 |
| Sulfate ($SO_4$) | 250 |
| Copper (Cu) | 3.0 |
| Zinc (Zn) | 15 |
| Phenols | 0.001 |
| **Total solids, desirable** | **500** |
| **Permitted** | **1,000** |

There are many environmental effects on groundwater quality. These effects are due to waste disposal such as sewage, industrial waste, cooling water, radioactive wastes, dump sites, watershed influences, connate waters, and sea water intrusion.

In formulating an effective plan for control and recovery of groundwater quality, full consideration must be given to geology, hydrology, and cultural development within the area of the groundwater reservoir.

## 7.2 ALTERNATIVE OF WASTE DISPOSAL

Environmental protection has taken its place beside efficient manufacturing and man's other activities. Pressures to handle wastes properly arise from several sources:

- legislative and regulatory actions
- concerns over known or suspected effects of a specific material, which is as-yet unregulated
- process economics improvement through waste reduction
- conservation of resources, including water
- protection of workers' health

Responses by all concerned people to these pressures largely demonstrate a high degree of responsibility. However, of concern to all parties is a comparison of the cost of the environmental protection measures versus their true benefits to the environment.

Alternatives of waste disposal are to provide practical technology useful in selecting and designing various environmental protection operations and processes which economically and reliably fill identified needs. The achievement of cost-effective means of disposing of liquid and solid wastes[3] involves the systematic consideration of a range of alternative approaches. Table 7.2 provides various technologies available for treatment and/or disposal of waste. Figure 7.1 shows technologies used for 70 sites in 1989 to either treat or dispose of the waste. Figure 7.2 depicts an overview of the available alternative means of waste water disposal. A comprehensive list of treatment alternatives for dilute and concentrated wastes is presented in Table 7.3. A treatment/disposal schematic for solid wastes[4] is shown in Figure 7.3.

## 7.3 DISPOSAL AND CONTROL

This section covers the disposal and control of wastes which have an unfavorable impact to the groundwater aquifers. Liquid wastes can easily migrate through soils causing contamination to the aquifers. Therefore, this section is restricted to disposal and control of all wastes which may produce contaminants in liquid state.

**Table 7.2 Treatment Technologies**

| **Physical treatment** | **Chemical treatment** |
|---|---|
| Sedimentation | Neutralization |
| Centrifugation | Chemical Precipitation |
| Flocculation | Chemical Hydrolysis |
| Oil/Water Separation | Ultraviolet Photolysis |
| Dissolved Air Floatation | Chemical Oxidation (Chemical Reduction) |
| Heavy Media Separation | Oxidation by Hydrogen Peroxide ($H_2O_2$) |
| Evaporation | Ozonation |
| Air Stripping | Alkaline Chlorination |
| Steam Stripping | Oxidation by Hypochlorite |
| Distillation | Electrolytic Oxidation |
| Soil Flushing/Soil Washing | Catalytic Dehydrochlorination |
| Chelation | Alkali Metal Dechlorination |
| Liquid/Liquid Extraction | Alkali Metal/Polyethylene Glycol (A/PEG) |
| Supercritical Extraction | |
| Filtration | |
| Carbon Adsorption | **Biological** |
| Reverse Osmosis | |
| Ion Exchange | Aerobic Biological Treatment |
| Electrodialysis | Activated Sludge |
| | Rotating Biological Contractors |
| | Bioreclamation |
| **Fixation/Stabilization** | Anaerobic Digestion |
| | White-rot Fungus |
| Lime-Based Pozzolan Processes | |
| Portland Cement Pozzolan Process | |
| Sorption | **Thermal Destruction** |
| Vitrification | |
| Asphalt-based (thermoplastic) | Liquid Injection Incineration |
| Microencapsulation | Rotary Kiln Incineration |
| Polymerization | Fluidized Bed Incineration |
| | Pyrolysis |
| | Wet Air Oxidation |
| **Potential Source** | Industrial Boilers |
| **Control Stratgies** | Industrial Kilns (Cement, Lime, Aggregate, |
| | Clay) |
| Recycling | Blast Furnaces (Iron and Steel) |
| Resource Recovery | Infrared Incineration |
| Materials Recovery | Circulating Bed Combuster |
| Waste-to-Energy Conversion | Supercritical Water Oxidation |
| Encapsulation | Advanced Electric Reactor |
| Waste Segregation | Molten Salt Destruction |
| Co-Disposal | Molten Glass |
| Leachate Recirculation | Plasma Torch |

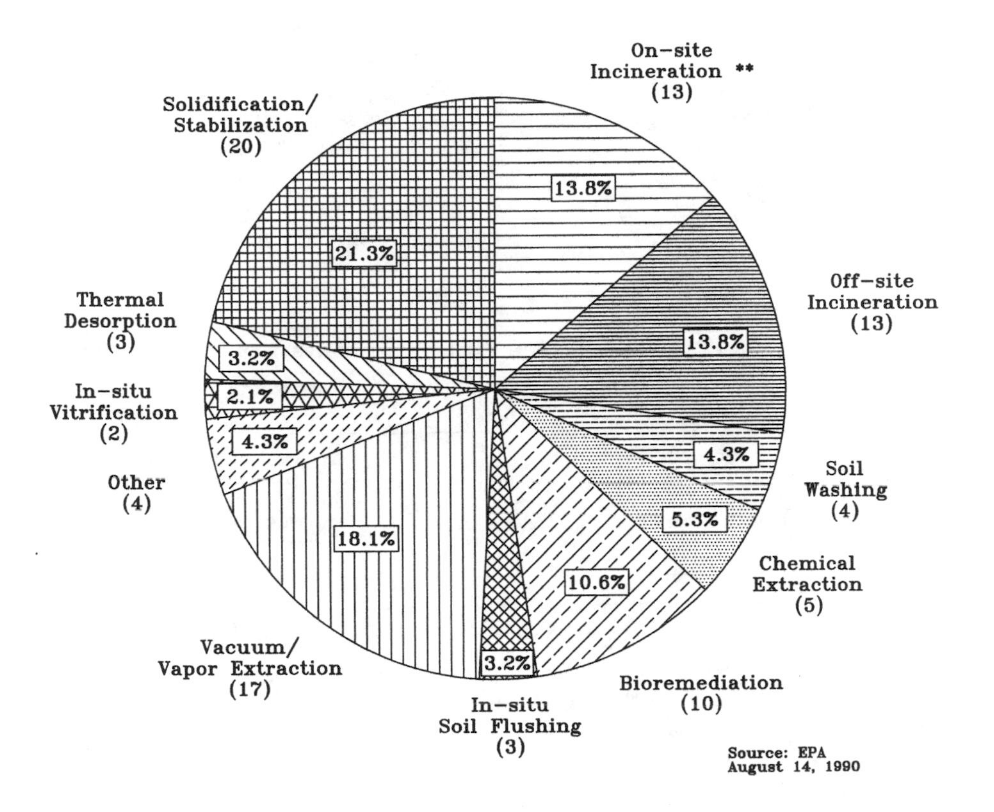


* Sources include solids, soils, sludges and liquid wastes; waste sources do not include ground water or surface water

** Also includes sites where location of incineration is to be determined

**Figure 7.1 Treatment technologies for 70 sites.[4]**

## 7.3.1 Types of Disposal

Manufacturing is the leading source of controllable man-made water pollutants; domestic waste is second. The industrial wastes are more likely to contain substances which will resist the normal treatment procedures. However, except for the types of industries which generate large amounts of incompatible wastes, many industries take advantage of the local municipal treatment facilities for some or all of their waste waters. It is often necessary to pretreat some of the industrial wastes before introducing them into the municipal treatment facility, but, most of the time, there is no difficulty in

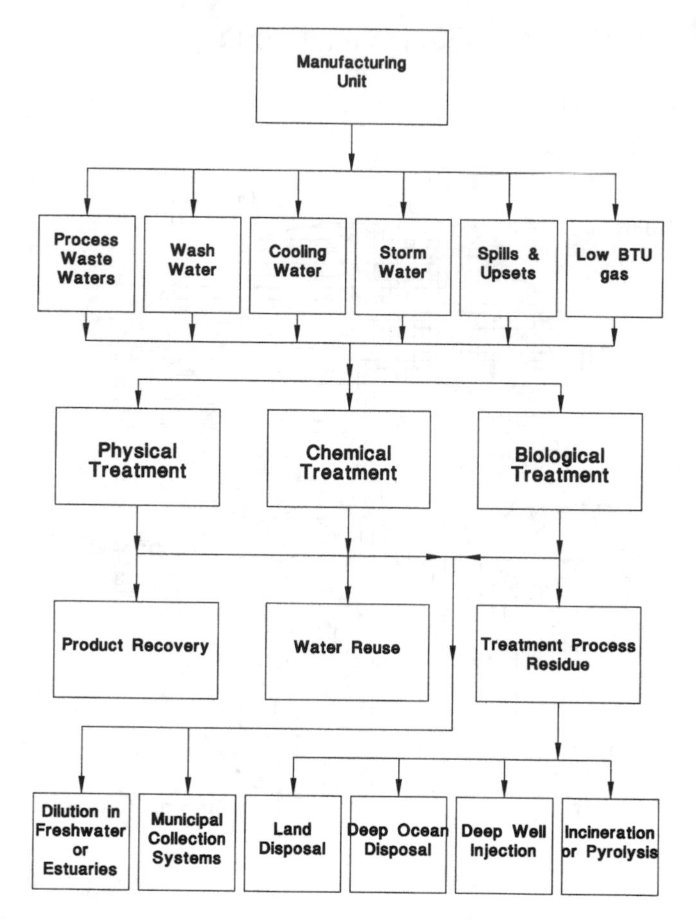


**Figure 7.2 General strategies in waste water disposal.**

handling the industrial wastes by the normal municipal technologies. Industrial wastes frequently constitute a large percentage of the volume treated by municipalities. Even if the industry has its own treatment facility, it usually operates on the same principles and employs many of the same techniques as the municipal systems.

There are three broad classes of treatment methods which are employed:

1. *Primary treatment.* Primary treatment removes from the waste water those substances which float or settle out. All processes in this category concentrate on removing the pollutants by physical means; hence, this is the mechanical treatment stage. Techniques included are grit removal, screening, grinding, and sedimentation.
2. *Secondary treatment.* This is based on biological oxidation; thus, it is a reproduction of the degradation processes which occur in nature.

**Table 7.3 Alternative Treatment for Water Wastes**

| |
|---|
| **Physical** |
| Equalization, adsorption, sizing, phase change, ion exchange, membrane process, force field separation, surface methods (foaming, skimming, etc.), extraction |
| **Biological** |
| Activated sludge (completely mixed, oxygen-based, etc.), fixed film processes (trickling, etc.), aerated stabilization, anaerobic (lagoons, etc.), algae stabilization ponds, balanced ecosystems (aquatic plants, animals), enzyme conversions (immobilized enzymes/cells) |
| **Chemical** |
| Acid-base treatments, chemical precipitation (coagulation, flocculation, etc.), oxidations (chlorination, ozonation, etc.), reduction reactions, complexation, photochemical reactions, hydration and clathrates, electromotive displacement, thermal decomposition |
| **Electrical and Electromagnetic** |
| Ultraviolet irradiation, electrolysis, magnetic separation, electrodialysis, electron beam radiation |
| **Acoustical** |
| Ultrasonic |
| **Nuclear** |
| Irradiation |

After Conway and Ross.[5]

The major purpose is to remove the soluble BOD (biological oxygen demand), as well as the suspended solids which were not removed in the primary treatment. The BOD is the amount of oxygen required to biologically oxidize the water contaminants to carbon dioxide. The three common methods used are activated sludge, trickling filters, and oxidation ponds (lagoons). All are based on having various microorganisms feeding on the organic impurities in the presence of oxygen, at a favorable temperature and for a sufficient time period.

3. *Tertiary treatment.* This primarily includes the various chemical treatments of waste water.

Most treatment facilities include primary and secondary treatment; some, particularly those associated with an industry, also include some form of advanced treatment. Figure 7.4 shows a typical system component for treating dilute process wastes.

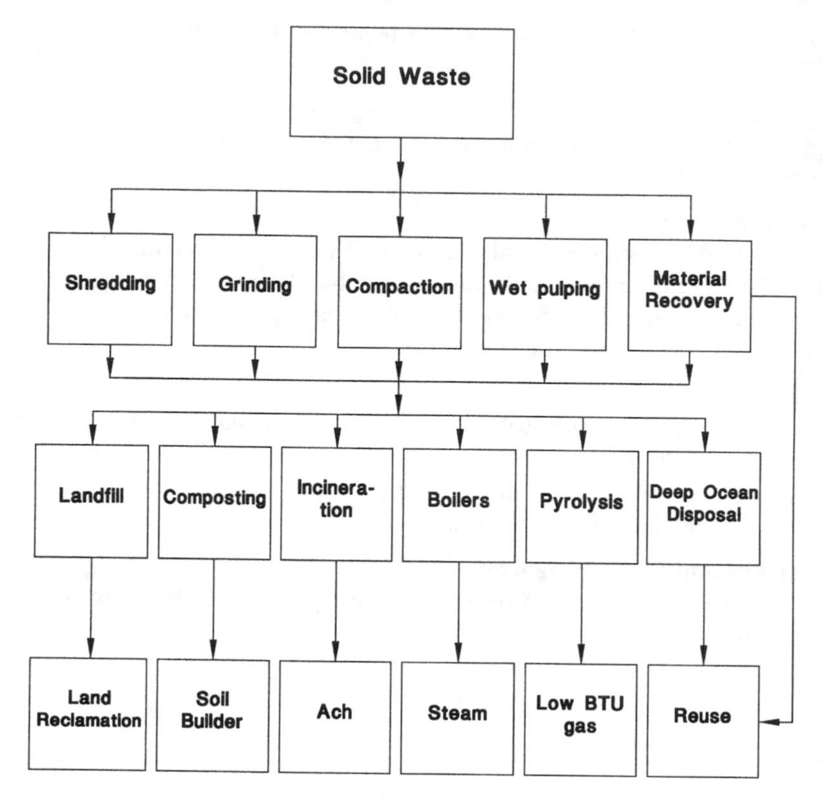


**Figure 7.3 Treatment/disposal sequence for solid waste.**

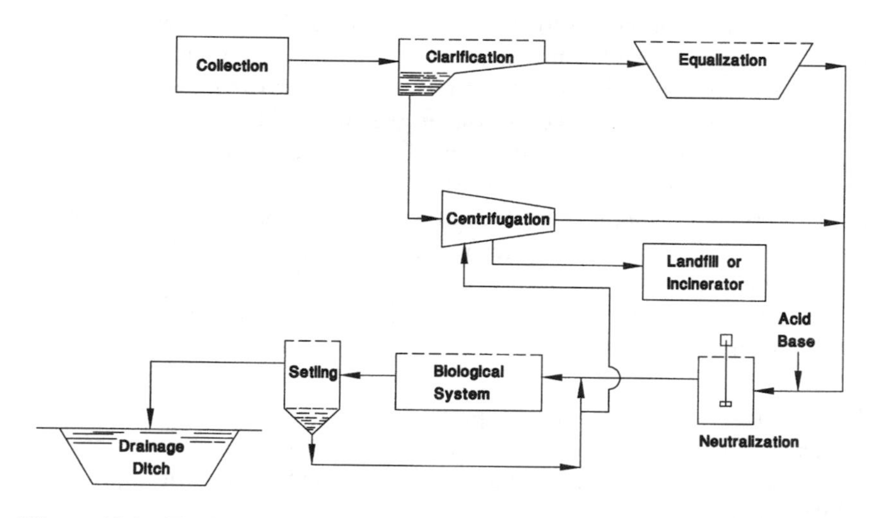


**Figure 7.4 Typical system components for treating dilute process wastes.**

There have recently been many projects devoted to studying the effects of using municipal sludge as a soil conditioner; as a liquid, dewatered or dried.

Landfilling is the most common method used for sludge disposal. Usually before the sludge can be landfilled, digestion is required to avoid odors, insects, and water pollution. All types of dewatered sludges can be disposed of by landfill, as can many other types of solid wastes. The sludges can be transported to the landfill site by truck, train, pipelines, or barges.

Severe measures have recently been taken in disposing and discharging the primary treated industrial wastes and sewage to deep-ocean outfalls. This is done to prevent any undesirable impacts on the marine environment.[6]

### 7.3.2 Disposal of Hazardous Waste

There has been much interest in hazardous waste disposal since the early 1960s. Hazardous wastes are those which are ignitable, corrosive, chemically reactive, or toxic.[7] Radioactive wastes are considered the most dangerous of hazardous wastes. Recently the catastrophe caused by a nuclear power plant in the USSR[8] had a disastrous impact on the environment. This includes the environmental impacts on the water resources and even the hydrological cycle.

In the past, hazardous wastes were often simply drummed, then disposed of in a landfill area. This was generally unsatisfactory. Several approaches are currently being taken for appropriate disposal techniques. Secured landfills are available in some locations and can be used to dispose of toxic wastes. These landfills are located in thick natural clay deposits or engineered with various layers of specially designed liners and leachate collection systems (Figure 7.5). Often, however, appropriate sites are too far from the plant which generates the waste to make them a logistically and/or economically viable disposal method. Therefore, other methods such as incineration are to be investigated and used for disposal of toxic waste.

Liquid waste incinerator is a type of incinerator which generally consists of a waste-fed system to deliver the liquids, slurries, sludges, and/or solids of thermal destruction chambers. There are usually two thermal chambers: one for oxidation and a separate one for gasification. Typical incinerators operate at 540–1400°C with residence times of the waste of 1–3 seconds or perhaps longer. Inclusion of heat recovery units can make the systems more economically feasible.[9]

Another potentially useful disposal technique for hazardous wastes consists of combined stabilization and solidification of these wastes. If properly done, the hazardous waste can then be disposed of in any well-designed landfill.

The technique used is dependent upon the nature of the waste. In the future, it is likely that processes will be changed to generate fewer toxic wastes, and more wastes will be converted to useful by-products or recycling may be increased.

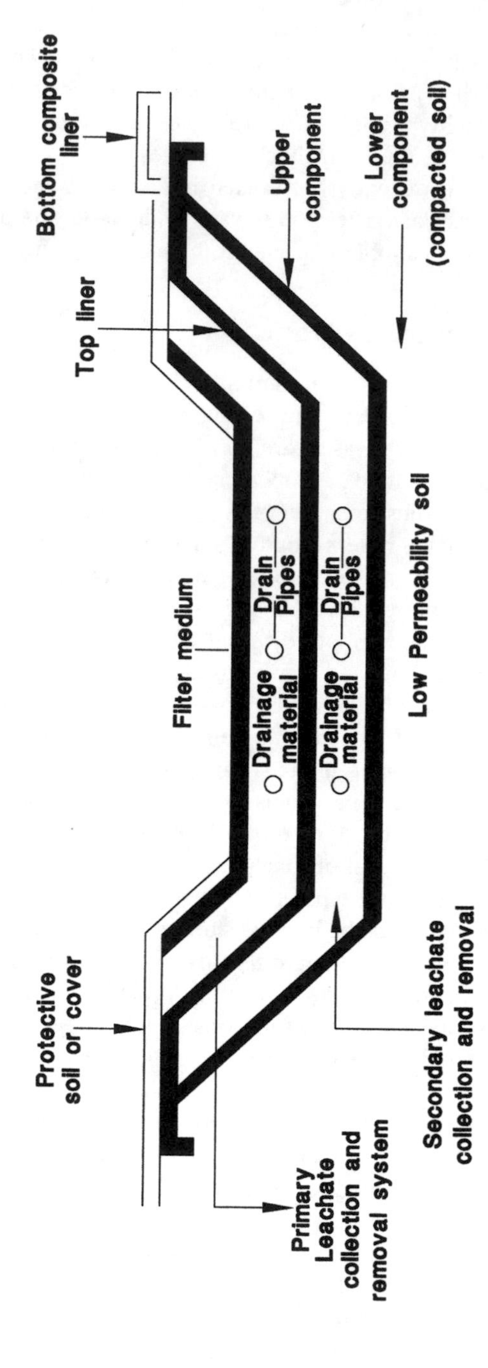


Figure 7.5 Linear leachate collection system for landfill.[10]

### 7.3.3 Salt Caverns for Disposal

Use of salt caverns for disposal of hazardous wastes has proved to be feasible in some locations worldwide. The solution mining method[11] provided the means for the creation of large storage capacities at economical costs. Moreover, underground sites in salt are safer from an environmental point of view, when compared to conventional shallow disposal sites.

Rock salt is practically impermeable to gas and liquids. This impermeability is due to the tightness of the structure and the absence of open natural joints and fissures which exist in many other types of rocks. Moreover, the high plastic deformality of rock salt hinders the development of artificial fissures through which liquids and gasses could leak out. The safe sealing of liquids leaks in a cavern in a rock-salt formation has two geochemical aspects:[12-14] the stability of the cavern and the tightness of the closure and the surrounding formations.

The stability of a cavern, at a particular depth and with predefined geometrical dimensions, is decisively affected by the geological situations and the geological properties of the rock salt, the rock mass temperature, the in-situ stress, and the pressure conditions in the cavern. Thus, the engineer and/or scientist has access to all data required for prediction models which allow realistic statements to be made and the time-dependent, load-bearing behavior of caverns for storage and waste disposal.[15,16]

In order to seal off a liquid-filled cavern from the biosphere for a long period of time, the following points should be noted:

1. Demonstration that cavern will be permanently sealed. This is primarily an engineering problem, namely a problem involving the sealing material.
2. Demonstration that the natural increase of pressure in the sealed cavern will not cause cracking in the surrounding rock formation.
3. The size of the critical pressure determined in the frac test depends on natural conditions, such as the mechanical properties of the surrounding halite and existing stress patterns, and on the test method itself, specifically on the time at which the test is performed and on the rate of pressure increase.
4. The results obtained from frac tests in boreholes cannot be directly applied to the case of a closed cavern with a natural pressure increase.
5. The previously permitted maximum pressure gradient for closed caverns should be subject to extensive testing.
6. Regarding a permanently sealed cavern, an accompanying computer analysis is urgently needed in order to correctly interpret the results of the frac tests and to evaluate correctly the danger of fractures in the surrounding salt formation.

One example of disposal of hazardous waste in salt caverns is the Waste Isolation Pilot Plant (WIPP) located in Southwestern New Mexico which was selected by the Department of Energy (DOE) for deep geologic depository for permanent disposal of radioactive wastes in the United States. The repository is located in bedded salt of Permian Age at a depth of 2150 feet below groundwater. Bedded salt is a preferred medium for permanent emplacement of wastes because of the favorable physical, thermal, and mechanical properties of halite. However, bedded salt does undergo different degrees of dissolution; therefore, it is necessary to have a thorough evaluation and understanding of the horizontal and vertical extent as well as the time and rate of salt dissolution.

The WIPP site has been investigated over a period of eight years to achieve a desirable level of assurance about its integrity and to ensure that there will not be a breach and leakage of the waste to either freshwater aquifer or the nearby Pecos River.

Prior to deposit of hazardous wastes in salt caverns, it is important to have a complete knowledge and thorough understanding of the geologic and hydrogeologic characteristics of the site which relate to the transport of waste to the biosphere in the event of a breach.

## 7.4 GROUNDWATER PROTECTION

### 7.4.1 Damage Prevention to Water Resource System

The prevention of damage to water resources is preferable and less costly than treatment. This applies only to man's activities such as mining and other constructions that affect the environment. However, many practical problems arising from dewatering by mines have been observed in different parts of the world.[17] Included are degrading of water and aquifer quality, reduction in yield or drying of wells, additional pumping costs for water supplies as a result of deeper pumping levels, intrusion of sea water, pollution of surface water by degraded groundwater, and land subsidence, especially sinkhole development.

Aquifers can be contaminated directly from mining wastes or as a result of rerouting of degraded waters due to mining activities. Contaminated surface streams may also affect local aquifers especially where heavy mine pumping has lowered the water table, which encourages recharge from the contaminated stream.

The extent of the problem in the USA is highlighted by the U.S. Geological Survey Water Summary for 1984,[18a] 1988,[18b] 1991,[18c] and 1993.[18d] Streamwater and groundwater quality were studied in 50 states, the District of Columbia, U.S. Trust territories, and the Virgin Islands; 32 states had surface water problems from such sources as toxic contaminants, pesticides, herbicides, acid precipitation, and bacteria; whereas 30 states showed groundwater pollution from causes such as hazardous waste sites, seepage from septic systems, landfill leachates, intensive pumping at coastal regions, and/or high concentrations of salt because of recirculation from irrigation water. Of these, 17 states showed both surface and groundwater problems from acid mine drainage and other mining activities.

Since mining is the most prevalent means of deteriorating the hydrogeological system, the prevention of damage to water resources due to mining is discussed in more detail. There are different methods to prevent this damage:

1. Pre-mining studies should investigate the impact of the proposed mine on the hydroenvironment and provide detailed baseline studies of the district to: (a) identify all aspects of the proposed mining activities liable to cause damage to water resources, (b) draw up plans to deal with the identified problems using the best approved methods, (c) make provisions for on-going impacts that may occur after the mine has started working and also after mining operations have ceased, and (d) prepare advanced plans to deal with the main types of possible accidents during mining which are likely to seriously damage the hydroenvironment.
2. A monitoring program should be carried out to provide adequate data on all relevant aspects of water quality in the existing mine area and in the district of the proposed mine. The monitoring system should be designed to provide an early warning system in the event of unforeseen changes in conditions which would be harmful to water resources.
3. Movement of polluted water from old workings should be controlled. This can be reduced by grouting of old boreholes, shafts, and fissures. Subsurface dams and grout curtains are also used to contain polluted waters.
4. Rehabilitation of site by remedial action can reduce damage to water. Compaction and landscaping with gently sloping topography can reduce the polluting potential of mining wastes to the hydroenvironment. Highly acidic waste heaps may be moved to an unexposed location. Revegetation by covering with metal-tolerant gasses and plants can help stabilize old tips and tailing ponds. It helps bind the surface and reduces both wind and water erosion. Various patent coagulants are also used to try to bind and stabilize tips and tailings. In many sites, a combination of different types of actions may be required to achieve maximum benefit.
5. If water resources are being polluted by mine waters and no ready means is available to prevent it at source, treatment may be necessary. The most commonly used methods of treating mine waters include neutralization, coagulation, and aeration.

Limestone and its derivatives are the most frequently used substances for neutralizing acid mine waters. Unslaked lime, hydrated lime, and, most recently, limestone have been widely used, primarily because of their low cost.[19,20] Besides adjusting the pH, the increased alkalinity also enables other methods of treatment to work more effectively especially with regard to the removal of metals from solution in the water.

A number of standard water treatment procedures involving the addition of chemicals, such as aluminum sulphate, are used to help remove fine particles. The coagulant binds together minute particles which then settle or can be filtered out.

The addition of oxygen is particularly helpful in facilitating the precipitation of the soluble forms of iron and manganese. Other polluting substances including hydrogen sulphide and cyanide, found in mine and industrial waters can also be removed by aeration.

Several other methods of treatment have been tried in different countries but are more limited in application, of less certain effectiveness, or controversial because of side effects.

Methods for waste control in industry include: minimizing use of water, controlling losses in processes, extraction by recycling water and recovery of products, fine tailings filtration in mineral industry, floatation in mineral industries, bioremediation/biological treatment, air stripping, and carbon bed adsorption.

### 7.4.2 Remediation of Groundwater Aquifers

Groundwater contamination by petroleum hydrocarbons, organic solvents, and other toxic nonaqueous phase liquids poses a serious threat to the groundwater resources worldwide. The processes that govern the behavior and fate of contaminants have been investigated by many researchers through mathematical models with various degrees of complexity.[21,22]

The goal of site remediation is to restore soil and groundwater quality to precontamination conditions, as much as possible. The remedial or cleanup standards are based on numerous criteria, including the site information, contamination type and extent, potential threat to human health, and protection of future soil and groundwater quality.[23]

Costs for site remediation can quickly escalate into considerable sums. Considerable progress has been made recently in regulation, engineering practice, environmental awareness, source control, and waste management procedures. This progress enables more reasoned decisions to be made regarding contaminant migration and threat to human health.

Numerous technologies to remediate sites exist. New technologies are being developed and tested. A brief review of some of the more common technologies follows:

1. *Excavation.* The removal of the contaminated materials for disposal at a hazardous waste or other disposal landfill site. Newer regulations favor alternative waste treatment technologies at the contaminated site; however, municipal refuse will continue to be placed at

landfills for some time to come in spite of the fact that it may contain small quantities of hazardous materials. Attempts must be made to remove hazardous materials from domestic refuse to keep the sanitary landfills from becoming hazardous waste landfills. Excavation projects must include the determination of volume of material to be removed, equipment to be used, source and type of clean backfill, compaction specifications, and sidewall stability. Excavation of hazardous waste for disposal is costly. Finally, post-excavation sampling and testing are required to determine the effectiveness of the excavation cleanup. Even though the cost of excavation is high, it is still the most effective way of site remedial.

2. *Air stripping*. Using the volatilization characteristics of the contaminant to separate it from groundwater. The contaminated water is pumped into a tank filled with a packing material to enhance aeration and slow the water movement. Air is blown into the hose of the tank as the water moves through it, and the volatile compounds are stripped and entrained to the atmosphere. This technology requires the contaminant to be highly volatile (such as solvents or light hydrocarbon fuels).
3. *Bioremediation.* There are several methods for bioremediation of groundwater. These include that used by Hazen et al.[24] who injected methane mixed with air into the contaminated aquifer via a horizontal well and extraction from the vadoze zone via a parallel horizontal well (Figure 7.6). The indigenous microorganisms were stimulated to degrade trichloroethylene (TCE), tetrachloroethylene (PCE), and their daughter products in-situ. Hazen et al. recorded in their test that all of the wells in the zone of effect showed significant decreases in contaminants in less than 1 month. Four of five vadose zone piezometers (each with three sampling depths) declined from concentrations as high as 10,000 ppm to less than 5 ppm in less than 6 weeks. A variety of other microbial parameters increased with methane injection indicating the extent and type of stimulation that had occurred.

   A bioremediation using a recirculation well is still in practice. Lang et al.[25] designed and showed that their model of in-situ bioremediation can be fairly cost effective.

   Bioremediation requires recirculating contaminated groundwater with treatment that contains an oxygen (such as peroxide) and nutrients. This delivers nutrients to allow biologic action to react on contaminants, cleaning the aquifer matrix. The groundwater is pumped by extraction wells, moving the treated waste through the contaminated zone and allowing biological respiration of the indigenous microfauna to further metabolize remaining contaminants. The contaminant plume must not move offsite, and the site geology must be acceptable (sandy, using methanotrophic bacteria for deg-

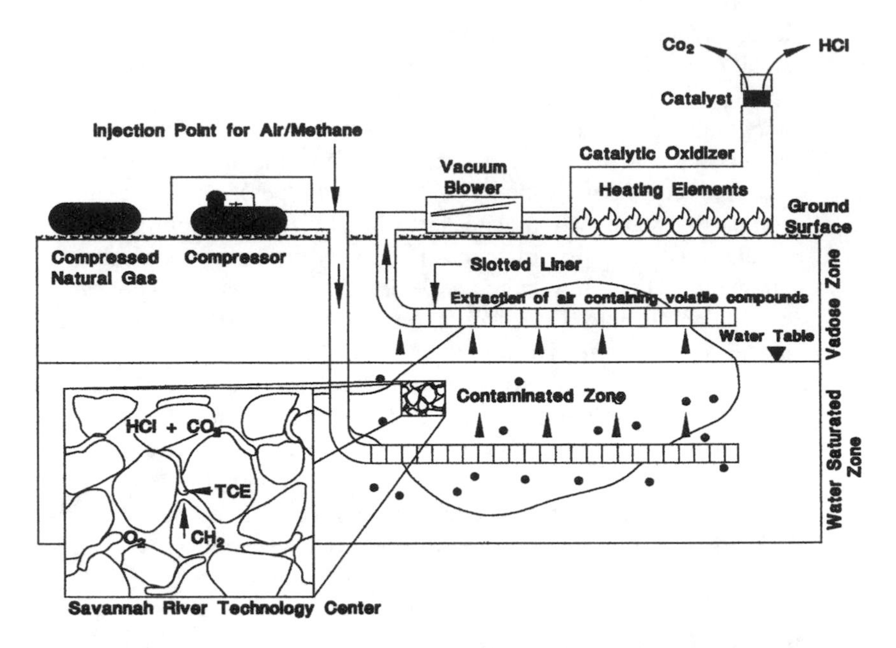


**Figure 7.6 In-situ bioremediation via horizontal well (after Hazen[24]).**

radation of volatile organic compounds). Methanotrophic bacteria produce the enzyme methane monooxygenase (MMO) in order for the microorganisms to utilize the supplied electron donor (methane). Lang's work investigated the use of a vertical recirculation well (Figure 7.7) to promote cometabolic transformation of vinylchloride (VC) using methanotrophic bacteria. The vertical recirculation well recirculated the target contaminants, thus increasing their contact time with the biologically active zone and allowing a greater extent of contaminant removal. The benefit of using recirculation wells is the elimination of the cost of pumping groundwater to the surface for above ground treatment.

Site remediation can be compromised based on the applicable regulations, hydrogeologic data, data interpretation, the final cleanup concentrations of contaminants, negotiation of the remedial action plan alternatives, the cost, and the remediation plan execution. The ability to estimate the safety risk and cost/benefit ratio of the remediation effort must be used.

Problems still exist with these methods as the release of the contaminant into the atmosphere is being discouraged in some areas, and a second treatment is often required to capture the airborne contaminants (such as carbon adsorption).

4. *Carbon bed adsorption.* An established technology used to treat contaminated groundwater. The contaminant becomes adsorbed to

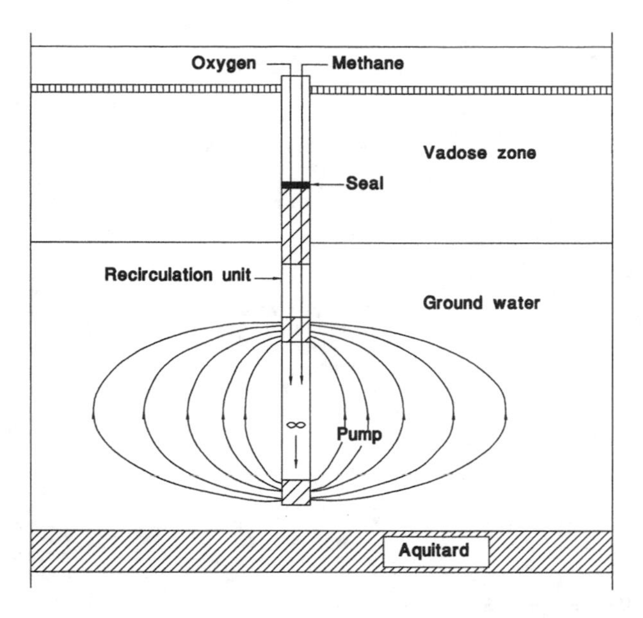


**Figure 7.7 In-situ bioremediation using a recirculation well (after Lang[25]).**

the carbon, which removes weakly polar molecules. The concentration of the contaminated influent directly affects the retention time of carbon adsorption before all the adsorption sites are filled and break through. The spent carbon can then be replaced with unused carbon, and the adsorption process repeated. Carbon is versatile and useable for many organic contaminants. Costs can become high, depending on how often the carbon must be regenerated or replaced.

## 7.5 RISK AND LEGAL ASPECTS OF WASTE DISPOSAL SITES[26]

Chemical solvents are potential threats to human health and the environment. The sequence of activities that can lead to releases of chemical solvents into the environment is shown in Figure 7.8. This section focuses on the human health effects that can occur at the point of ultimate disposal as related to landfills as the final disposal method.

In the past, landfilling has been a major disposal option for solvents. The U.S. Environmental Protection Agency (USEPA) has estimated that about 22.4 million metric tons of hazardous waste are disposed of on land each year.[27] The risks from land disposal of chemicals originate from the potential for chemicals to migrate into water supplies or soils and to evaporate into the atmosphere. Because of these problems, landfilling of solvents has become a major public issue. The Hazardous and Solid Waste Amendments of 1984 (P.L.

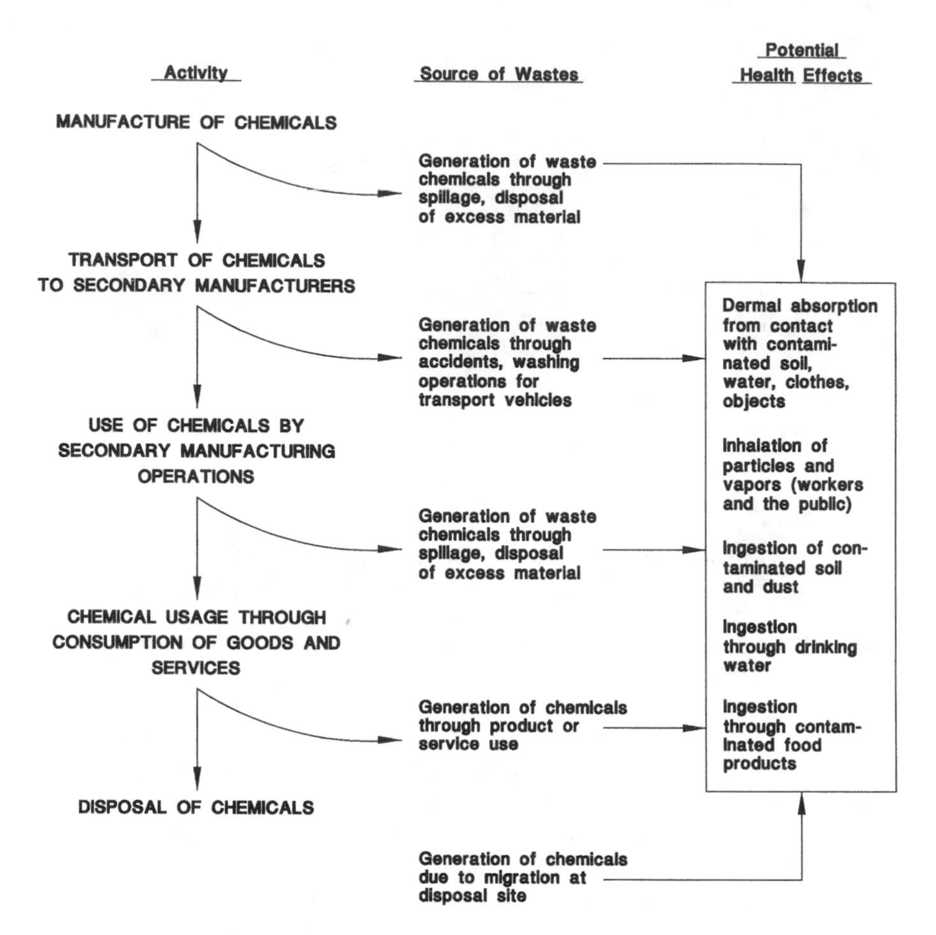


**Figure 7.8 Generalized framework for the origins of risks from organic chemical solvents.**

98-1134, Section 201 (e)) prohibit the land disposal of certain liquid hazardous wastes unless the EPA can demonstrate that there is no threat to human health.

Risk assessment is used widely to evaluate environmental health risks. The application of risk assessment to the land disposal of organic solvents has been applied to a limited extent in connection with site investigations or remedial cleanup actions under the Comprehensive Environmental Response Compensation and Liability Act (CERCLA or "Superfund"). The application of risk assessment to evaluate which chemicals should be disposed of on land and in what concentrations has received less attention. The procedures for risk assessment, the application of these procedures to chemical disposal decisions, and the uncertainties that occur in the assessment process are outlined with emphasis on the effect that these uncertainties can have on the level of risk that is calculated.

Risk assessment techniques are used to evaluate the use of land disposal for 24 organic solvents. Risk acceptability, risk evaluation, and risk management are important dimensions of the use of risk assessment, and perception literature and the public's perception and acceptability of the risks from organic solvents in landfills must be determined. A framework follows for translating physical, chemical, biological, and health-related attributes of the solvents into dimensions of public perception of risks from organic solvents.

### 7.5.1 Definition of Risk and Risk Assessment

Risk has been defined as "the potential for realization of unwanted, negative consequences of an event"[28] and as "the complete description of possible undesired consequences of a course of action, together with an indication of their likelihood and seriousness."[29] Risk also has been defined in more quantitative terms as follows:

> The probability per unit time of the occurrence of a unit cost burden [and it] represents the statistical likelihood of a randomly exposed individual being adversely affected by some hazardous event. Thus, risk involves a measure of probability and severity of adverse impacts.[30]

Quantitative risk assessment is a procedure for estimating the probability of an adverse health effect occurring in a population from some event and the probability of the occurrence of the event. A level of risk associated with the use of a chemical solvent is calculated as the combined probability (or product of the probabilities) of the hazard existing, the chemical's release into the environment occurring, population exposure, and potential health effects.[31, 32]

Risk assessment is most accurate when decision situations, causative events, and effects are specific. In the context of siting a waste disposal facility, this means that accuracy is greatest for a site where geological, hydrological, and meteorological characteristics of proposed sites, as well as the physical, chemical, and biological characteristics of the waste material, are known. When risk assessment is used to evaluate an existing waste disposal facility, results are the most accurate when applied to the behavior of individual chemicals known to produce or suspected of producing adverse health effects.

### 7.5.2 Application of Risk Assessment in the Context of Waste Disposal

Risk analysis or assessment has received attention from the scientific community for more than two decades. Between 1973 and 1982, a number of Federal agencies proposed carcinogen assessment policies or guidelines that incorporated risk assessments. Risk assessment was used extensively by the EPA in 1980 for the development of water quality guidelines. Its importance for the hazardous and toxic waste legislation of the late 1970s and early 1980s has been recognized by regulatory agencies.[33,34] Regulations were proposed by the EPA to establish and systematize methods for conducting risk assessments for a variety of its environmental programs. In the regulations, risk assessment is a method of quantifying the carcinogenic, mutagenic, and development effects of toxic substances[35-37] and the health effects associated with complex chemical mixtures.[38]

Risk assessment can be applied to several kinds of waste disposal situations, including abandoned waste disposal sites, operation and maintenance of an existing facility, closure of an existing facility, and the siting and construction of a new disposal facility.

In each of these four situations, risk assessment can be used for the following purposes (some of these have been suggested by Anderson et al.[39] and by Russell and Gruber):[40]

- setting priorities among existing waste sites or potential waste disposal site locations as a basis for regulatory action (e.g., the Hazard Ranking System used to establish the National Priorities List under CERCLA)
- setting standards or guidelines for the release of contaminants to air, water, and land in both the general environment and occupational settings
- evaluating the residual risk to human health and the environment of alternative technologies
- evaluating the suitability of alternative sites for waste disposal

Of these uses of risk assessment, the most common use is for the development of environmental standards. Many existing environmental standards are directly applicable to assessing the impacts of solvent disposal in landfills. In fact, regulations for the National Contingency Plan under CERCLA require that, at least for remedial action for uncontrolled waste sites, these standards be reviewed prior to undertaking a formal risk assessment.[41] A summary of programs under which standards are developed is given in Table 7.4 for the chemicals under review here.

*Drinking water standards.* The EPA's Office of Drinking Water makes extensive use of risk assessment in the development of "Recommended Maximum Contaminant Levels" (RMCLs) for drinking water under the Safe Drinking Water Act.[42] The most recent version of these RMCLs for the organic solvents under study is given in Table 7.4. Prior to the EPA's work, the National Academy of Sciences (1977 and ongoing) published the extensive drinking water limits based on risk assessments in its "Drinking Water and Health" series. A number of states, such as New York, have also been active in the area of developing drinking water standards using risk assessment techniques.

*Ambient water quality standards.* One of the earliest applications of risk assessment to standards development was for water quality criteria. In Novem-

**Table 7.4 Regulations Applicable to Chemical Solvents**

| | Clean Water Act | | Safe Drinking Water Act Proposed | | | Toxic Substances Control Act | | | | | |
|---|---|---|---|---|---|---|---|---|---|---|---|
| | Sec. 311 reportable quantities (in lbs.)[a] | Priority pollutants[b] | TTHMs[c] | VOCs[d] | OSHA | 4(e) | 8(b) | 8(e) | ITCL list | Clean Air Act 112-NESHAPs candidates | RCRA/ CERCLA listed hazardous waste |
| Chlorobenzene | B | | | | x | x | x | x | | x[e] | x |
| o-Dichlorobenzene | B | x | | | x | | x | | | | x |
| Cresol | C | | | | x | x | x | | x | | x |
| Ethylbenzene | C | x | | | x | | x | | | | x |
| Nitrobenzene | C | x | | | x | x | x | | x | | x |
| Pyridine | | | | | x | x | x | | x | | x |
| Toluene | C | x | | | x | x | x | | x | x[e] | x |
| Xylene | C | | | | x | x | x | | x | x[f] | x |
| Cyclohexanone | | | | | x | x | x | | x | | x |
| Acetone | | | | | x | | x | | | | x |
| Methyl Ethyl Ketone | | | | | x | x | x | | x | | x |
| Methylisobutylketone | | | | | x | x | x | | x | | x |
| N-Butyl Alcohol | | | | | x | | x | | | | x |
| Isobutyl Alcohol | | | | | x | | x | | | | x |
| Methanol | | | | | x | | x | x | | | x |
| Carbon Disulfide | D | | | | x | | x | | | x | |
| Carbon Tetrachloride | D | x | x | x | x | | x | | | x[g] | x |
| Ethyl Acetate | | | | | x | | x | | | | |
| Diethyl Ether | | | | | x | | x | | | | x |

**Table 7.4 Regulations Applicable to Chemical Solvents** *(continued)*

| | Clean Water Act | | Safe Drinking Water Act Proposed | | | Toxic Substances Control Act | | | | | |
|---|---|---|---|---|---|---|---|---|---|---|---|
| | Sec. 311 reportable quantities (in lbs.)[a] | Priority pollutants[b] | TTHMs[c] | VOCs[d] | OSHA | 4(e) | 8(b) | 8(e) | ITCL list | Clean Air Act 112-NESHAPs candidates | RCRA/ CERCLA listed hazardous waste |
| Methylene Chloride | | | | | x | | x | | | x[g] | x |
| Tetrachloroethylene | | x | x | x | x | | x | | | | x |
| Trichloroethylene | C | x | x | x | x | | x | | | x[g] | x |
| 1,1,1-Trichloroethane | | | x | x | x | x | x | x | | x[g] | x |
| Chlorofluorocarbons | | | | | x | x | x | | | | x |

[a] Reportable Quantities are defined under 40 CFR 117.3, and are applicable to discharges of hazardous substances into navigable waterways, unless covered under other statutes. B = 100 lbs., C = 1,000 lbs., D = 5,000 lbs.

[b] Water-quality criteria listed in *Federal Register,* November 28, 1980, Part V EPA, which resulted from court case brought by NRDC.

[c] Promulgated drinking water standards under Safe Drinking Water Act. TTHMs = Total trihalomethanes.

[d] Proposed limits for Volatile Organic Compounds, *Federal Register,* Vol. 50, No. 219 (November 13, 1985) and required to be finalized under Phase I of the new regulatory program of the Safe Drinking Water Act Amendments of 1986. Ethylbenzene, chlorobenzene, o-Dichlorobenzene, xylene, and toluene are to have drinking water standards developed under the Phase II program.

[e] Under the Clean Air Act, the decision has been made not to regulate the chemical, though it was once a candidate.

[f] A preliminary health screening only has been conducted.

[g] Intent-to-list under National Emission Standards for Hazardous Air Pollutants.

The items under the Clean Air Act are as of 1/13/86 and as itemized in Cannon (1986).

**Table 7.5 Concentrations for Selected Organic Solvents Equivalent to a $10^{-5}$ Risk Level**

| | Concentrations in μg/L | | | |
|---|---|---|---|---|
| | | Drinking water concentrations Equal to $10^{-5}$ risk[b] | | |
| Chemical | Ambient water quality criteria[a] | NAS | CAG | RMCLs[c] |
| Trichloroethylene | 27 | 45 | 26 | 5 |
| Tetrachloroethylene | 8 | 35 | 6.7 | NA |
| Carbon tetrachloride | 4 | 45 | 2.7 | 5 |
| 1,2-Dichloroethane | 9 | 7.0 | 3.8 | 5 |
| Vinyl chloride | 20 | 10 | 0.15 | 1 |
| 1,1-Dichloroethylene | 0.3 | NC | 0.61 | 7 |
| Benzene | 7 | NC | 13 | 5 |
| 1,1,1-Trichloroethane | | | | 200 |
| p-Dichlorobenzene | | | | 750 |

[a] From Anderson et al. (1984, p. 288), updating U.S. EPA *Federal Register* (November 28, 1980, pp. 79318–79379).

[b] Cited in 50 *Federal Register* (November 13, 1985, p. 46883). *Abbreviations:* CAG, Cancer Assessment Group in the EPA; NAS, National Academy of Sciences; NC, Not calculated.

[c] RMCLs or Recommended Maximum Contaminant Levels are from U.S. EPA (November 20, 1985).

ber 1980, the EPA proposed water quality criteria in the form of water concentrations assuming risk levels ranging from $10^{-5}$ to $10^{-7}$. These levels are given in Table 7.5 for the solvents under review.

*Air quality standards.* The EPA is drawing upon risk assessments in the development of National Emission Standards for Hazardous Air Pollutants (NESHAPs) under the Clean Air Act. These risk assessments are appearing in health assessment documents conducted on a chemical-by-chemical basis by the EPA Office of Health and Environmental Assessment. Health effects assessments have been conducted for many of the solvents covered in this study: acetone; carbon tetrachloride; tetrachloroethylene; toluene; 1,1,1-trichloroethane; trichloroethylene; and xylene. The EPA has issued "intent-to-list" notices under NESHAPs for the following solvents under review here: carbon tetrachloride, trichloroethylene, methylene chloride, and 1,1,1-trichloroethane.[43] Decisions have been made not to regulate toluene and chlorobenzenes. Xylenes are currently undergoing a preliminary health screening under NESHAPs. States are also taking the lead in using risk assessments for the development of emission limits (called performance standards) for new sources of air pollutants under the Clean Air Act. As of mid-1986, 17 states had developed air toxic programs, and 19 more were in the process of doing so. Many of the chemicals selected for state action are organic solvents. Fourteen of the states use or plan to use risk assessments in the evaluation of air toxic pollutants.[43]

*Occupational health standards.* The Occupational Safety and Health Administration (OSHA) and its research division, the National Institute of Occupational Health and Safety, develop criteria and standards primarily for air pollutants in occupational settings under the OSHA Act. Risk assessments are used in the development of many of these guidelines. Some of the pollutants are organic solvents.

*Consumer protection standards.* The Consumer Product Safety Commission and the Food and Drug Administration also set limits for contaminants, including some solvents, in certain settings that affect the general public. Risk analyses are used for some of these limits.

*Hazardous waste disposal and toxic standards.* Chemicals, including solvents, are listed for regulation under the Comprehensive Environmental Response Compensation and Liability Act and its 1986 Amendments, the Resource Conservation and Recovery Act, and the Toxic Substances Control Act.

### 7.5.3 An Outline of the Risk Assessment Process

Risk assessment is a useful tool for evaluating the potential threats that chemical solvents pose to human health and the environment. It provides a unified analytical framework that integrates the sources of chemical releases, environmental fate and transport, exposure, and health effects of such solvents. A risk assessment provides an estimate of the lifetime risk to an individual or to a population of the exposure to a hazard. Individual risk is the probability that an individual will experience a risk of death or disease in the course of a lifetime. Population risk is the number of excess cases of disease or death that could appear as a result of lifetime exposure of the population to a particular hazard.

While the concept of risk assessment is straightforward, its application depends upon a variety of complex models whose mathematical formulations are far from refined and whose data requirements can be enormous. While debates exist about the definition of the overall process of risk analysis or risk assessment, several components are generally accepted as critical to the process. These components include:

1. *Risk or Hazard Identification.* A qualitative identification of the likelihood that hazards can result in risk and the nature of the risk agents
2. *Risk Estimation.* A quantification of the types and magnitudes of the risks to humans, i.e., how bad they are and the probability of the risks occurring

3. *Risk Acceptability.* A determination of the acceptability of risk levels to society and to individuals, taking into account perceptions of and attitudes toward risks
4. *Risk Evaluation.* An evaluation of the economic and social ramifications of the risks associated with an activity including, but not limited to, an assessment of the costs and benefits of the risks
5. *Risk Management.* The design and implementation of a management and decision-making system that unites analytical elements of the risk assessment process with judgmental and institutional factors

A schematic illustration of the risk-assessment process is shown in Figure 7.9. This overall process has been called by various names: risk analysis or assessment, risk control, and risk management. Alternative frameworks for the process have been proposed by Lowrance,[44] Rowe,[28] and Lave.[45] The concept of risk control was discussed by Davies.[46] The term *risk assessment* will be used primarily to characterize the risk identification and estimation process as it applies to chemical solvents.

## 7.6 COMPONENTS OF THE RISK ASSESSMENT PROCESS

### 7.6.1 Risk or Hazard Identification

*(Note: Hazard identification should not be confused with hazard assessment. Hazard assessment is a term often used to describe dose-response estimation.)*

Hazards are threats to human health, and risks are the probability of harm occurring from such hazards. The first step in risk assessment is to identify hazards and risks through a qualitative or semi-quantitative screening process. Hazard identification is a form of problem identification. In this step, a potential linkage between a risk agent and an adverse effect is identified.[47] Identified are those substances that have a high probability of being harmful to human health. The decision-making process might never culminate if the sources and consequences of risks are not identified adequately at the beginning of a risk assessment. In fact, decision-making could proceed incorrectly.

#### *7.6.1.1 Methods for Hazard Identification*

Lave[45] identified five methods for hazard identification which are applicable to chemical solvents. These five methods have also been incorporated in proposed federal regulations for risk assessment[35] and developed in the National Academy of Science.[47]

1. The identification of abnormal disease patterns through an analysis of case clusters;

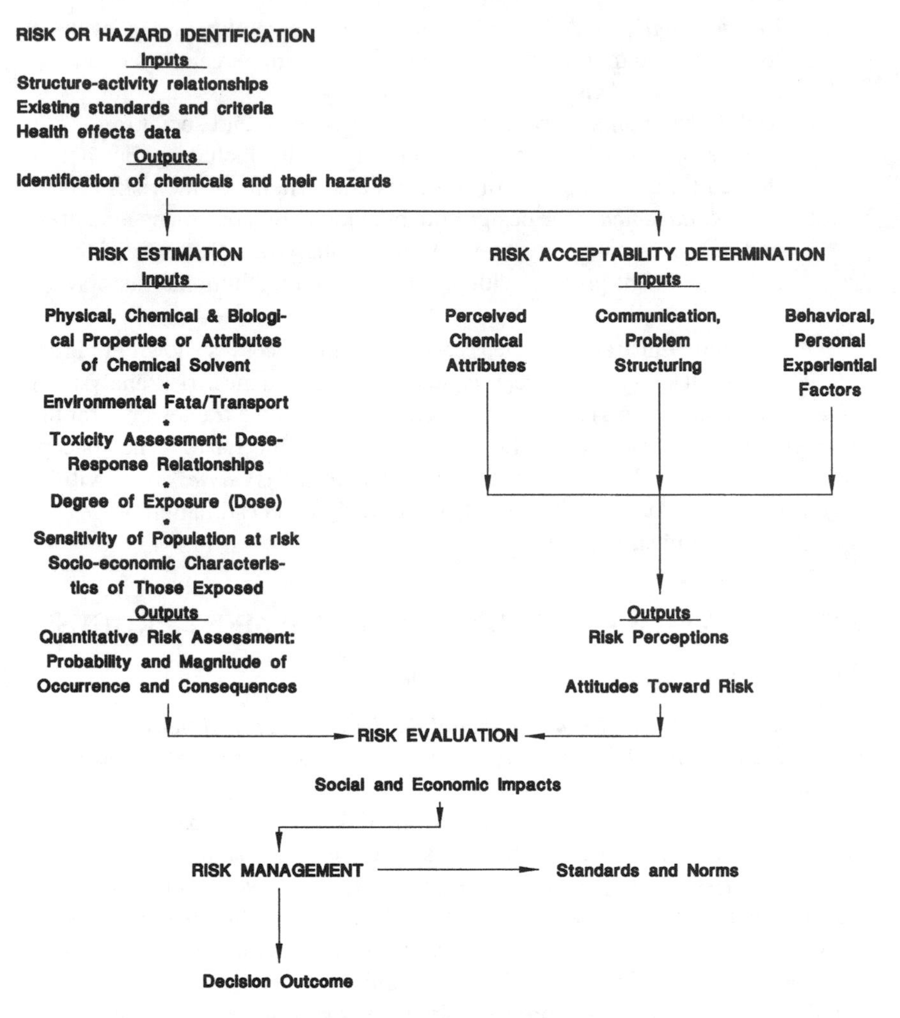


**Figure 7.9 A risk assessment and management framework for chemical solvents.**

2. The identification of structural and toxicological similarities (known as structure-activity relationships) among chemicals through structural toxicology;
3. The use of data from experimental tests of chemicals on simple organisms whose response time is rapid;
4. The use of data from experimental tests through animal bioassays, and
5. The creation of inferences from epidemiological studies.

Many of the above methods have been incorporated into hazard identification procedures for uncontrolled hazardous waste or "Superfund" sites under CERCLA and analogous state legislation. The guidance document for conducting public health assessments at Superfund sites, for example, uses hazard identification procedures to obtain a list of "indicator" chemicals. These indicators are considered representative of a larger number of chemicals found in a particular landfill. The physical, chemical, and biological attributes of chemicals are used to identify and select these indicator chemicals and determine their health risks. Once the chemicals are identified, the chemical attributes are used as a basis for quantifying risks more precisely.

The output of hazard identification for an existing landfill is an identification and ranking of chemicals likely to be found in the vicinity of the landfill. According to the EPA health assessment manual, the ranking of the chemicals is based upon the product of two chemical attributes: the toxicity and the concentration of each chemical found at a particular landfill site.[48] The concentration is either measured directly or estimated from a wide range of chemical attributes such as solubility, permeability in soil, and volatility. The set of indicator chemicals is considered representative of different chemical groups. In the case of Superfund sites, these indicator chemicals are used as the basis for analyzing the relative risks from alternative remedial actions. An important input into the toxicity component for the ranking of chemicals is a "weight of evidence" determination. Such a determination rates the quality of the toxicity data from high to low. The criteria developed by the International Agency for Research on Cancer (IARC) are commonly used to characterize the weight of evidence for carcinogens. The IARC categories are: sufficient evidence, limited evidence, inadequate evidence, no evidence, and no data. This is used to group carcinogens into five categories: human carcinogens, probable carcinogens, possible carcinogens, not classified, and no evidence.[35,49]

Since the parameters and methodologies used in hazard identification are similar to those used for certain steps in risk estimation, they will be discussed in detail below in Section 7.6.2. The major difference between the two steps is that hazard identification uses data qualitatively, whereas risk estimation uses data more quantitatively. The objective of hazard identification is to obtain a subset of chemicals for more detailed investigation.

### *7.6.1.2 Uncertainties and Errors in Hazard Identification*

Errors and uncertainties that occur in hazard identification can adversely affect the accuracy of identifying chemicals which have a high potential for entering the environment from a landfill. When sampling data is used to identify chemicals and their prevalence in a landfill, a number of potential sources of error can occur. First, analytical techniques can miss chemicals. Analytical techniques used under the Superfund (CERCLA) program, for example, are designed to analyze for a number of compounds simultaneously. They rely on scanning devices that do not spend enough time to identify particular chemicals. It is, therefore, not an optimum technique when the detection of individual compounds is important. Isaacson, Eckel and Fish[51] recently reanalyzed 3,000 samples from Superfund sites known to contain both organic and inorganic compounds. They detected 28 organic compounds that had been undetected in a previous analysis oriented toward identifying multiple compounds simultaneously.

A second source of error can be the choice of an improper indicator chemical to characterize the risks from a group of chemicals. Often, the most toxic compounds in a group are used to characterize the group. This can cause an overestimation of risk. Sometimes, a chemical is chosen to characterize a group of chemicals with risks ranging over several orders of magnitude. In such cases, additional chemicals should be chosen to characterize the group.

Errors can result from exclusive reliance on tests with lower organisms as screens for potential carcinogens. Evidence of mutagenicity is often used as an indicator of potential carcinogenicity since the two often show strong correlations. There have been significant exceptions; that is, false negatives occur. For example, carbon tetrachloride is a chemical which has not been shown to be mutagenic in bacteria or cultured liver cells, yet it is carcinogenic in animals.[51]

### 7.6.2 Risk Estimation

Once chemicals and their hazards are identified, estimates of the magnitude of risks from those hazards and the consequences of the risks can be made. This estimation procedure consists of the following steps:

1. *Exposure assessment* is "the determination of the estimation (qualitative or quantitative) of the magnitude, frequency, duration and route of exposure."[52] Exposure assessment includes a knowledge of the following:
   - The sources of the chemical.
   - The rates of release of the chemicals into the environment and their fates and transport in the environment. The rates are related to the physical, chemical, and biological attributes of the chemicals and the way in which these attributes are transformed under environmental conditions.
   - The routes or pathways of exposure from environmental end points to human organisms via oral, inhalation, and dermal pathways.
2. *Health effects assessment* primarily uses health data to determine the likelihood that, once exposed, a given individual or population will actually experience the risks. Where health data are available, socio-economic characteristics of the population would be used as

an indirect measure of health sensitivity, but these relationships are not well established. The assessment consists of the following:

- Estimation of intake levels.
- Absorption of the body.
- Toxicity of the risk agent, once in the body.
- State of health of the organism (though usually not feasible to include this).

3. *Dose-response relationships (or toxicity assessment)* relationships are expressed as the responses of different organisms to varying doses of the chemicals. The responses are a function of the health-related characteristics of the chemicals (for example, chronic or acute toxicity, carcinogenic potential).
4. *Risk calculation (or risk characterization),* the output of a quantitative risk assessment, is expressed as a risk to an individual or population for a lifetime of constant exposure. The individual risk is the chance of contracting a disease or dying from it. The population risk is the increased number of disease cases or deaths resulting from the exposure. Once that risk level is known, measures can be designed to reduce the level of exposure to some acceptable level.

These steps are portrayed in a simple form in Figure 7.10.

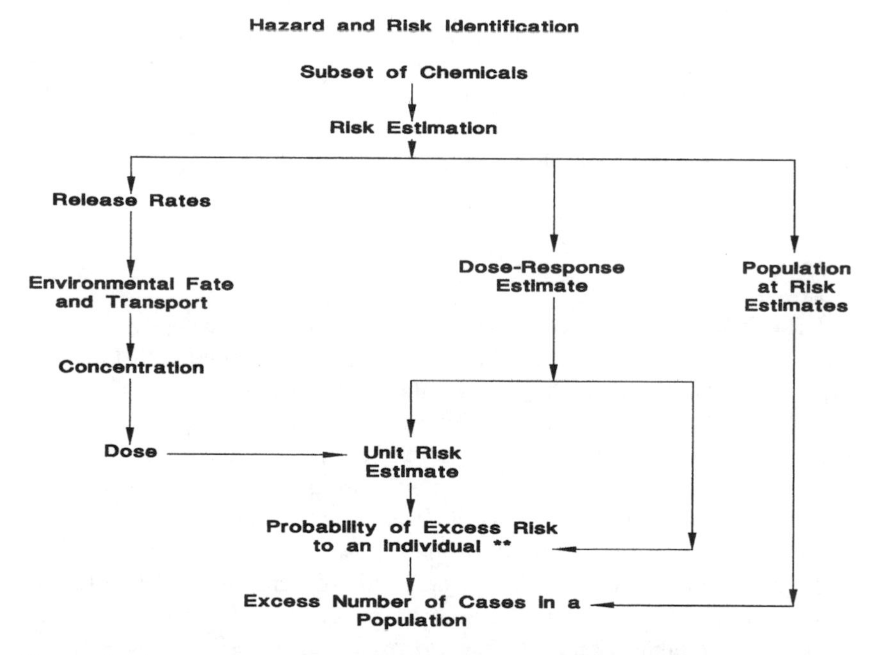


**Figure 7.10 Elements in risk estimation and their relationships to the risk calculation. *, suspected carcinogens or reference doses for threshold effects. **, non-threshold effects.**

To envision the interrelation of the above steps, one can begin at the end of the risk-estimation process: the risk calculation.

#### *7.6.2.1 The Risk Calculation*

The formulation of the equations for the risk calculations depends on whether: (1) the risk agent is a carcinogen and (2) a threshold exists; that is, a dose of the chemical below which no health effect is observed. Carcinogens are considered to have no threshold, and most noncarcinogens are assumed to have a threshold.

#### *7.6.2.2 Carcinogens (No Threshold)*

A simplified version of the calculation of the individual lifetime risk of exposure to a carcinogen (assuming no threshold) is as follows:

$$T_i = R_i * D_i \tag{7.2}$$

where i = the specific chemical or chemicals
T = the probability that a human response such as the development of cancer will occur over a lifetime of exposure to a constant dose, D, of a chemical, i
R = the unit risk factor in terms of number of excess cases of cancer that develop in a given population after a lifetime exposure to a chemical. It is expressed as the inverse of milligram (mg) of the chemical consumed per kilogram (kg) of body weight per day
D = the dose of the chemical in terms of mg/kg/per day

The risk, Z, to an entire population, P, is:

$$Z_i = T_i * P = R_i * D_i * P \tag{7.3}$$

The simplicity of the equation conceals a number of complexities. The mathematical formulation in Eq. 7.2 is based on the assumption that a linear relationship exists between the dose and the response. This assumption is often made for low doses or concentrations of a substance at which no dose-response data exists. This assumption may not always be valid. At higher doses common formulations have the probability, T, of increasing exponentially with dose. Such an exponential relationship could also be occurring at low doses.

The dose, D, is the end result of the exposure assessment and is derived from one of the outputs of environmental fate and transport modeling for air, water, and soil-concentration. The unit risk factor, R, is derived from experimental work relating chemical doses to responses in test organisms (dose-response curves). The unit risk factor for carcinogens is the response level

**Table 7.6 Daily Ingestion Rates**

| | Adult | Child |
|---|---|---|
| Water | 2 l | 1 l |
| Air | 20 $m^3$ | 5 $m^3$ |
| Fish | 6.5 g | — |

*Note:* Average body weight of an adult is assumed to be 70 kg and average weight of a child is assumed to be 10 kg. There is some variation in the rates used for quantitative risk assessments. For example, air intake rates of 17 $m^3$ have been used rather than 20 $m^3$ (U.S. EPA, December 27, 1985, p. 26).

Summarized in ICF (1985, p. 75) from water-quality criteria and National Academy of Sciences, *Drinking Water and Health* series.

associated with a dose of 1 mg/$m^3$ (via inhalation) and 1 mg/l (via water ingestion) for an adult whose body weight is 70 kg when exposed over a 70-year lifespan. When using a linearized multistage model to extrapolate potency of chemicals down to low doses, the unit risk is the upper bound of the 95% confidence limit of the maximum linear term derived from the dose-response data, i.e., the steepest slope.

To obtain acceptable concentrations as a goal for mitigation, one can use Eq. 7.2 to back calculate to a desired concentration reflecting some desired risk level (e.g., $1 \times 10^{-6}$ chance of getting cancer). To do so, some response or "acceptable risk" level must be defined.

Utilizing assumptions about human body weight and the amount of a given chemical consumed per day (Table 7.6), one can calculate the "acceptable" concentration from the dose that corresponds to the acceptable risk level. For a body weight of 70 kg and a consumption rate of 2 liters (l) of water per day, Eq. 7.3 is:

$$C_i = D_i * 70 \text{ kg}/2 \text{ l/day} \qquad (7.4)$$

where C = the concentration in mg/l.

This "acceptable" concentration can then be compared with measured concentrations, i.e., at a disposal site, or with concentrations from a series of environmental fate and transport models for water, soil, and air. Alternatively, this "acceptable" concentration can be used as an input into environmental fate and transport models to obtain an acceptable discharge of the substance into the environment.

#### *7.6.2.3 Non-Carcinogens*

Acceptable risk levels for non-carcinogens are generally expressed in terms of acceptable daily intake (ADI) levels. The expression for ADIs usually takes the following form:

$$ADI = (B/U) * (A/C) \quad (7.5)$$

where ADI = acceptable daily intake (in mg/l per day)
A = the weight of the consuming organism (e.g., 70 kg for humans)
B = the amount sufficient to produce an unwanted health effect, which has been determined empirically (expressed in mg/kg of body weight). Lethal doses that will kill 50% of a population ($LD_{50}$) could be used as one extreme. Concentrations at which no effect is observed (No Adverse Effect Levels or NOELs) could be used as the other extreme.
U = an uncertainty factor which, for drinking water, has ranged from 10 to 1,000 to introduce conservatism into the risk estimate.
C = the amount of water/air/food consumed daily (e.g., the consumption of 2 liters of water per day or inhalation of 20 $m^3$ per day by humans).

ADI and B are usually subdivided into subchronic and chronic effects. Chronic levels are preferable to acute levels when chronic estimates are available. Eq. 7.5 can be reformulated with B as the dependent variable. This allows one to calculate a targeted dose once an acceptable level has been established.

Figure 7.10 above illustrates the relationship among the parameters in the risk calculation. If the doses needed to compute the risk level, T, are not available from reliable monitoring data, they can be calculated from the amounts of a given chemical released into the environment and from environmental fate and transport models. Unit risk estimates are obtained from dose-response relationships. The product of doses and unit risk estimates gives the excess risk to an individual. Multiplying this product by total population gives the risk level in terms of the number of excess cases in a population expected from exposure to the given dose of the chemical.

### 7.6.3 Exposure Assessment: Identification of Sources of Chemicals

There are a number of ways of defining the source of chemical solvents in landfills. On the broadest level, the source can be defined in terms of the origin of the chemical in the economy. The source can also be defined as the mixture of chemicals that enter or are inputs to a landfill.

This is a difficult task even though operating landfills are required to document and control incoming wastes through a manifest system. A large variety of chemicals arrive at a landfill and in widely varying concentrations.

Solvents, for example, typically arrive at landfills in drums. An analysis of solvents in over 1200 drums nationwide revealed that the mean and maximum concentrations of organic solvents in the drums often differed by several orders of magnitude.[53] The mean concentration of carbon tetrachloride was 342 parts per million (ppm) and the maximum was 400,000. The mean concentration for chlorobenzene was 85 ppm and the maximum was 57,000. Once the solvents are dispersed in a landfill, their concentrations can become even more variable. This uncertainty is particularly acute for existing landfills. Operators of new landfills are required by RCRA Permit to determine the correlation of effluent in incoming wells prior to accepting the waste thereby controlling this source of uncertainty. Operators are also required to categorize waste to prevent their liability from accepting questionable and/or banned waste. New landfills can presumably control this source of uncertainty by sampling incoming wastes and adjusting the amounts accepted for disposal.

A third definition of the source is in terms of the outputs from the landfill (i.e., the chemical mixtures that leave the landfill). This probably has a greater degree of variation than estimates from sources.

## 7.6.4 Exposure Assessment: Chemical Releases/Environmental Fate and Transport

### *7.6.4.1 General Considerations*

Release or emission estimates are the first step in evaluating the fate and transport of chemicals in the environment. Releases of chemicals into the air, land, and water occur in a variety of ways. Releases to water occur via leaching (or percolation) and runoff. Releases to the air occur via direct air emissions from soil, fugitive dust emissions, and evaporation from land and water surfaces. The amount of material stored in a landfill is a major factor in predicting discharges from the landfill. As discussed above, uncertainties occur in estimating these amounts since the quantity of materials in landfills is highly variable. This can cause emission and effluent rates to vary by several orders of magnitude.

The transformation and migration of released chemicals can be estimated via mathematical models. These models are specific to the particular medium in which chemicals are found. These media are soil, groundwater, surface water, air, foods (vegetables, fish, shellfish), and objects with which humans come into contact. Some of the major input parameters to models used to characterize the fate and transport of chemicals in the environment are listed in Table 7.10. The values assigned to many of these parameters for organic solvents are given in Tables 7.7, 7.8 and 7.9.

Onishi[54] distinguished between primary and secondary release mechanisms of contaminants. A primary release, he noted, is directly from the contaminant source. A secondary release occurs from a location or a source that has become contaminated from the primary source. For example, a primary

**Table 7.7 Environmental Release/Transport Potential into Air**

| | Ignitability | | Explosivity limits (% by volume) | | |
|---|---|---|---|---|---|
| | Flash point °F | Vapor pressure mmHg 20°C | Lower | Upper | Relative volatility (evaporation rate[a]) |
| Chlorobenzene | 82 | 8.8 | 1.3 | 7.1 | 1.07[b] |
| o-Dichlorobenzene | 160 | 0.348 | 2.2 | 9.2 | 0.15[b] |
| Cresol, Cresylic acid | 86 | 0.15 | | | |
| Ethylbenzene | 20 | 7.1 | 1.0 | 6.7 | |
| Nitrobenzene | 88 | 0.15 | | | |
| Pyridine | 20 | 18.00 | 1.8 | 12.4 | |
| Toluene | 45 | 38 | 1.2 | 7.0 | 1.5[b] |
| Xylene | 80 | 9.5 | 1.0 | 7.0 | 0.75[b] |
| Cyclohexanone | 129 | 7.0 | 1.1 | 8.6 | 0.31 |
| Acetone | –4 | 185.0 | 2.6 | 12.8 | 7.7 |
| Methyl Ethyl Ketone | 16 | 70.6 | 1.8 | 10.0 | 4.6 |
| Methylisobutylketone | 60 | 16.0 | 1.2 | — | 1.6 |
| N-Butyl alcohol | 97 | 4.39 | 1.2 | 10.9 | 0.46 |
| Isobutyl alcohol | 85 | 8.8 | 1.45 | 11.25 | 0.63 |

| | | | | | |
|---|---|---|---|---|---|
| Methanol | 54 | 96 | 6.7 | 36.0 | 3.5 |
| Carbon Disulfide | –30 | 300 | | | |
| Carbon Tetrachloride | — | 90.0 | — | — | 6.0[b] |
| Ethyl Acetate (99%) | 24 | 76 | 2.2 | 11.0 | 4.1 |
| Diethyl Ether | –45 | 440 | | | |
| Methylene Chloride | — | 340 | — | — | 14.5[b] |
| Tetrachloroethylene | — | 13.0 | — | — | 2.1[b] |
| Trichloroethylene | — | 59.0 | 8.0 | 10.5 | 4.46 |
| 1,1,1-Trichloroethane | | 100 | | | |
| Chlorofluorocarbons | [c] | 284–502 | — | — | — |

[a] Given for acetone = 1 unless indicated otherwise.

[b] Given for n-butyl alcohol = 1.

[c] No flash points or explosive limits are available for most chlorofluorocarbons; practically all evaporation rates are unspecified.

ChemCentral. *Physical Properties of Common Organic Solvents and Chemicals*. Chicago, IL: ChemCentral, 1980. The reader is referred to these tables for detailed notes.

**Table 7.8 Environmental Release/Transport Potential through Soil and Water**

| | Mobility in soil, based on $K_{oc}$[a] | Dielectric constant, 25°C | Solubility, mg/L, 30°C in water[b] |
|---|---|---|---|
| Chlorobenzene | 2 | 5.65 | 448 L |
| o-Dichlorobenzene | 1 | 6.83 | 150 L |
| Cresol, Cresylic acid | 4 | 11.8 (m) | 21,800(m)VS |
| | | 11.5 (o) | 24,500 (o) VS |
| | | 9.9 (p) | 19,400 (p) VS |
| Ethylbenzene | 1 | 2.41 | 150 L |
| Nitrobenzene | 3 | 34.8 | 2,000 MS |
| Pyridine | 4 | 12.3 | CM |
| Toluene | 2 | 2.44 | 500 S |
| Xylene | 1,2 | 2.44 (m) | 175 SS |
| | | 2.57 (o) | |
| | | 2.27 (p) | |
| Cyclohexanone | 4 | 18.3 | 50,000 VS |
| Acetone | 4 | 20.7 | —[c] CM |
| Methyl Ethyl Ketone | 4 | 18.5 | 270,000 VS |
| Methylisobutylketone | 4 | 15.0 | 19,000 VS |
| N-Butyl Alcohol | 4 | 17.7 | 70,800 VS |
| Isobutyl Alcohol | 4 | 17.9 | 87,000 VS |
| Methanol | 4 | 32.6 | —[c] CM |
| Carbon Disulfide | 3 | 2.64 | 2,940 MS |
| Carbon Tetrachloride | 2 | 2.205 | 800 L |
| Ethyl Acetate | 4 | 6.02 | 85,300 VS |
| Diethyl Ether | 4 | 4.2 | 60,050 VS |
| Methylene Chloride | 4 | 9.1 | 13,200 VS |
| Tetrachloroethylene | 2 | 2.35 | 150 L |
| Trichloroethylene | 2 | 3.42 | 1,000 MS |
| 1,1,1-Trichloroethane | 2 | 7.1 | 700 L |
| Chlorofluorocarbons | 1,2,3,4 | | |

[a] $K_{oc}$ scale: 1 = low mobility; 2 = medium mobility; 3 = high mobility; 4 = very high mobility.

[b] Solubility scale: L = low; SS = slightly soluble; S = soluble; MS = moderately soluble; VS = very soluble; CM = completely miscible.

[c] Completely miscible.

Summarized from R.A. Griffin and W.R. Roy, Interaction of Organic Solvents with Saturated Soil-Water Systems. Open File Report, University of Alabama, Environmental Institute for Waste Management Studies, Tuscaloosa, AL, 1985.

**Table 7.9 Environmental Release/Transport Potential: Biodegradability**

| | Biodegradability[a] | | | | |
|---|---|---|---|---|---|
| | % Removed | Rate of degradation (mg COD/g · VOS-hr) | Half life in soil[b] | Bioconcentration | Bioaccumulation |
| Chlorobenzene | | | H | 2 | 450 |
| o-Dichlorobenzene | | | H | 2 | 89 |
| Cresol, Cresylic Acid | 95–96 | 54–55 | H | | |
| Ethylbenzene | | | | | |
| Nitrobenzene | 98 | 14 | M | 1 | 15 |
| Pyridine | | | H | 0 | |
| Toluene | | | H | 1 | 15–70 |
| Xylene | | | | | |
| Cyclohexanone | 96 | 30 | | | |
| Acetone | | | | | |
| Methyl Ethyl Ketone | | | M | 1 | |
| Methylisobutylketone | | | M | | |
| N-Butyl Alcohol | 99 | 84 | | | |
| Isobutyl Alcohol | | | | | |
| Methanol | | | | | |
| Carbon Disulfide | | | L | 1 | |

**Table 7.9 Environmental Release/Transport Potential: Biodegradability** *(continued)*

| | Biodegradability[a] | | | | |
|---|---|---|---|---|---|
| | % Removed | Rate of degradation (mg COD/g · VOS-hr) | Half life in soil[b] | Bioconcentration | Bioaccumulation |
| Carbon Tetrachloride | | | L | | 17–30 |
| Ethyl Acetate | | | | | |
| Diethyl Ether | | | | | |
| Methylene Chloride | | | MH | 1 | |
| Tetrachloroethylene | | | MH | 3[c] | 49 |
| Trichloroethylene | | | L | 3[c] | |
| 1,1,1-Trichloroethane | | | | 1 | 9 |
| Chlorofluorocarbons | | | | | |

[a] From W.W. Eckenfelder, Jr., *Principles of Water Quality Management,* pp. 279–281.

[b] Key to half-life in soil: H = High, less than 7 days; MH = Moderately high, 1–4 weeks; M = Moderate, 1–5 months; ML = Moderately low, 5–9 months; L = Low, greater than 9 months. From Berkowitz, J.B., Harris, J.C., and Goodwin, B., Identification of hazardous waste for land treatment research, in *Proceedings of the 7th Annual Research Symposium,* Schultz, D.W. and Black, D., Eds., U.S. EPA, Philadelphia, 1981, pp. 168–177.

[c] Bioconcentration in marine life. Source of bioconcentration and bioaccumulation data: GCA Corp., Disposal Alternatives for Certain Solvents, U.S. EPA, Washington, DC, January 1984, p. 77.

**Table 7.10 Summary of Input Parameters for Risk Estimation for Chemical Solvents**

**Release Rates**

Amount of surface exposed or disrupted
Chemical and environmental factors enhancing release rates (see parameters below)

**Environmental Fate and Transport**

*Environmental Factors*

Generation rate of methane gas (if it is present)
Wind speed
Wind direction
Atmospheric stability
Temperature
Precipitation, moisture
Surface area
Gradients or slope
Soil type
Soil moisture
Type of soil cover
Depth of soil cover
Soil density and porosity
Water flow rate
Hydraulic conductivity

*Chemical Factors*

**Soil:**
- Soil adsorption ($K_{oc}$)
- Persistence (biodegradability)
- Amount of the chemical (concentrations in soil pores and at the air-soil interface)
- Vapor pressure of the chemical
- Diffusivity of the chemical

**Water:**
- Solubility
- $K_{ow}$
- Henry's Law Constant

**Air:**
- Vapor Pressure
- Flash Point
- Explosivity

source would be a spill or discharge of chemicals to soils surrounding a factory. The secondary source would be the release of those chemicals to groundwaters or surface waters driven by flood waters. This aspect of environmental fate and transport is often ignored. Monitoring data used as a basis for calculating doses often does not include a record of the weather conditions at the time of the sampling. Heavy rains and floods can substantially affect chemical detection in both soil and water.

An important aspect of the fate of organic chemicals in the environment is the extent at which they disappear altogether. Processes of degradation are hydrolysis, reduction, oxidation, photolysis, or biodegradation. A measure of the resultant of all of these forces is persistence. It is measured in terms of the "half-life" of a chemical. The half-life of a chemical is the amount of time it takes for half of the chemical to disappear, regardless of the mechanisms of degradation. While this is an extremely useful measure, data are not always available for every chemical, and values vary with environmental conditions.

### *7.6.4.2 Air*

Evaporation is a major mechanism by which organic chemicals leave landfills. A recent review of the literature on organic chemical emission rates at landfills was conducted by Bennett.[55] Experiments and models that estimate air emissions from landfills reveal that the rate of air emissions is a function of the following (often interrelated) variables:[55,56]

**Environmental Conditions:**
- Wind speed
- Wind direction
- Atmospheric stability
- Temperature
- Precipitation, moisture

**Landfill Attributes and Surrounding Conditions:**
- General rate of methane gas (if it is present, it can physically drive chemicals out of a landfill)
- Surface area
- Soil type
- Soil moisture
- Type of soil cover
- Depth of soil cover
- Soil disturbance (vehicles, other human activity, animals)
- Soil porosity/permeability

**Chemical Attributes:**
- Amount of the chemical (its concentration in soil pores and at the air-soil interface)
- Vapor pressure of the chemical

Diffusion rate of the chemical
Solubility

Emission rates are negatively correlated with the amount of soil moisture, depth of soil cover, the solubility of the chemical in water, and the weight of the chemical. Emissions are positively correlated with the diffusion rates of the chemicals, the porosity and permeability of the soil, amount of exposed surface area, vapor pressure, Henry's Law constant, temperature, the velocity of the methane gas (if it is present), and wind velocity.[55]

The quantification of atmospheric concentrations of organic chemicals via air quality monitoring is limited by several factors. First, the scope of existing air quality monitoring networks is limited in terms of the coverage of organics and the quality of the data. Only 368 chemicals are covered by the EPA's National Emissions Inventory and adequate monitoring information is available for only about 10% of these.[43]

Second, chemical transformations in the atmosphere are not easy to capture through a monitoring network. The fates and ultimate concentrations of emissions into the atmosphere are usually estimated via a variety of air-quality simulation models. These models tie together many of the variables listed above. They have been developed by or under the auspices of the EPA for compliance with the requirements of the Clean Air Act. The most popular series of models in the User's Network for Applied Modeling of Air Pollution (UNAMAP). These models use emissions and meteorologic data as inputs and provide concentrations of selected air pollutants as outputs. The applicability of any given model to landfill emissions depends upon the terrain, the physical setting of the facility, the type and frequency of the emissions, and the distance of the receptors. For the purpose of modeling emissions from landfills, landfills are generally treated as emitters located at ground level.

A major factor in determining the degree of model accuracy in reflecting the dispersion and deposition of emissions from a given source to a given receptor is the choice of coefficients that represent how fast a plume will rise and fall. Another major factor is the source emission rate. Garavanos and Shen[57] have demonstrated that wind speed is a major factor in influencing the rate of emissions of carbon tetrachloride and trichloroethylene from landfills. Emissions vary more according to wind speed than they do by soil types. When estimated and actual emission rates are compared, the variation often ranges between 50% and 150%, unless corrections are made using wind speed data.[57] Thus, both of these factors can cause outputs to vary by several orders of magnitude.

### *7.6.4.3 Groundwater Soil/Surface Water*

The parameters that influence the migration of organic chemicals into groundwater are:

**Landfill/Environmental Conditions:**
Hydraulic Conductivity (water flow rate)
Net infiltration from precipitation, runoff, and evapotranspiration
Soil density and porosity
Thickness of soil layer
Organic content

**Chemical Attributes:**
Soil-water partition coefficient or organic carbon-water partition coefficient ($K_{oc}$)
Persistence (biodegradability)
Solubility
Octanol-water partition coefficient ($K_{ow}$)
Henry's Law Constant

Empirical work indicates that leaching to groundwater is inversely proportional to the Henry's Law Constant, diffusion rate, organic content of the soil, soil adsorption rates, runoff of water over land surfaces, and evapotranspiration. Leaching is directly proportional to infiltration and rainfall, soil permeability, solubility of the chemical in water, and the size of the site.

Four parameters to estimate the mobility of nonpolar organic solvents in soils co-vary with one another.[58] These parameters are the octanol-water partition coefficient, $K_{ow}$; the organic carbon-water partition coefficient (or the ability of chemicals to adsorb to organic carbon), $K_{oc}$; solubility; and the retardation factor.[58, 48] Since each of these four parameters represents the mobility of a chemical in soil in approximately the same manner, any one parameter can be used to represent chemical mobility in soil. This is only plausible if there is roughly a linear correlation among them.

The advantage of using $K_{oc}$ is that it is available for specific chemicals. There are several drawbacks, however, to the use of $K_{oc}$ as an indicator of chemical mobility. A first drawback is that the value of $K_{oc}$ varies with the organic carbon content of the soil (especially when the organic carbon content is less than about 0.1%). Organic content in soils is often difficult to estimate and is known to vary between 0.1 and 3%.[59, 60] Second, the prediction is unreliable for soils with large particles (greater than 50 micrometers in diameter). Third, adsorption rates for organic solvents to clay and soil change because of interaction or reaction effects between the chemical and the soil or clay; i.e., the presence of chemical solvents can change the affinity of the soil or clay for the solvent. Affinity is not necessarily linearly proportional to the concentration of the solvent.[58] Fourth, adsorption is influenced by the rate of flow of water through the soil; that is, non-equilibrium conditions may exist at fast flow rates. Finally, the value of $K_{oc}$ varies from one to one million.[61] Other variables have wide ranges as well. Vapor pressure, for example, ranges from 0.001 to 760 millimeters of mercury for liquids. Any error or variability of these parameters could distort computations in which the parameter is used by several orders of magnitude unless an index with a narrower range is constructed from those values.

The mobility of organic chemicals in soil is inversely proportional to $K_{oc}$; however, there are several qualifications. A compound that normally has a high affinity for soil has low mobility, but it can still migrate due to runoff or soil erosion.

The variation in the range of parameter values (other than $K_{oc}$) used to estimate the migration of contaminants in groundwater can be quite substantial. For example:

- Groundwater flow velocities can range from 1 to 100 meters per year[62] with a velocity of 9,000 meters per year representing an extreme situation (glacial outwash conditions).[63] A velocity range of 10 to 100 meters is typical.
- Retardation factors affect the flow of chemicals relative to groundwater. Retardation factors for organic solvents (e.g., 1,1,1-trichloroethane; trichloroethylene; and tetrachloroethylene) range from one to ten. Thus, the estimate of the velocity of the solvent relative to groundwater flow ranges from 10% to 100%.[62]
- The storage capacity of an aquifer is another parameter that has a considerable range of variation. McKown, Schalla, and English[64] quote a range of 0.05 to 0.25 for storativity in porous unconfined aquifers. The range they quote for confined aquifers is even greater: 0.00001 to 0.001.

There are many groundwater models that can be used to predict the patterns and rates of movement of groundwater and associated chemicals.[65] A series of groundwater models is commonly combined to estimate the movement of contaminants through the saturated and unsaturated zones. For a review of these models see Sternberg,[66] Onishi,[65] and Javandel, Doughty, and Tsang.[67] The model that the EPA has recommended for estimating movement in the saturated zone is the Vertical-Horizontal Spread (VHS) model. A source of uncertainty in the modeling of the transport of landfill contaminants in groundwater arises from assumptions about whether the zone of subsurface contamination beneath a landfill is portrayed as a single plume or many plumes.[62] Mackay et al. noted that plumes are often delineated with inorganic nonreactive tracers. Organic contaminants, which are reactive, do not follow the same pattern. In fact, they estimated that plumes of trichloroethylene and 1,1,1-trichloroethane arrive sequentially rather than simultaneously. Arrival times are much sooner than predicted from groundwater velocity and retardation factors alone.

A back calculation can be used to derive the concentration of chemicals in groundwater from their concentration in soil using the following simple model.[68]

$$C_{water} = C_{soil}/[K_{oc} * (OC)] \tag{7.6}$$

where $C_{water}$ = concentration in water (μg/l)
$C_{soil}$ = concentration in soil (μg/kg)
$K_{oc}$ = adsorption coefficient (μg/kg organic carbon/μg/l water)
OC = fractional organic carbon content (grams of carbon per gram of soil)

The rate of movement of water into the soil can be estimated from the product of the area and the infiltration rate of water. Infiltration, in turn, is computed by subtracting runoff, evapotranspiration, and evaporation from precipitation.[61]

The concentrations of chemicals in surface water can be estimated from surface-water flow rates and the rate of seepage of contaminants into the surface waters. Many of the same parameters that influence the migration of pollutants into groundwater also affect their migration into surface water. In addition, the following variables influence surface water migration of contaminants:

- Sewage or leaching rate from soil or groundwater into surface water
- Surface-water flow
- Distance or distances from point of seepage

Ultimately, the outputs from water-resource modeling that are used as inputs into risk-assessment models are the steady-state concentrations of contaminants at the following locations:[61]

- within the landfill
- under the site
- at the boundary of the site
- in adjacent streams
- in aquifers, and
- in environmentally sensitive areas, such as wetlands and floodplains, that are hydrologically contiguous with the landfill

### *7.6.4.4 Food*

The characteristics that relate water concentrations of chemicals to concentrations in biota consumable by humans are bioaccumulation, bioconcentration, and biomagnification. Bioconcentrations in foods are estimated in a variety of ways. For fish and shellfish, bioconcentration factors usually are derived from concentrations of the chemicals in the water. Thus, the concentration in micrograms per kilogram weight of fish (kg) of a particular chemical in fish or shellfish (in μg/kg) is calculated as the product of the bioconcentration factor and the concentration (μg/l) in water.[68] The tendency of chemicals to accumulate in the food chain is directly correlated with $K_{oc}$. Bioaccumulation is more likely to occur at higher values of $K_{oc}$.[48]

### *7.6.4.5 Areas of Uncertainty in Fate and Transport Estimates*

Areas of uncertainty in environmental fate and transport estimates occur anywhere between the initial measurement of parameters characterizing chemical transport and the underlying assumptions and simplifications in the models. Of the many examples given above, a few are particularly significant.

One of the most important parameters is the estimation of the tendency of a chemical to migrate in soil and water, $K_{oc}$. There are also several related parameters that describe similar tendencies. Errors in both laboratory and field measurements for these parameters can be substantial (several orders of magnitude). The range in the value of $K_{oc}$ for chemicals can be as large as six orders of magnitude. Since $K_{oc}$ is used directly in risk estimation, this error becomes a part of the risk estimate.

The rate of flow of groundwater is another parameter with a high degree of uncertainty. Errors of a few orders of magnitude can result from improper field measurements alone, for example, where monitoring wells have been improperly drilled or installed.[64]

In environmental fate and transport modeling, uncertainties exist in the structure of both air and water quality models. Estimates of the deposition and emission of materials can be a source of considerable error in air quality models. The accuracy of both deposition and emission estimates depends on the accuracy of estimates for wind speed and direction. Deposition estimates are affected further by the existence of barriers to uniform dispersal and by chemical degradation and transformation processes in the atmosphere.

## 7.6.5 Exposure Assessment: Routes of Exposure

Exposure pathways are the routes by which chemicals come into contact with human receptors. The importance of the route depends on the type of waste disposal activity that is being conducted and the type and magnitude of population exposed (e.g., the general population versus workers).

The exposure pathways typical of land-disposal facilities for hazardous wastes are inhalation, ingestion, and dermal adsorption:

- Inhalation of air-borne soil particles or chemical vapors as they volatilize from soil
- Ingestion of soil particles directly or from clothing, skin, animals, or plants
- Dermal absorption (and absorption by other body fluids such as blood or tears) from direct contact with contaminants in soil particles and water or airborne vapors and particulates

- Ingestion of foods that have become contaminated via the air or water such as fish, shellfish, vegetables, and dairy products
- Ingestion of drinking water that is obtained from contaminated surface or groundwaters
- Inhalation, dermal contact, and ingestion from bathing in waters that are contaminated

The importance of each of these routes of exposure varies by type of chemical, by type of environmental condition, and by proximity of individuals to the land-disposal facility. For example, contact through ingestion of soil particles that become airborne from the land or from the movement of vehicles is a common route of exposure for workers at a landfill.

#### *7.6.5.1 Exposure Points*

Exposure points are locations where human contact occurs with the contaminants. One can identify exposure points from the routes by which people can be exposed to a chemical from a landfill. While exposure points are unique to specific sites, the following general categories of exposure routes for chemicals from landfills can be identified:

- Groundwater: ingestion of water from a well
- Surface water: dermal contact from swimming or ingestion of water from a surface water supply
- Soil: ingestion of or dermal contact with windblown soil
- Air: inhalation at points of population concentration, such as a school, shopping center, or office building

Uncertainties that can occur in identifying routes of exposure include omission of important routes of exposures and overemphasis or underemphasis of routes of exposure that have been identified. These uncertainties result because human behavior cannot be predicted exactly. For example, not all private wells in use can be identified; children might be playing in contaminated areas where such activity is prohibited.

#### *7.6.5.2 Health Effects*

Human health effects from chemicals are related to doses of the chemicals that are taken into the body (intake rates), the amounts that are absorbed or retained by the body, the size and nature of the population at risk, the state-of-health or resistance of individuals to the effects, and the effects that the absorbed chemicals are likely to have on health (dose-response relationship).

#### *7.6.5.3 Dose Estimation*

Dosage is a function of the amount of air, water, and food and the concentration of a chemical that is taken into the body. Typical intake rates are given in Table 7.6. Considerable variation in rates obviously exists. These variations result from differing assumptions about the sizes and weights of individuals, their levels of activity, and their tastes and preferences for food and water.

Given these intake rates, daily human doses can be calculated per unit of body weight according to the following equation:

$$D = (C * R)/W \tag{7.7}$$

where D = daily intake in mg/kg/day
C = concentration of the chemical in mg/L (water) or mg/m$^3$ (air)
R = daily intake rate in units comparable to C, i.e., L/day (water) or m$^3$/day (air)
W = body weight in kg

#### *7.6.5.4 Absorption*

Absorption of a substance by the human body is one component of the effect of exposure. The level of exposure in mg/day is equivalent to the product of the concentration (in mg/L) of the chemical taken in, the weight of the medium by which it is taken in (e.g., food), and the rate of absorption. The amount of a substance that is actually absorbed by the human body (as distinct from the exposure to the substance) is dependent upon the route by which the substance enters the body, the length of exposure, type of species exposed, and the behavior of the substance in general and relative to the body's chemistry. In addition, the rate depends on the state of health of the individual, degree of starvation, and whether the individual is exercising during exposure. When information is lacking on absorption rates, the ratio of the concentration of substances absorbed to the concentration of the substance in the exposure medium is often set equal to one.[39]

The variation in absorption rates of organic solvents is illustrated by some empirical findings reported by the EPA for a few selected chemicals. For trichloroethylene, absorption ranged from about 50–65% in humans via inhalation and 95–98% in rats and mice via ingestion.[69] For carbon tetrachloride, absorption ranged from 65–85% in rats via inhalation and 30–60% via ingestion.[50] For xylene, absorption via ingestion was 85–90% and 55–70% via inhalation.[70] Toluene exhibited approximately 100% absorption via ingestion and 30–55% via inhalation.

#### *7.6.5.5 State of Health*

The health of individuals in a population can determine the actual toxic effect that is felt. Animal and human experimental data giving the effects of doses under varying health states is an important input into these determinations. In reality, such information is rarely available. Most experiments used as the basis for establishing dose-response curves are conducted on healthy organisms.

#### *7.6.5.6 Size and Nature of the Population at Risk*

The degree to which sectors of the population are at risk depends on the proximity of people to the chemical releases, the time spent near sources, the level of physical exertion, resistance to the effects of exposure, and the extent to which chemicals are transported to people via wind, water flow, etc. One can work back from an acceptable dose level and the behavior or migration of the chemical along each of the routes of exposure to obtain a radius from the source within which populations can be considered at risk. However, the populations at risk may not always be easily defined as a direct function of distance from a source of chemical emissions. Various concentration effects that are not linear with distance can occur. For example, dispersed chemicals can be concentrated in water through entrainment in sediment and through turbulence, in air from downwash effects and eddys, and in soil.

The sectors of the population that are typically considered to be at higher risk from chemical exposures than other sectors of society are defined in terms of age, activity, or state of health. The higher risk groups typically are children at play or in school, nursing mothers, the infirm and chronically ill, the elderly, athletes, trespassers, workers in proximity to the chemical disposal site (e.g., landfill operators, truckers), and persons that typically spend a lot of time outdoors (e.g., maintenance workers and the homeless). Certain populations defined by environmental or behavioral factors (e.g., smoking) may experience greater toxicity once exposure has occurred.

### 7.6.6 Dose-Response Estimation*

Experimentally derived data on the responses of bacteria, animals, or humans to doses of a particular chemical are used to create dose-response curves. These curves can only be constructed in ranges where chemical concentrations and the responses are easily measured. The dose-response relationship is used to develop the unit risk estimate (R in Eq. 7.2 above). The unit risk estimate is defined as "the increased individual lifetime risk" for an individual of a given weight (usually 70 kg) exposed continuously to a unit amount of a chemical (e.g., 1 $\mu g/m^3$ in air or 1 mg/l in water) over a lifetime (usually 7 years).[39] Unit risk is computed for acute and chronic toxicity and for more long-term health effects such as carcinogenesis.

* Dose-response estimated is considered a distinctly separate step from exposure assessment.

## 7.7 HYDROGEOLOGICAL SYSTEMS AND MONITORING[71-74]

Acceptable waste management for groundwater protection must be based on an adequate and detailed monitoring system. It must be representative of the geology, stratigraphy, structure, depositional environment, and water-bearing beds. It must provide adequate information about base level conditions in the aquifer or aquifers involved with regard to recharge, storage and discharge conditions, and water quality parameters. The monitoring system must reflect the conditions of contamination or pollution as well as the dynamic aspects of these conditions, e.g., the natural impacts of climate, rainfall, snowfall, freezing, and evaporation/transpiration.

A list of federal laws and regulations has been provided in Table 7.4; however, six of those are of most importance regarding groundwater protection and should be emphasized as follows:

1. The Clean Water Act of 1972 gives the EPA authority to protect surface and groundwater.
2. The Safe Drinking Water Act of 1974 sets drinking water standards that are used to protect groundwater. A provision of the act allows the EPA to designate an aquifer as the sole source of drinking water for an area and denies federal money to water projects that threaten to contaminate the aquifer. The EPA has set maximum contaminant level standards for public drinking water supplies.
3. The Federal Insecticide, Fungicide, and Rodenticide Act of 1974 (FIFRA) assigns to the EPA control over availability and use of pesticides that may leach into groundwater.
4. The Toxic Substances Control Act of 1976 gives the EPA authority to limit certain uses of chemicals, to require warning labels, and to reduce risks from chemicals that have the potential to contaminate groundwater.
5. The Resource Conservation and Recovery Act of 1976 gives the EPA authority to set up programs to prevent hazardous wastes from leaching into groundwater from landfills, surface impoundments, and underground tanks.
6. The Comprehensive Environmental Response, Compensation, and Liability Act of 1980 is called the Superfund bill, because it set up a fund to support federal and state responses to hazardous waste problems. The law gives the EPA authority and money to clean up hazardous waste sites that are a threat to human health and the environment.

It must be recognized that federal regulations and rules must in turn be implemented at state level; therefore, within each state there is a companion organization to the U.S. Public Health Service and U.S. Environmental Protection Agency. In Alabama, it is the Alabama Department of Environmental Management and, in Georgia, the Department of Environmental Regulations and Department of Environmental Resources. The state of Florida passed legislation to be applied to the development of groundwater. These regulations require strict adherence to defining the impact on the surface water, the shallow surficial aquifer, and the deeper intermediate and Floridan aquifers. Regulations require the Development of a Regional Impact Statement or DRI. Extensive pumping tests, surface-water and groundwater studies, and monitoring for discharge, water levels, and quality of water are required to identify these impacts.

The Southwest Florida Water Management District (SWFWMD) in its Code Section 16, CR-0.15 (5) (A) states, "The water crop, in the absence of data to the contrary, is 1,000 gallons per day per acre." The Cooperative's Project tract is 7,810 acres, and the water crop established legally for the acreage involved is more than needed for the proposed projected mining operations. However, "the 5-3-1 Criteria," which also applies, requires that a determination be made to show that there will not be more than a 5-foot average decline in water level in the Florida Aquifer at the boundary of the property, not more than a 3-foot decline in the Surficial Aquifer at the boundary, and no more than a 1-foot decline in the nearest water body (pond, lake, etc.). In addition, surface-water flow in streams of the area must not be decreased more than 5% unless a variance to the rule is obtained.

Surface-water flow is affected by differences in soils, geology, vegetation cover, altitude, elevation, and precipitation intensities for the various surface-water basins within the tract. Each year, within the project area, streams recede to low flows from April to June. Therefore, a seasonal distribution of average monthly flow must be determined. The annual minimum instantaneous or daily flow is subject to alterations by transient, natural, or man-made causes; therefore, the lowest 7-day average flow each year is used as the reference period for low flows. The yearly minimum 7-day low flows are determined from data collected by the U.S. Geological Survey at gauging stations strategically located over the state. One or more of these long-term gauging station records provide the 7-day minimum flows. Regression models are used to obtain a site-specific extension of the annual flows to an equivalent 40-year period.

Environmental monitoring strategy must achieve goals established by a great variety of governmental legislation and regulations pertaining to a wide diversity of uses. Specifications for monitoring are contained in a series of federal laws that address the need for protection of groundwater quality.

Table 7.4 lists the laws enacted by Congress and summarizes the applicable groundwater activities associated with each law. Of the 16 statutes listed in Table 7.4, 10 statutes have regulatory programs which establish groundwater monitoring requirements for specific sources of contamination. Table 7.5 summarizes the objectives and monitoring provisions of the federal acts. While the principal objectives of the laws are to obtain background water quality data and to evaluate whether groundwater is being contaminated, the monitoring provisions contained within the laws vary significantly. Acts may mandate that groundwater monitoring regulations be adopted, or they may address the need for the establishment of guidelines to protect groundwater. Further, some statutes specify the adoption of rules that must be implemented uniformly throughout the United States, while others authorize adoption of minimum standards that may be made more stringent by state or local regulations.

Specific groundwater monitoring recommendations can be found in the numerous guidance documents and directives issued by agencies responsible for implementation of the regulations. Examples of guidance documents include the Office of Waste Programs Enforcement Protection Agency, the Office of Solid Waste Documents SW-846 and SW-611 (United States Environmental Protection Agency), and CERCLA and RCRA documents.

Purpose and importance of proper groundwater monitoring well installation must have as a primary objective a monitoring well that will provide an access point for measuring groundwater levels and permit the procurement of groundwater samples that accurately represent in-situ groundwater conditions at the specific point of sampling. To achieve this objective, it is necessary to fulfill the following criteria:

1. Construct the well with minimum disturbance to the formation;
2. Construct the well of materials that are compatible with the anticipated geochemical and chemical environment;
3. Properly complete the well in the desired zone;
4. Adequately seal the well with materials that will not interfere with the collection of representative water quality samples; and
5. Sufficiently develop the well to remove any additives associated with drilling and provide unobstructed flow through the well.

In addition to appropriate construction details, the monitoring well must be designed in concert with the overall goals of the monitoring program. Key factors that must be considered include the following:

1. Intended purpose of the well;
2. Placement of the well to achieve accurate water levels and/or representative water quality samples;
3. Adequate well diameter to accommodate appropriate tools for well development, aquifer testing equipment, and water quality sampling devices; and
4. Surface protection to assure no alteration of the structure or impairment of the data collected from the well.

There are many excellent references for well construction, spring development and sampling, well testing and sampling, and systematic monitoring programs. Each, however, must be based on a detailed knowledge of the geology, stratigraphy, structure, and depositional environment, as well as all the man-made factors involved.

## REFERENCES


1. American Society of Civil Engineers, *Manual on Groundwater Management*, ASCE Mann (40), 1972.
2. APHA, *Standard Methods for the Examination of Water and Waste Water*, 14th Ed., American Public Health Association, New York, 1976.
3. Panoni, J.L., Heer, J.E., Jr., and Hagerty, D.L., *Handbook of Solid Waste Disposal; Materials and Energy Recovery*, Van Nostrand Reinhold, New York, 1975.
4. Hazardous Waste, Proposed Guidelines and Regulations and Proposal on Identification and Listing, *Federal Register*, 43, 243, 58945-59028.
5. Conway, R.A. and Ross, R.D., *Handbook for Industrial Waste Disposal*, Van Nostrand Reinhold Co., New York, 1980, chap. 1.
6. Sell, N.J., *Industrial Pollution Control, Issues and Techniques*, Van Nostrand, New York, 1981, chap. 5.
7. Hazardous waste disposal, *Science News* 114, 26, Dec. 23 and 30, 1978, p. 440.
8. *Chernobyl, Special Programs of United Nations System Organizations Relating to the Aftermath of the Accident at the Nuclear Power Plant*, UNESCO, 25 October 1990.
9. The fire next time, *Environmental Science and Technology* 12, 2, Feb. 1978, p. 134.
10. Environmental Institute for Waste Management Studies, *Disposal of Solvents and Solvent-Contaminated Wastes to Land—A Position Paper*, The University of Alabama Press, Tuscaloosa, December 1985.
11. Kozlovsky, E.A., Ed., *Geology and the Environment, An International Manual*, 3 Vol., UNESCO-UNEP, 1989, pp. 100–120.
12. Langer, M., *Geotechnical Investigation Methods for Rock-Salt*, Paris, Bull. I AEG, 25, 1982, pp. 155–164.
13. Lux, K.M., *Gebirgsmechanischer Entwurf und Feldefahrungen in Salzkavernenbau*, Enke-Verlag, Stuttgart, 1984.
14. Schmidt, M.W. and Quast, P., Disposal of MLW and LLW in Leached Cavern, Braunschweig, Report GSF-T, 166, 1984.
15. Langer, M., Rheological Behaviour of Rock Masses, Int. Association of Rock Mechanics Engineers, Proc. 4th Int. Congress on Rock Mechanics, Montreaux, Vol. 3, 1979, p. 29–96.
16. Hardy, H. and Langer, M., *The Mechanical Behaviour of Salt*, Proc. 1 Conference, Penn State University, Trans Tech Publ. Clausthal, 1984.
17. Libby, et al., *The Taa Mines Story*, Transactions of the Institute of Mining and Metallurgy, London, 1985.

18a. U.S. Geological Survey, National Water Summary, Hydrologic Events and Issues, Water Supply Paper 2250, 1984, pp. 79–239.

18b. U.S. Geological Survey, Hydrologic Events and Groundwater Quality, Water Supply Paper 2325, 1988, pp. 143–546.

18c. U.S. Geological Survey, Hydrologic Events and Floods and Droughts, Water Supply Paper 2375, 1991, pp. 163–582.

18d. U.S. Geological Survey, Hydrologic Events and Stream Water Quality, Water Supply Paper 2400, 1993, pp. 155–576.
19. Williams, R.E., *Waste Production and Disposal in Mining, Milling and Metallurgical Industries*, Miller Freeman Publication, 1975, pp. 38–83.
20. Down, C.G. and Stocks, L., *Environmental Impact of Mining*, Applied Science Publishers, London, 1977, pp. 131.
21. Tyagi, A.K. and Turner, W., *Underground Storage Tanks, Spill Prevention and Clean Up*, Engineering Extension, OSU, Stillwater, OK, 1987.
22. Tyagi, A.K. and Martell, J., *Fate/Transport Modelling of Imissible LNAPL in Unsaturated Aquifers*, Int. Symposium on Engineering Hydrology Proc. ASCE, 1993, pp. 701–705.
23. Palmer, C.M., *Principles of Contaminant Hydrogeology*, Lewis Publ., Boca Raton, FL, 1992, p. 161.
24. Hazen, J.C. et al., *In-Situ Bioremediation Via Horizontal Wells*, Int. Symp. on Engin. Hydrology Proc. ASCE, 1993, pp. 862.
25. Lang, M.M. et al., *In-Situ Bioremediation Using a Recirculation Well*, Int. Symp. on Engin. Hydrology Proc. ASCEP, 1993, p. 880.
26. Zimmerman, R., *Risk Assessment Procedures for Organic Solvent Disposal in Landfills (Procedures and Their Uncertainties)*, Open File Report prepared for the Environmental Institute for Waste Management Studies, University of Alabama, June 1987.
27. Office of Technology Assessment, Superfund Strategy, Washington, DC, OTA, 1985.
28. Rowe, W.D., *An Anatomy of Risk*, John Wiley & Sons, New York, 1977.
29. Vlek, C. and Stallen, P., Rational and personal aspects of risk, Acta Psychologica, 45, 1980, pp. 273–300.
30. Sage, A.P. and White, E.B., Methodologies for risk and hazard assessment: a survey and status report, *IEEE Transactions on Systems, Man and Cybernetics,* 10, August 1980, pp. 425–446.
31. Sloan, III, A. and Sepesi, J.A., Public health: a quantitative way of determining threat, Proceedings of the Management of Uncontrolled Hazardous Waste Sites, Hazardous Materials Control Research Institute, Silver Spring, MD, 1984.
32. Ess, T. and Shih, C.S., Perspectives of risk assessment for uncontrolled hazardous waste sites, Proceedings of the Management of Uncontrolled Hazardous Waste Sites, Hazardous Materials Control Research Institute, Silver Spring, MD, 1983.
33. Ruckelshaus, W.D., Science, risk, and public policy, *Science,* 221, September 9, 1983, pp. 1026–1028.
34. U.S. Environmental Protection Agency, *Risk Assessment and the U.S. EPA*, Washington, DC, U.S. EPA, December 1984.
35. U.S. EPA, Guidelines for Carcinogen Risk Assessment, *Federal Register,* 51, September 24, 1986a, 33992–34003.
36. U.S. EPA, Guidelines for Mutagenicity Risk Assessment, *Federal Register,* 51, September 24, 1986b, 34006–34012.
37. U.S. EPA, Guidelines for the Health Assessment of Suspect Developmental Toxicants, *Federal Register,* 51, September 24, 1986d, 34028–34040.
38. U.S. EPA, Guidelines for the Health Risk Assessment of Chemical Mixtures, *Federal Register,* 51, September 24, 1986c, 34014–34025.

39. Anderson, E.L. et al., Quantitative Approaches in Use to Assess Cancer Risk, *Journal of Risk Analysis,* 3, 1983, pp. 277–295.
40. Russell, M. and Gruber, M., Risk assessment in environmental policy-making, *Science,* 236, April 17, 1987, pp. 286–290.
41. U.S. EPA, National oil and hazardous substances pollution contingency plan; final rule, *Federal Register,* 50, November 20, 1985, pp. 47912–47979.
42. U.S. EPA, National primary drinking water regulations; volatile synthetic organic chemicals; final rule and proposed rule, *Federal Register,* 50, November 13, 1985, pp. 46080–46933.
43. Cannon, J.A., The regulation of toxic air pollutants — a critical review, *Journal of the Air Pollution Control Association,* 36, 5, May 1986, pp. 562–573.
44. Lowrance, W.W., *Of Acceptable Risk,* William Kaufman, Los Altos, CA, 1976.
45. Lave, L.B., Methods of risk assessment, in *Quantitative Risk Assessment in Regulation,* Lave, L.B., Ed., Brookings, Washington, DC, 1982.
46. Davies, J.C., Science and policy in risk control, in *Risk Management of Existing Chemicals by the Chemical Manufacturers Association,* Government Institutes, Rockville, MD, June 1984.
47. National Academy of Sciences, *Risk Assessment in the Federal Government, Managing the Process*, National Academy Press, Washington, DC, 1983.
48. ICF, Inc., *Draft Superfund Public Health Evaluation Manual,* U.S. Environmental Protection Agency, Office of Emergency and Remedial Response and Office of Solid Waste and Emergency Response, Washington, DC, December 18, 1985.
49. International Agency for Research on Cancer (IARC), IARC Monographs on the Evaluation of the Carcinogenic Risk of Chemicals to Humans, Supplement 4, IARC, Lyon, France, 1982.
50. U.S. EPA, Drinking Water Quality Criteria Document for Carbon Tetrachloride (Final Draft), National Technical Information Service, Springfield, VA, January 1985a.
51. Isaacson, P.J., Eckel, W.P., and Fisk, J.F., Low occurrence compounds: analytical problem or environmental process?, Proceedings of the Management of Uncontrolled Hazardous Waste Sites, Hazardous Materials Control Research Institute, Silver Spring, MD, 1985, pp. 130–135.
52. U.S. EPA, Guidelines for exposure assessment, *Federal Register,* 51, September 24, 1986e, pp. 34042–34054.
53. Blackman, W.C., Jr. et al., Chemical composition of drum samples from hazardous waste sites, Proceedings of the Managmenet of Uncontrolled Hazardous Waste Sites, Hazardous Materials Control Research Institute, Silver Spring, MD, 1984, pp. 39–44.
54. Onishi, Y., Chemical transport and fate models, in *Principles of Health Risk Assessment,* Ricci, P., Ed., Prentice-Hall, Englewood Cliffs, NJ, 1985, pp. 155–234.
55. Bennett, G.F., *Fate of Solvents in a Landfill,* Open File Report of the University of Alabama Environmental Institute for Waste Management Studies, Tuscaloosa, AL, August 1985.
56. Thibodeaux, L.J., Estimating the air emission of chemicals from hazardous waste landfills, *Journal of Hazardous Materials,* 4, 1981, pp. 235–244.

57. Garavanos, J. and Shen, T.T., The effect of wind speed on the emission rates of volatile chemicals from open hazardous waste dump sites, Proceedings of the Management of Uncontrolled Hazardous Waste Sites, Hazardous Materials Control Research Institute, Silver Spring, MD, 1984, pp. 68–71.
58. Griffin, R.A. and Roy, W.R., Interaction of Organic Solvents with Saturated Soil-Water Systems, Open File Report, University of Alabama Environmental Institute for Waste Management Studies, Tuscaloosa, AL, August 1985.
59. Tucker, W.A., Gensheimer, G.J., and Dickinson, R.F., Coping with uncertainty in evaluating alternative remedial actions, *Proceedings of the Management of Uncontrolled Hazardous Waste Sites,* Hazardous Materials Control Research Institute, Silver Spring, MD, 1984, pp. 306–312.
60. Freeze, R.A. and Cherry, J.A., *Groundwater*, Prentice-Hall, Englewood Cliffs, NJ, 1979.
61. ICF-Clement, *Risk Assessment of the Tyson's Dump Site, Montgomery County, PA,* Final Report, ICF-Clement, Arlington, VA, February 21, 1986.
62. Mackay, D.M., Roberts, P.V., and Cherry, J.A., Transport of organic contaminants in groundwater, *ES&T,* 19, 1985, pp. 384–392.
63. Guven, O., Mo, F.J., and Melville, J.G., An analysis of dispersion in a stratified aquifer, *Water Resources Research,* 20, 10, 1984, pp. 1337–1354.
64. McKown, G.L., Schalla, R., and English, C.J., Effects of uncertainties of data collection on risk assessment, *Proceedings of the Management of Uncontrolled Hazardous Waste Sites,* Hazardous Materials Control Research Institute, Silver Spring, MD, 1984, pp. 283–286.
65. Onishi, Y., Chemical transport and fate in risk assessment, in *Principles of Health Risk Assessment,* Ricci, P., Ed., Prentice-Hall, Englewood Cliffs, NJ, 1985, pp. 117–154.
66. Sternberg, Y.M., Mathematical Models of Contaminant Transport in Ground Water, Open File Report, University of Alabama Environmental Institute for Waste Management Studies, Tuscaloosa, AL, 1985.
67. Javendal, I., Doughty, C., and Tsang, C.F., *Groundwater Transport: Handbook of Mathematical Models,* American Geophysical Union, Washington, DC, 1984.
68. A.D. Little, Inc., *Quantitative Risk Assessment for Bloody Run Capping and Excavation Options,* A.D. Little, Cambridge, MA, October 5, 1984.
69. U.S. EPA, Drinking water quality criteria document for trichloroethylene (draft), National Technical Information Service, Springfield, VA, January 1985b.
70. U.S. EPA, Drinking water criteria document for ethylbenzene (final draft), National Technical Information Service, Springfield, VA, March 1985a.
71. Aller, L. et al., *Handbook of Suggested Practices for the Design and Installation of Ground-Water Monitoring Wells,* National Water Well Association, Dublin, OH, EPA 600/4-89/034, 1989.
72. National Research Council, The Management of Radioactive Waste at the Oak Ridge National Laboratory: A Technical Review, Panel for Study of the Management of Radioactive Waste at the Oak Ridge National Laboratory, National Academy Press, Washington, DC, 1985.
73. Davis, S.N. and DeWiest, R.J.M., *Hydrogeology,* John Wiley & Sons, New York, 1966.

74. Meinzer, O.E., The Occurrence of Ground Water in the United States, with a Discussion of Principles, Department of the Interior, U.S. Geological Survey Water Supply Paper 489, Washington, DC, 1923.

# Chapter 8
# Case Studies

# 8.1 THE NUBIAN SANDSTONE AQUIFER SYSTEM IN EGYPT

## 8.1.1 INTRODUCTION

The Nubian Sandstone aquifers in Egypt are mostly located in the Western Desert of Egypt (Figure 8.1.1), an area of 750,000 sq. km., representing 75 percent of the whole area of Egypt. Groundwater of good chemical quality occurs in the Nubian Sandstone aquifer system in the Western Desert. The Nubian aquifers also occupy a vast area in the northern region of Africa. This vast aquifer system extends from northeastern Sudan and Chad through eastern Libya and Egypt west of the Nile River, nearly to the Mediterranean Sea. It covers an area of 2 million square kilometers, of which about one-third lies in the Egyptian territories.

While the tableland areas in the Western Desert are characterized as low productivity soils, the depression areas proved to contain arable land resources. Agricultural development activities have been started in the last 20 years in the two southern depressions, Kharga and Dakhla, as pilot development projects. These developments were associated with drilling of a prolific system of deep wells which were originally flowing. The production of the wells has been decreasing year after year with a continuous decline of the water levels. The majority of the wells in Kharga Oasis ceased flowing and have been pumped. Figures 8.1.2 and 8.1.3 show the general physical features of the study area. The Western Desert consists of large plateaus which rise 300 to 700 m above mean sea level. From south to north these are the southern El-Gilf El-Kebir Sandstone Plateau, the central Eocene-Cretaceous Plateau, and the northern Miocene Marmaricau Plateau. Such plateaus are interrupted by a series of low-lying depressions, namely, from south to north: South Kharga, Kharga, Dakhla, Farafra, Bahariya, Siwa, and Qattara depressions. The southern plateau extends from the Nile River in the east to Tibesti and Ennedi highlands (Chad Republic) in the west and from the highlands of Kordofan (Sudan) in the south to the Eocene-Cretaceous Plateau in the north. The land

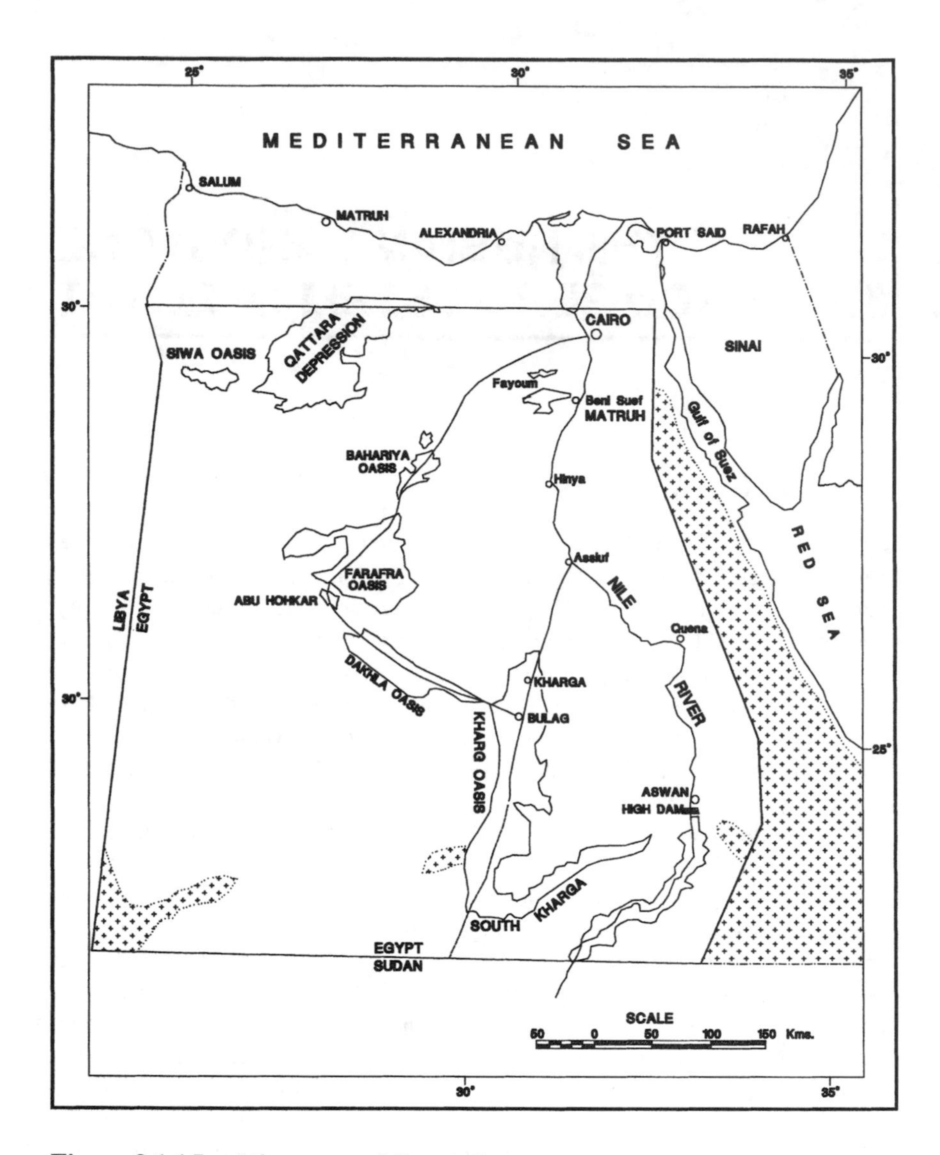


**Figure 8.1.1 Location map of the study area.**

surface is mainly underlain by the Nubian Sandstone series with some inlands of Pre-Cambrian crystalline basement complex. The central plateau covers an extensive area west and east of the Nile, with Eocene limestone underlying its land surface. In the south, it is defined by a steep escarpment overlooking the sandstone lowlands and is interrupted by the Nile Valley in the middle part. The northern boundary is characterized by a steep scarp on the northern

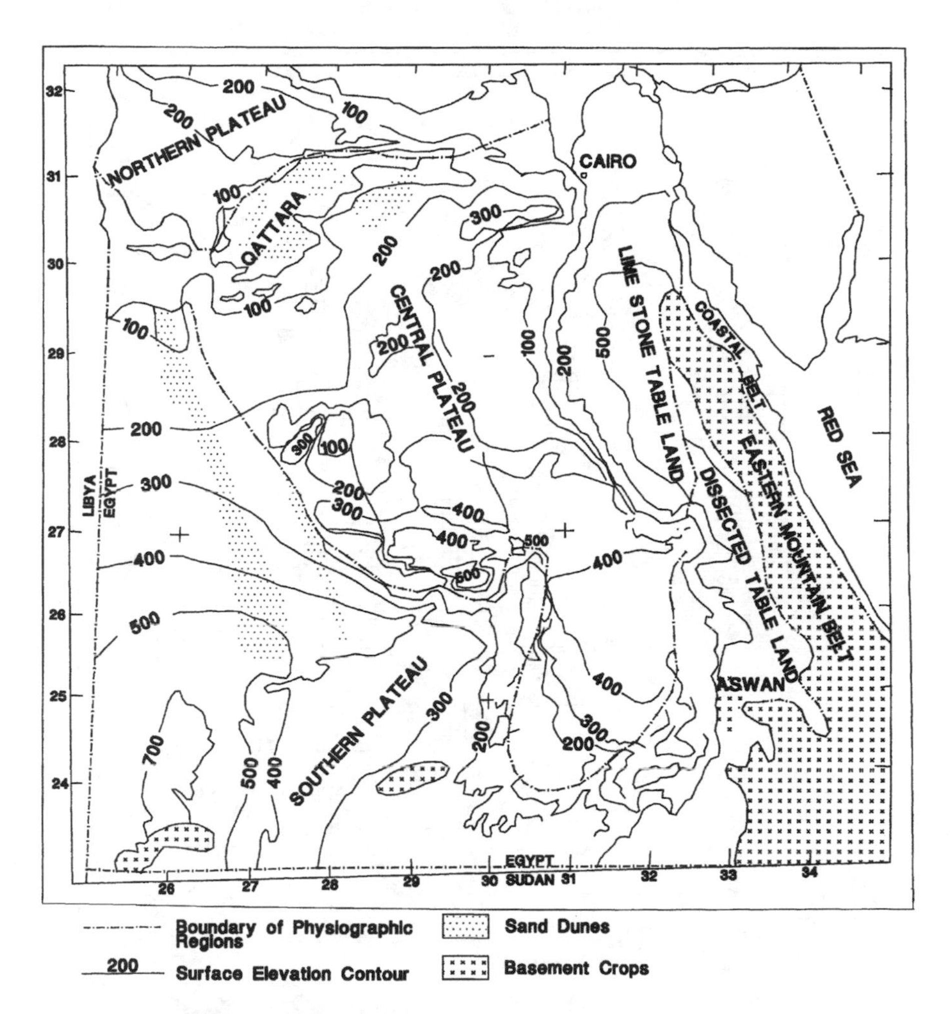


**Figure 8.1.2 Physical features of the study area.**

side of Siwa Oasis and Qattara depression. The plateau is covered on its western side by the Great Sand Sea.

The northern plateau extends from the bold escarpment at Siwa and Qattara depressions northward to the coastal plain along the Mediterranean Sea and from the Nile delta westward to Cyranaica (Libya). The lithologic character of the formations of the plateau are limestone and sandstones of Miocene age.

In the Western Desert, the maximum temperature ranges from 48°C during summer to about 7°C during winter nights. The rainfall is sparse in the Western Desert regions (annual average of 9.6 mm at Siwa and 0.3 mm in Kharga Oasis). The average humidity ranges between 30 and 60%.

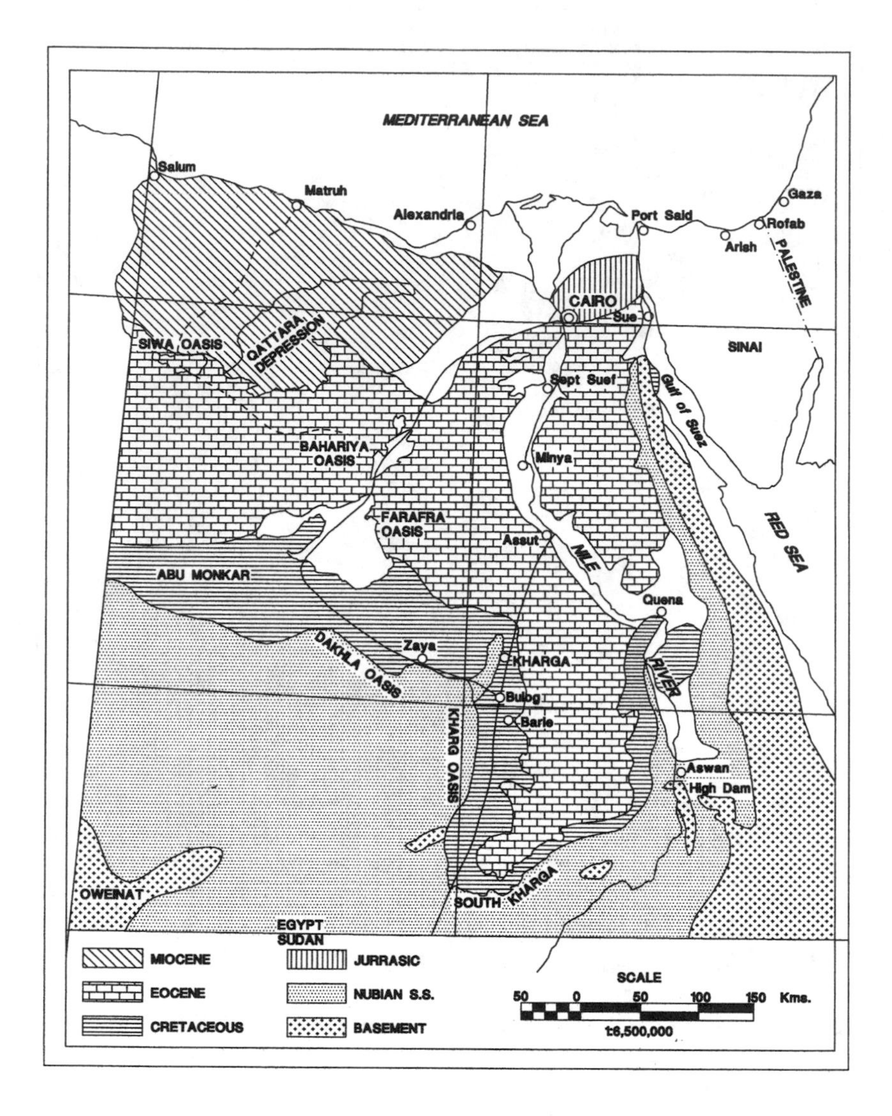


Figure 8.1.3 Schematic geological map of the study area.

## 8.1.2 GEOLOGICAL AND HYDROGEOLOGICAL CHARACTERISTICS

### 8.1.2.1 The Basement Complex

The total sedimentary successions in the Western and Eastern Deserts have been deposited on the eroded surface of the crystalline basement complex of Pre-Cambrian age. The basement complex crystalline rocks largely crop out in the Eastern Desert, where they form the backbone of the elevated Red

Sea Mountains, extending continuously from latitude 28° to the Sudan. The basement surface dips progressively from the Red Sea Mountains belt toward the west, to an elevation of –1,600 m near El-Monya to –4,000 m east of Assuit and increases in elevation to the north of Aswan.

In the Western Desert, basement rocks are prominently to the southwest in Gebel Oweinat, while smaller exposures occur in the area between Oweinat and Aswan and to the south of Kharga Oasis in the vicinity of Abu-Bayan.

Figure 8.1.4 shows a compiled basement relief map. A series of structural trends affecting the basement surface were traced (Ministry of Land Reclamation in Egypt, 1979), forming discontinuities in its general north-dipping trend. The map indicates that, in the area south of latitude 25°N, shallow basement occurs at elevations ranging between –500 m and –1,000 m but some occur at mean sea level.

The elevation of the top of the basement ranges between –500 m at the southern periphery of Kharga Oasis, –1,000 m in northern Kharga, –2,000 m in Baharia, –3,300 m at Sirva, and –4,000 m to the east of Assuit.

The sandstone system consists of alternating beds of sandstone, shale, and clay. Shaly and clayey beds occurring within the sandy sequence of the

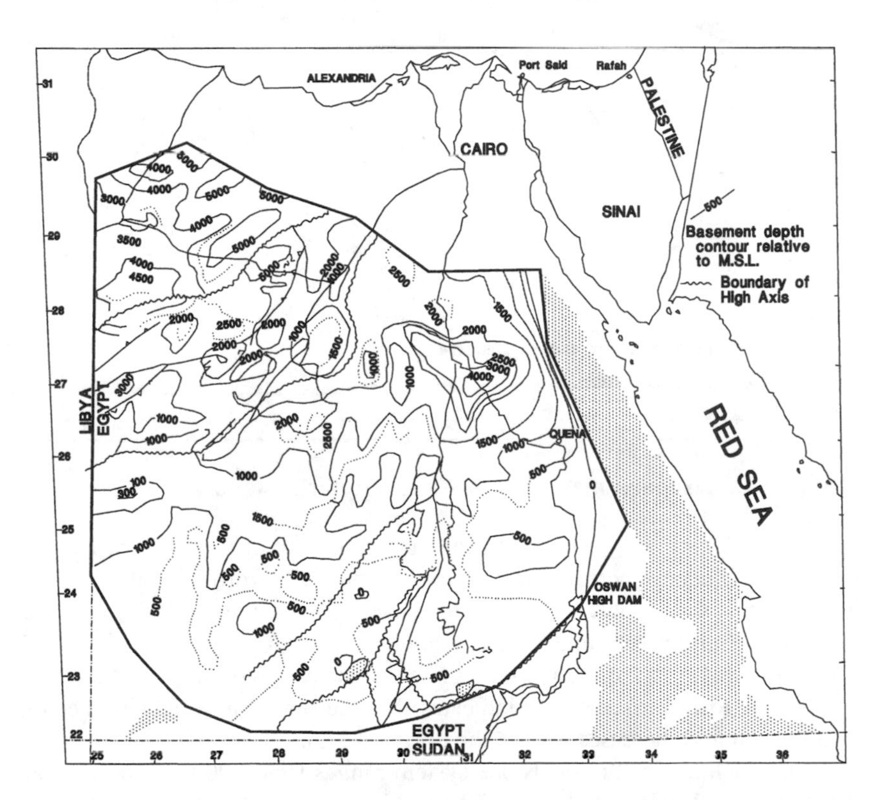


**Figure 8.1.4 Structure contours on the surface of the basement complex.**

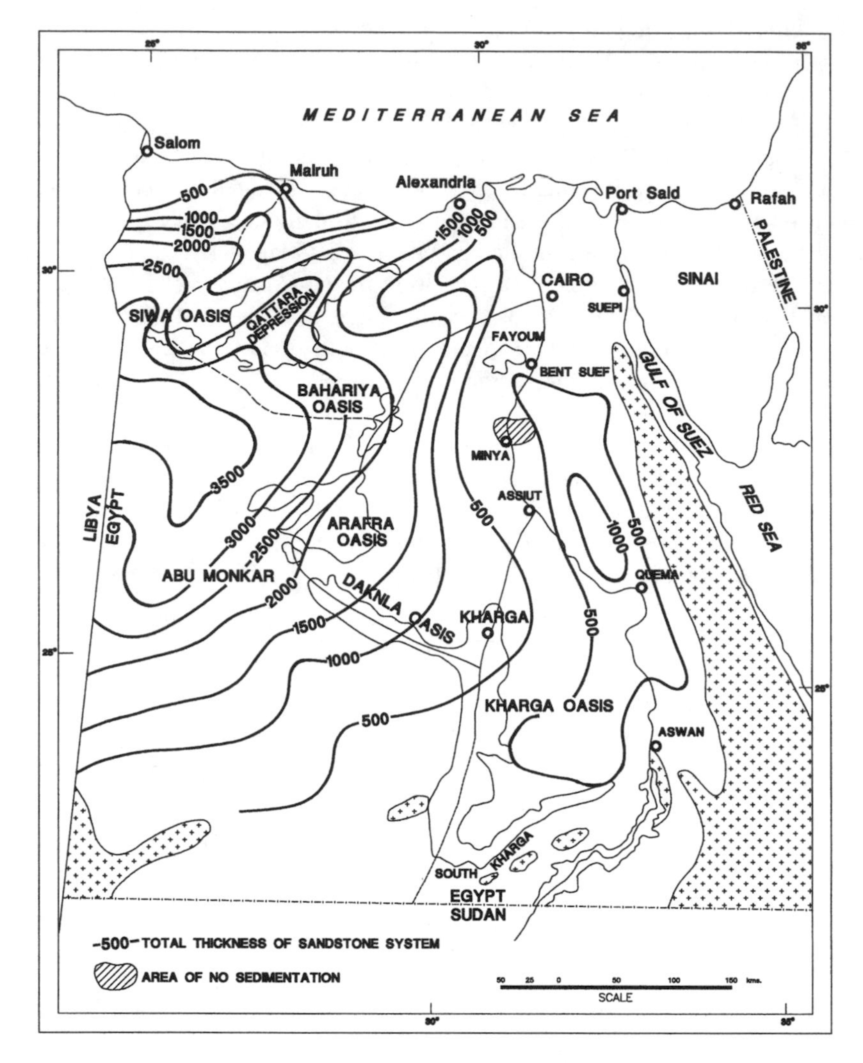


**Figure 8.1.5 Isopach map of Sandstone aquifer system.**

system is of minimum proportion in the southern part of the Western Desert, but their frequency increases to the north. Analysis of drilled wells records indicates that the proportion of sandstone averages about 40 to 90%, with the grain size ranging from fine to coarse-grained.

Figure 8.1.5 is an isopach map of the sandstone system from the upper Cretaceous to the basement. It indicates that the minimum thickness occurs in the south and increases toward the north and northwest.

The thickness of the sandstone system ranges from 500 to 1,000 m near the Sudanese borders and El-Gilf El-Kebir area (with some exception of reduced thickness in the vicinity of basement outcrops), 1,000 m north of

Kharga Oasis, 2,000 m in Farafra, and 3,500 m south of Sirva Oasis, where it forms the most conspicuous, thick sedimentary basin.

In the Eastern Desert east of the River Nile, the sandstone system forms a North/South elongated belt of outcrops directly overlying the basement that plunges westward. It forms the eastern continuation of the immense Pre-Cretaceous sandstone system of the Western Desert. The sandstone system in the Eastern Desert attains a maximum thickness of 1,000 m east of Assuit and decreases gradually toward the Red Sea Mountains where it is 100 m thick.

The sandstone system is overlain by the Upper Cretaceous-Eocene shale-carbonate complex in the northern part of the study area, while in the southern part (south of latitude 25°30′N), Toshka Basin and the central part of the Eastern Desert, the system crops out.

Basement uplifts in the southern and southeastern parts of the Western Desert are at elevations higher than the water table and are considered as the lateral boundary to the system.

The top of the sandstone system is capped by a thick sequence of low permeable shale carbonate complex of the Cretaceous-Eocene age in the north and central parts of the Western Desert. Under these caps rocks, the aquifer system is highly confined. In the southern part of the study area and the central part of the Eastern Desert, where the sandstone system crops out, water table conditions prevail.

The intercalated shale and clay beds within the sandstone system are lateral extents. Although individual beds or groups of beds may be continuously traced in some local areas within the system, on a regional basis, the entire sandstone aquifer system is to be hydraulically regarded as one single aquifer system.

The geologic framework of the area was described in early publications of Zittle,[1] Beadnell,[2] Little,[3] Ball,[4] Hellstrom,[5] Little and Attia,[6] Caton-Thompson,[7] Murray,[8] Paver and Pretorius,[9] Shukri,[10] Shazly et al.,[11] Shatta,[12] Ghobrial,[13] Barakat and Milad,[14] Grandic and Koscec,[15] Jacob,[16] Borelli and Karanjac,[17] Hammad,[18] and FAO.[19] Many additional studies include reports on gravity and magnetic surveys for most of the oasis and provide valuable information on the surface of the underlying crystalline basement rock.

## 8.1.3 HYDROGEOLOGY

The Nubian aquifer system (Figure 8.1.6), one of the largest groundwater systems of the Sahara, is formed by two major basins — the Kufra Basin in Libya, northeastern Chad, and northwestern Sudan and the Dakhla Basin of Egypt. The aerial extent of the aquifer includes the southernmost strip of the Northwestern Basin of Egypt and the Sudan Platform. The total area is about two million square kilometers.

The aquifer mainly consists of continental sandstone and intercalations of shales and clays of shallow marine and deltonic origin. To the south, east, and west, the aquifer is limited by basement outcrops. In the southwest, the

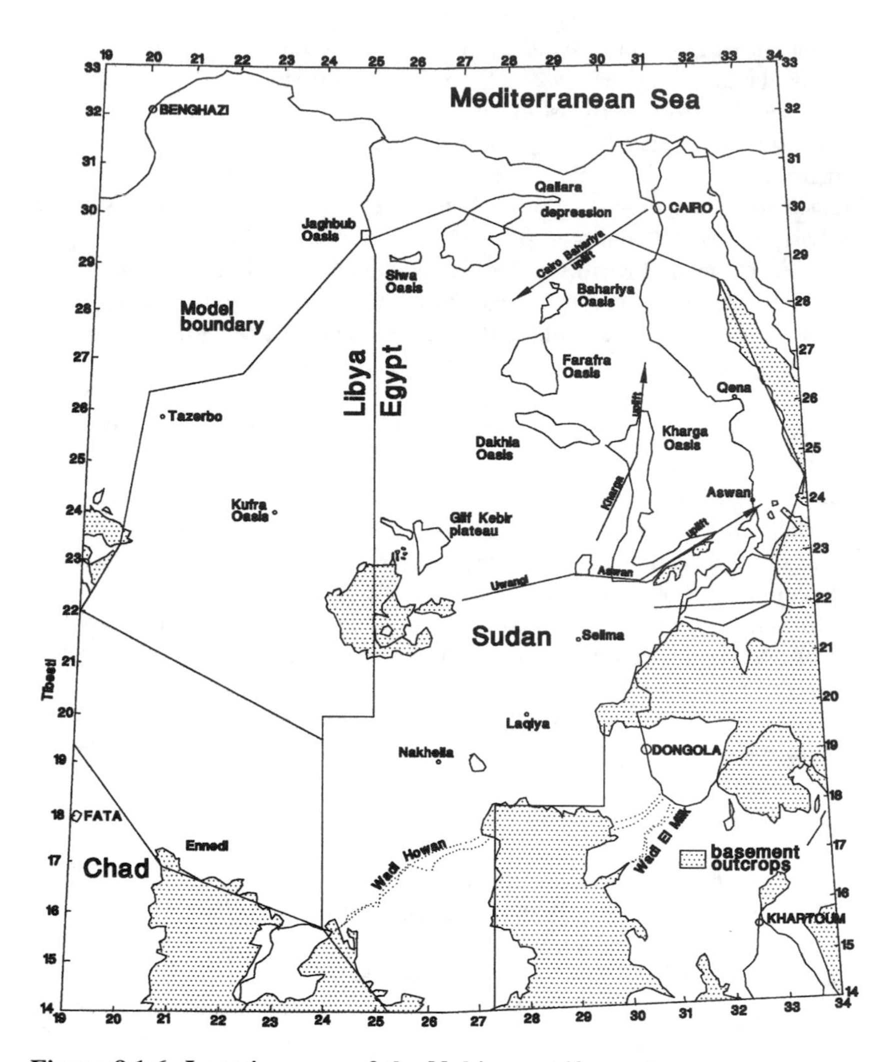


**Figure 8.1.6 Location map of the Nubian aquifer system.**

sandstone layers crop out at the rim of Chad Basin. In the northwest, the sandstones are connected to the Sirte Basin. In the north, a possible groundwater movement is limited by the freshwater–saltwater interface. North of the 25th parallel, the aquifer is confined under thick marine shales.

In spite of the hyperarid climate, these are huge groundwater reserves. In the center of the basin, where the average precipitation is less than 5 mm/year, there are several thousand meters of saturated sandstones. Obviously, there is no recent groundwater recharge in most parts of the system.

For the origin of the groundwater, two concepts have been under discussion. The first and older one is based on observations of groundwater levels and postulates a large-scale flow from mountainous recharge areas in the southwest (Tibesti and Ernedi) to discharge areas in the northeast. Present recharge and a more or less steady state prevail. Thus, groundwater can be considered a renewable resource. According to the second concept which is based on investigations of groundwater isotopes, groundwater had been formed locally in the surroundings of the present discharge areas during a more humid climate prevailing all over the present desert. If so, groundwater extraction must be regarded as mining of a nonrenewable resource under unsteady state conditions.

In order to clarify this point several groundwater models were constructed. A brief description of each model is given herein together with its achievement.

## 8.1.4 REGIONAL FLOW PATTERN

Piezometric mapping began with the earliest data collected on water levels from wells and springs in the oases. The first areal water-level maps providing the earliest concepts of groundwater movement were made by Ball (Figure 8.1.7),[4] Hellstrom (Figure 8.1.8),[5] and then by the Ministry of Land Reclamation in 1960 (Figure 8.1.9). The flow lines give evidence of some eastern flow.

Ezzat et al.[20] estimated the total groundwater inflow rate to the Western Desert Sandstone system to be about $1306.3 \times 10^6$ $m^3$/year. The groundwater outflow is mainly occurring through natural springs and drilled well discharges as well as natural losses by evaporation and/or evapotranspiration. Ezzat et al.[21] in their model study on the sandstone aquifer system south of Qattara area, concluded that the belt of the topographically low lands of Al-Eng, Bahrein, and Sitra, located at the southern periphery of the Qattara depression, represent the only northern natural discharging area for the aquifer system. Groundwater natural losses were estimated differently by several investigations (Barber and Carr[22] and Ezzat[21]).

## 8.1.5 GROUNDWATER MODELS

Several models of both analog and digital types have been constructed for the regional sandstone aquifer system in the Western Desert and for the Kharga-Dakhla Oases area.

Salem[23] constructed two regional one-layer R-C analog models for the sandstone aquifer system of the Western Desert. The first model was set to simulate the artesian condition, while the second one represented the water-table condition, with a vertical, low hydraulic conductivity value so that an elapsed period of time will occur between the earlier artesian effect and the development of water-table conditions. Cause-effect response of the aquifer system was studied under artesian and water table conditions as a result of two pumping programs at the different oases areas for a time period of 50 years.

Borelli et al.,[17] who made the two-layer model of the Kharga-Dakhla area, described the results of this R-C analogue model in which a regional one-layer model of the Western Desert was represented to define the boundary condi-

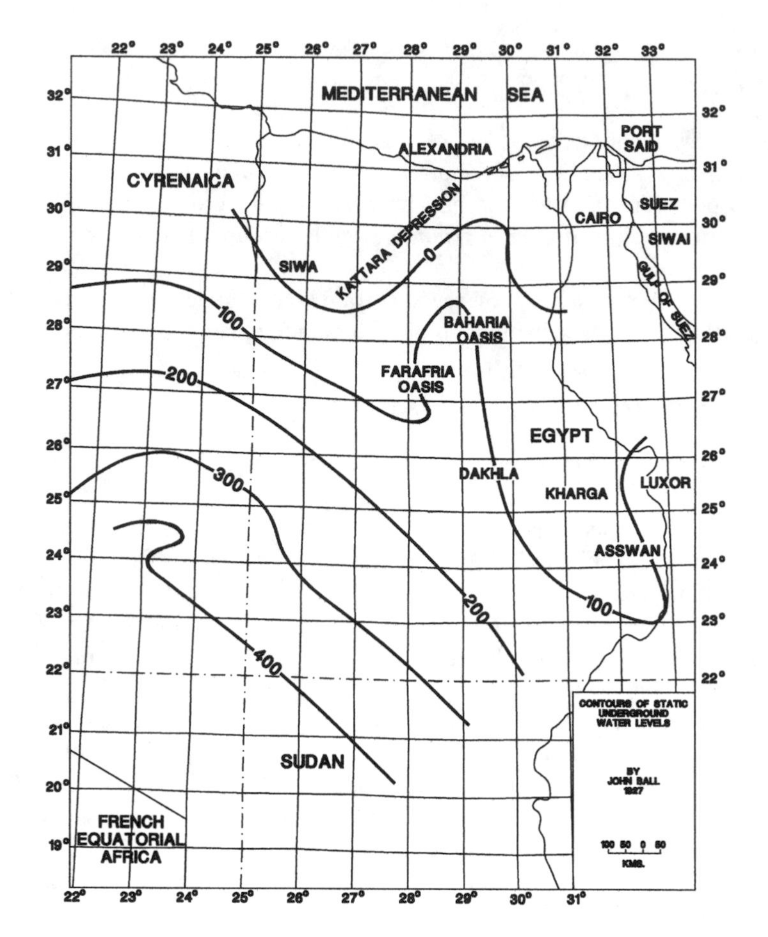


**Figure 8.1.7 Contours of underground static water levels. (From Ball, J., Problems of the Libyan desert, *Geog. J.*, London, 1927.)**

tions. Steady state simulations were achieved by invoking various conditions, both geometrically and hydraulically, which violates present understanding of the systems' hydraulic features.

Ezzat et al.[21] reported on a regional groundwater one-layer model of the Western Desert and a semi-detailed model of the Kharga-Dakhla Oases area. The models were carried out using the Honeywell "ECAP" program. The regional model output was used to control boundary conditions on the semi-detailed one, with a very simplified system geometry. Steady state simulation for the piezometer in the Kharga-Dakhla area was achieved with fair accuracy. Time-controlled calibration proved impossible; therefore, the forecasts of the system response to future extractions are questionable.

In the framework of the United Nations Development Program technical assistance to the New Valley project, Western Desert, evaluation of ground-

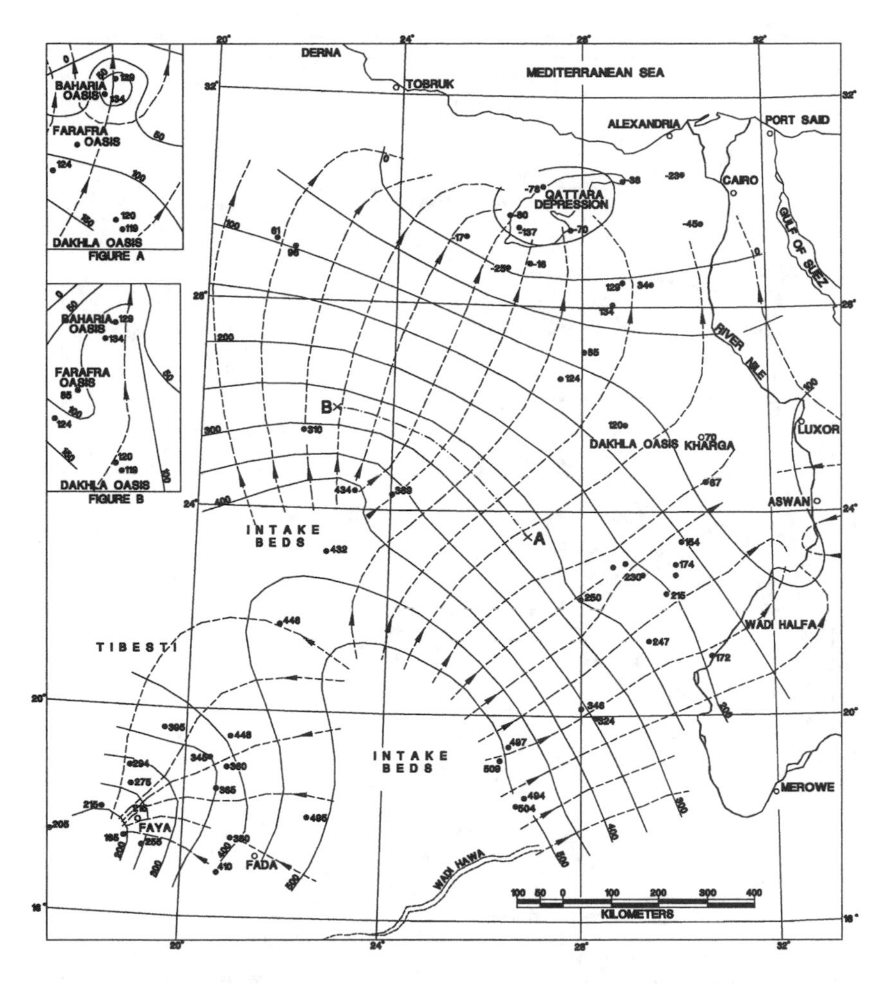


**Figure 8.1.8 Stream lines of equal pressure in the Libyan desert. (From Hellstrom, B., *Sartreyte Geogr. Annater.*, 34, 1940, pp. 206–239.)**

water resources in the Kharga-Dakhla Oases area was carried out.[22] A detailed numerical model, based on the integrated finite difference method, was made. The objective of the model was to determine the economic quantities of groundwater to be extracted to support long-term development of irrigated agriculture in the area. The model simulated two-layer aquifer system connected by vertical leakage. However, the drawbacks of the simulated input data considered a unique average value for the horizontal hydraulic conductivity all over the modeled area and setting of an arbitrary flow line outside the two oases area as western and northern boundary of the model. This boundary was considered a no-flow boundary in the steady state condition and

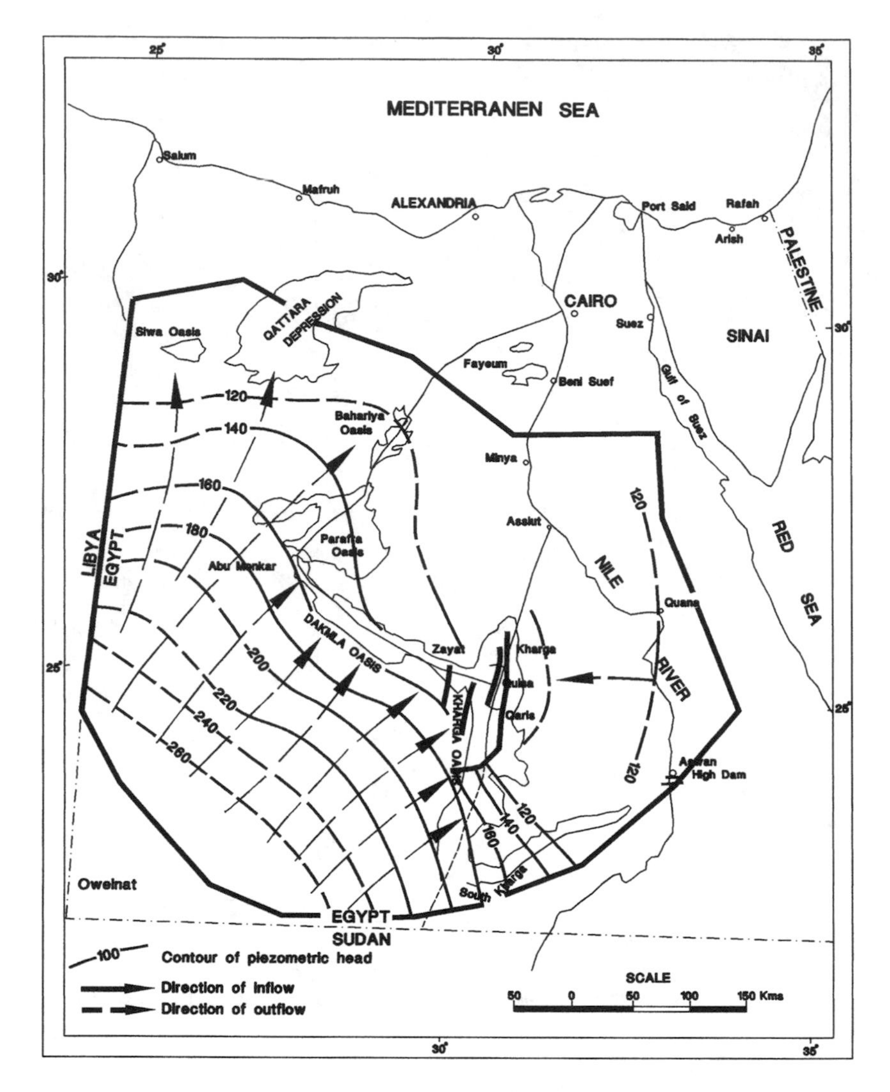


**Figure 8.1.9 Regional steady state piezometry of Sandstone aquifer system (1960) (Ministry of Land Reclamation).**

step-down boundary in the transient stage. Figure 8.1.9 shows the output of that assumption.

A regional numerical two-dimensional model, simulating the pre-upper Cretaceous sandstone aquifer system in the Western and Eastern Desert as one-layer system was made by Amer et al.[24] using the finite element technique. The model was prepared to forecast the aquifer system hydraulic response against future foreseen water extractions at the different development desert areas. Amer considered

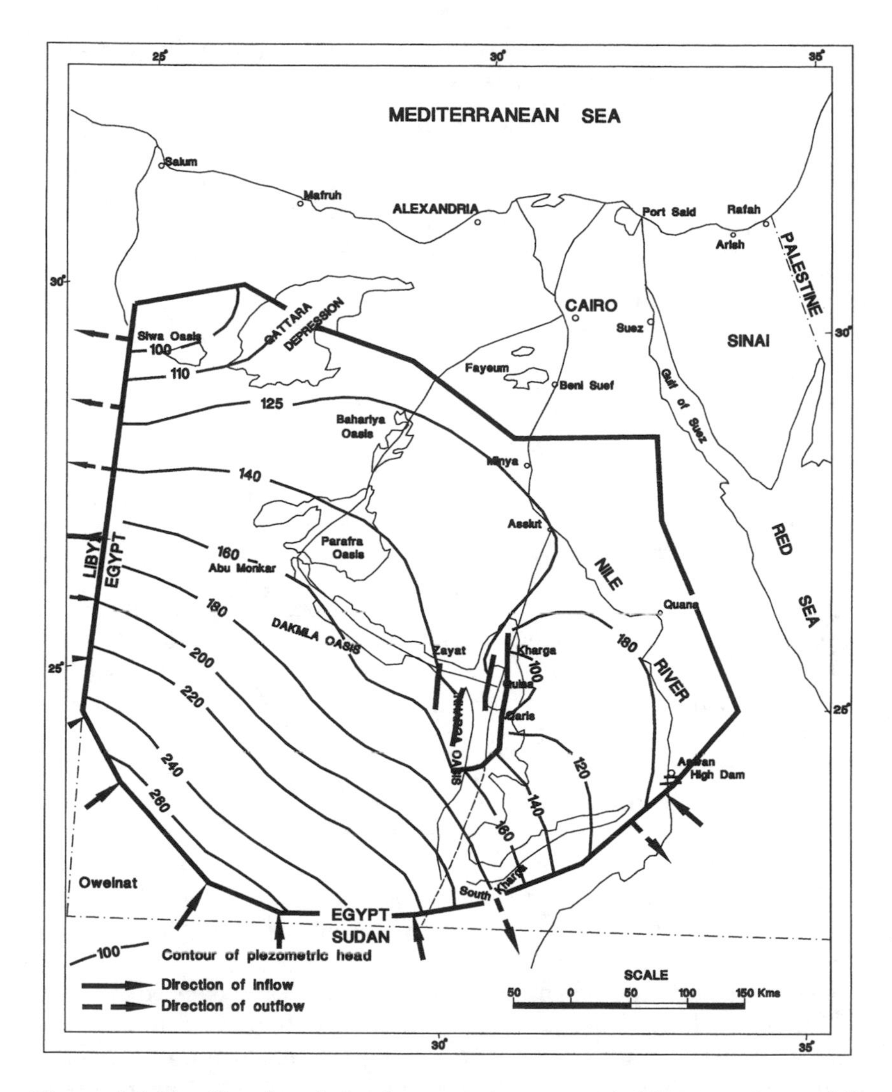


**Figure 8.1.10 Simulated steady state piezometry (with oases permeability and natural losses values adjustements — Run 8).**

also that the sandstone aquifer system was nonhomogeneous and anisotropic with essentially horizontal flow. The model output for a steady state condition is given in Figure 8.1.10. The final water balance of the aquifer system was also obtained from the model study as given in Table 8.1.1.

Heinl and Brinkmann[25] constructed a two-dimensional horizontal finite element model for the simulation of the Nubian aquifer system (1989). The finite element grid covered an area of two million square kilometers. Thus, large distance flow from the Chad to the Qattara depression was modeled as

**Table 8.1.1 Final Simulated Steady State Inflow — Outflow Groundwater Pattern Sandstone Aquifer System, Egypt**

| Area | Total Outflow (M · m³/year) | | | | Total inflow through boundaries M · m³/year |
|---|---|---|---|---|---|
| | By wells | Natural losses | Outflow through boundaries | Total | |
| Kharga | 51.80 | 54.12 | | 105.92 | |
| Dakhla | 114.20 | 36.73 | | 150.93 | |
| Farafra | 0.84 | 5.77 | | 6.61 | |
| Bahariya | 25.50 | 29.41 | | 54.91 | |
| Siwa | 54.03 | 15.00 | | 69.03 | |
| South Qattara | — | 87.25 | | 87.25 | |
| Western Front | | | 114.92 | 114.92 | 489.23 |
| Southern Front | | | 32.86 | 32.86 | 133.19 |
| Total | | | | 662.43 | 662.42 |

well as the transition of several thousand years from a semi-arid climate to the present hyper-arid conditions with a corresponding flow distance. The model was designed as a closed system, where reliable zero-flow boundary conditions could be identified at the outcrops of the basement, i.e., the natural boundaries of the system. All groundwater flow, recharge, and discharge occurred within the model.

The confined part of the system in the north was considered as a leaky aquifer, allowing vertical water exchange between the Nubian aquifer and overlying sediments, i.e., exfiltration to the large Egyptian depressions such as Kharga or Dakhla and possible infiltration from highlands. The hydrogeological system parameters, viz., transmissivity, storage coefficients, and leakage factors, were deduced from field data.[26]

The simulated extraction plans in Egypt and Libya are compiled in Figure 8.1.11. The total assumed extraction plan is $2800 \times 10^6$ m³/year in Egypt and $2200 \times 10^6$ m³/year in Libya. The total extraction imposed on the Nubian aquifer is 5 km³/year. The groundwater flow pattern at the beginning of the new projects in 1990 is still marked by natural flow to the oases.

In 2070, after 80 years of additional extraction, deep drawdown cones will have been formed (Figure 8.1.12). In the unconfined part of Egypt, the new Farafra projects are now in the center of a common drawdown cone between Bahariya and Dakhla. It will also include Sirva and Kharga and a small part of Libya. The model predicts a maximum drawdown of 130 m in Bahariya and Farafra. The extraction of E-Oweinat and Qena-Lagita will cause separate drawdown cones, also exceeding 100 m.

Heinl and Brinkmann[25] concluded that infiltration, supporting equilibrium conditions, stopped some 8000 years ago but continued on a minor scale in different areas and time intervals. The present recharge in Wadi Howar on the

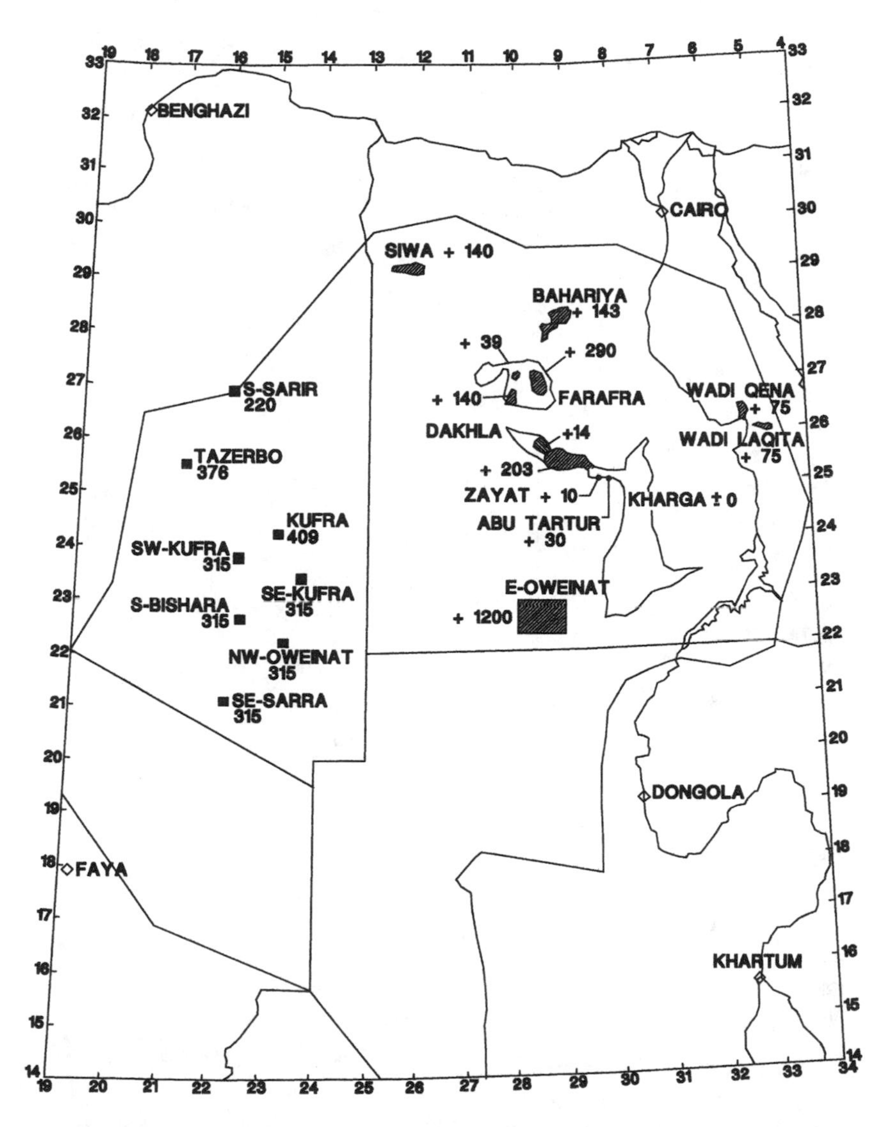


**Figure 8.1.11 Proposed future extraction in Libya (total) and Egypt (additional) in $10^6$ $m^3$ $year^{-1}$.**

Tibesti mountains is relevant only for natural flow conditions in geological time scales; for artificial extraction it is negligible. The River Nile, acting as a drainage channel, does not recharge the system. Heinl and Brinkmann also concluded that groundwater extraction in the Nubian aquifer is groundwater mining of a limited and non-renewable resource. Its costs and gains have to be carefully evaluated.

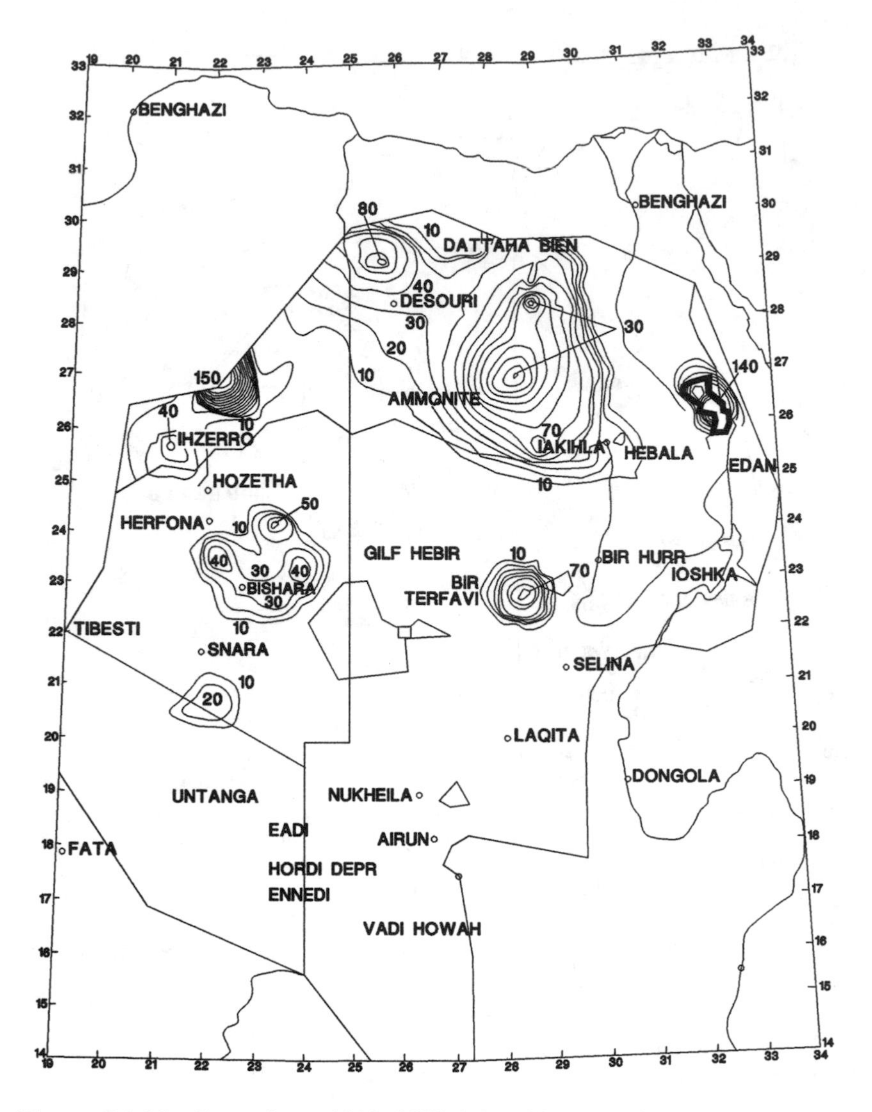


Figure 8.1.12 Drawdown 1990–2070 (after 80 years of additional extraction).

## 8.1.6 ENVIRONMENTAL PROBLEMS

Two environmental problems can occur from using the groundwater from the Nubian aquifer without good water management plans. The first problem is rapid groundwater depletion, while the second problem is increasing the existing water supply with eventual salinization of the soils.

## REFERENCES


1. Zittle, A.K., Beitrage Zur Geologic and Palaontologic der Libyschen Wuste und der Angrezenden gabiete von Aegypten, *Paleotographica,* 30, 1983.
2. Beadnell, H.J.L., Flowing wells and subsurface water in Kharga Oases, *Geol. Mag*., 5, 524, 1908, pp. 49–57.
3. Little, O.H., Preliminary Report on the Water Supply of Kharga and Dakhla Oases, Egyptian Survey Dept, 1931.
4. Ball, J., Problems of the Libyan Desert, *Geog. J*., London, 1927.
5. Hellstrom, B., The subterranean water in the Libyan Desert, *Sartreyte Geogr. Annater.*, 34, 1940, pp. 206–239.
6. Little, O.H., and Attia, M.I., The Deep Bores in Kharga and Dakhla Oases: Cairo Geological Survey, Report, 1942, 58 pp.
7. Caton-Thompson, G., *Kharga Oases in Prehistory with a Physiographic Introduction by E.W. Gardner*, University of London, Athlone Press, London, 1952.
8. Murray, G.W., The Artesian Water of Egypt: Survey Dept., Ministry of Finance and Economy, Paper No. 52, 1952, 20 pp.
9. Paver, G.L., and Pretorius, D.A., Hydrogeological investigation of Kharga and Dakhla Oases, *Bull. Egyptian Desert Institute*, 4, 1954, pp.
10. Shukri, N.M., Remarks on the geological structure of Egypt, *Bull. Soc. Geogr. d'Egypte*, 27, 1954, pp. 65–82.
11. Shazly, M.M. et al., Contribution of the study of the stratigraphy of el-Kharga Oasis, *Bull. Inst. Desert d'Egypt*, 10, 1, 1959, pp.
12. Shatta, A., Remarks on the regional geological structure of the groundwater reservoir of El-Kharga and Dakhla Oases, *Bull. Soc. Geogr. d'Egypt*, 34, 1961, pp. 177–186.
13. Ghobrial, M.G., The Structural Geology of the Kharga Oasis, Cairo, Geol. Survey, and Min. Research Dept., Paper No. 43, 1967.
14. Barakat, M.G., and Milad, G.S., Subsurface geology of Dakhla Oasis, *J. Geol.*, 10, 2, 1966, pp. 145–154.
15. Grandic, S., Koscec, B., Geological characteristics of sedimentary complex of the Nubia Formation: Report of Egyptian general desert development organization, INDUSTROPROJECT, Zagreb, Yugoslavia, 1968.
16. Jacob, C.E., Geology and Hydrology of Kharga Oasis, in Water Well Design, Western Desert, Egypt, Rosscoe Moss Company, Report to Egyptian E.G.D.D.O., 1964.
17. Borelli, M., and Karanjac, J., Kharga and Dakhla Oases, Determination of Hydrogeological and Hydraulic Parameters: Reports, presented by INDUSTROPROJECT, Zagreb, Yugloslavia, 1968.
18. Hammad, H.Y., Groundwater Potentialities in the African Sahara and the Nile Valley, Beirut Arab University, Beirut, 1970.
19. FAO, Development of the New Valley Region in the Western Desert, Fact-finding Mission's Report, Draft: Rome, EGY/8903 Terminal Report, 1970.
20. Ezzat, M.A., and Abuer, Atta, A., Regional hydrogeological conditions, Groundwater Series in A.R.E. Part I, Ministry of Land Reclamation, Cairo, 1974.
21. Ezzat, M.A., Nour, S.E., Morshed, T., and Mishriki, M., South Qattara Area Groundwater Model, General Petroleum Company, Cairo, 1977.

22. Barker, W.M. and Carr, D.P., Groundwater Model of the Kharga-Dakhla Area: Working Document No. 7, UNDP/FAO, AGON:EGY 71/561, 1976.
23. Salem, M.H., Study of the Hydrologic Parameters of the Nubian Sandstone Aquifer with Reference to the Productivity of Pattern for Well Development in Kharga Oasis, Egypt, 1970.
24. Amer, A.M., Nour, S.E., and Mishriki, M.F., A Finite Element Model of the Nubian Aquifer System in Egypt, Groundwater Seminar, Egyptian Ministry of Land Reclamation, 1979.
25. Heinl, M., and Brinkmann, P.J., A groundwater model of the Nubian aquifer system, *J. Hydrol. Sci.*, 34, 4, 1989, p. 8.
26. Hesse, K.H., Hissene, A., Kheir, O., Schnacher, E., Schneider, M., and Thorweike, U., Hydrogeological investigations in the Nubian aquifer system, *Berliner Geowissen-Schaftliche Abhandlungers (A)*, 75, 2, 1987, pp. 397–464.

# 8.2 SITING A SECURE HAZARDOUS WASTE LANDFILL IN A LIMESTONE TERRANE

## 8.2.1 INTRODUCTION

This study describes the geologic and hydrologic settings for a secured hazardous landfill in a limestone terrane in New York.

The site is located in a part of the Niagara River corridor which occurs adjacent to the Niagara River from the northern part of Buffalo to Lewiston, New York (Figure 8.2.1). The Niagara River corridor, in the town of Niagara and the city of Niagara Falls in the southwestern part of New York State, is highly industrialized as a result of the abundant water supply available for industrial processing, waste assimilation, and power generation. There are numerous disposal sites in the area (Figure 8.2.1).

The area was investigated to determine site suitability for a hazardous waste landfill. The results of geological and hydrological investigations together with proper design and engineering of the site and installation of double liner with leachate collection and adequate pre- and post-monitoring systems in place characterize this site suitable for landfill.

## 8.2.2 TOPOGRAPHIC AND GEOGRAPHIC SETTING

Niagara Falls is within a lowland bordered on the north by the Lake Ontario plain and to the south by the Lake Erie plain. The Niagara escarpment, which crosses the area along an east-west line, forms a 200-foot high cliff, which crosses north of the Niagara River but diminishes to a broad, sloping incline toward the south (falls). The Niagara River declines about 320 feet along its 30-mile length between Lake Erie and Lake Ontario, including the 160-foot drop at Canadian Falls.

The topography of the site is relatively flat, with the exception of the artificial mounds created by landfills in the area. Natural ground surface elevation ranges from 574 to 584 feet above sea level. Surface drainage is

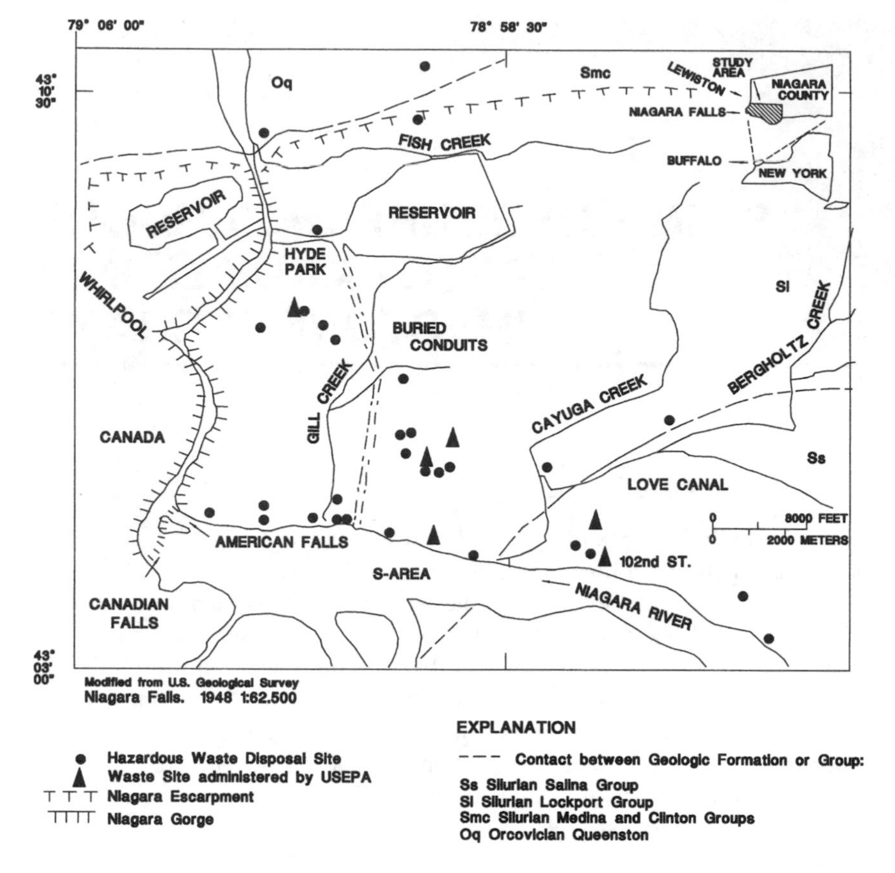


**Figure 8.2.1 Major geographic features and geologic setting of Niagara Falls area and location of hazardous waste sites.**

relatively poor due to the apparently undersized culverts immediately downstream of the site which cannot accommodate the high volumes of runoff. The runoff is in turn caused by the clayey glacial soils.

## 8.2.3 GEOLOGIC SETTING

Unconsolidated glacial deposits of till and lacustrine clay, silt, and sand overlie gently dipping sedimentary rocks throughout the Niagara Falls area.

The Middle Silurian Lockport Dolomite directly underlies the northern part of the area, and the younger Upper Silurian Salina Group underlies the southern part as a result of the bedrock's southward dip of about 30 feet per mile. In the Niagara Falls area, the Lockport Dolomite ranges from 130 to 160 feet in thickness and consists of five members that have been differentiated on the basis of lithologic characteristics and fossil evidence.

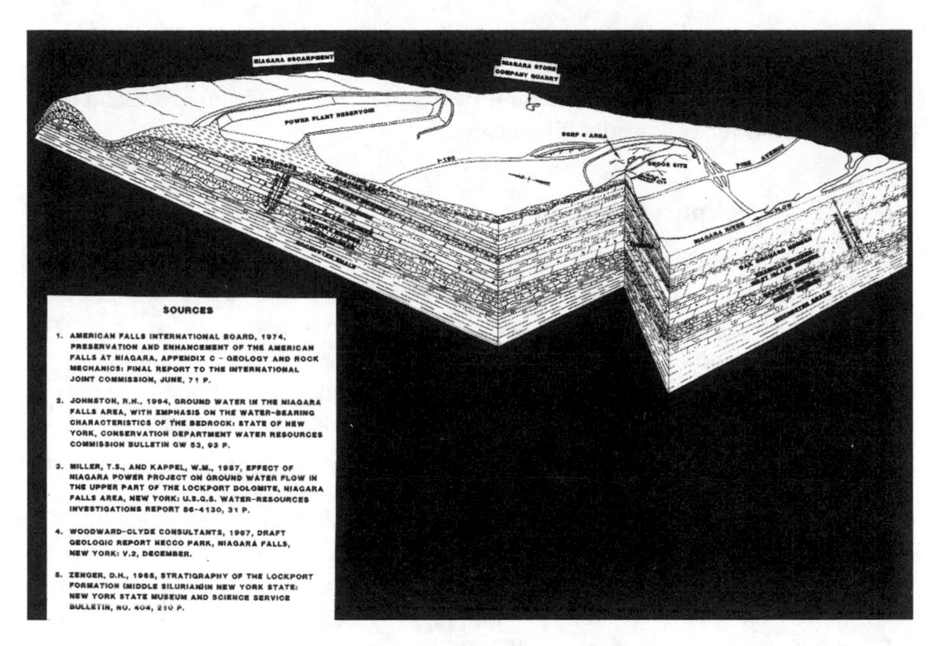


**Figure 8.2.2 Schematic block diagram.**

The overburden, Lockport Dolomite, and Rochester Shalc, considered to be the most important geologic units, are described below. Figure 8.2.2 is a schematic bloc diagram showing the geologic setting.

### 8.2.3.1 Overburden

The overburden materials in the Niagara Falls area consist of predominantly natural sands, silts, and clays and man-deposited miscellaneous fill.

A 1- to 5-foot thickness of glacial till generally occurs at the base of the overburden. Glacial till contains very poorly sorted sands, silts, clays, and gravels. The till in the Niagara Falls area was deposited near the end of the Wisconsinan Glaciation during the Pleistocene Epoch. The tills are characteristically stiff red clays with varying amounts of sand, silt, and gravel. Above the till there is usually a variable thickness of glaciolacustrine lake sediments consisting of sand, silt, and clay deposited about 12,000 years ago as the continental ice sheets retreated northward. These sediments, commonly represented as varved (banded) silts and clays, were deposited in temporary lakes which formed at the ice front (preglacial lakes). Additional sediments were later deposited when a large post-glacial lake formed on the flatland between the Niagara and Onondaga Escarpments. This lake (Lake Tonawanda) stretched for over 50 miles to the east of the Niagara Falls area.[1] A 1- to 2-foot thickness of topsoil overlies the glaciolacustrine sediments in undisturbed regions. Since much of the Niagara Falls area has been disturbed by human activities, many areas exist where sections of natural overburden have been removed and/or replaced with miscellaneous fill material.

### 8.2.3.2 Lockport Formation

The Middle Silurian Lockport Formation, consisting of approximately 140 feet of relatively competent dolomite, lies beneath the overburden in the Niagara Falls area. This unit thickens to the southeast and thins to the west toward the Niagara Gorge and to the north toward the Niagara Escarpment. The Lockport Formation, which has also been referred to as the Lockport Dolomite (or Dolostone), is subdivided into five principal members: the Oak Orchard, Eramosa, Goat Island, Gasport, and DeCew Members.[2]

The Lockport Formation is primarily dolomitic and characterized generally by brownish-gray to dark gray color, medium granularity, medium to thick bedded, stylolites, carbonaceous partings, vugs, and poorly preserved fossils. The Lockport is subdivided into its five principal members based on variations within this general description.[2] A stratigraphic column showing the Lockport Formation is provided on Figure 8.2.3. Figure 8.2.4, shows the locations of geologic cross sections. Figure 8.2.5 shows the northeast-southwest geologic cross section, and Figure 8.2.6 shows the northwest-southeast geologic cross section.

#### *8.2.3.2.1 Oak Orchard Member*

The Oak Orchard Member, the uppermost and thickest member of the Lockport Formation, ranges from approximately 80 to 120 feet in thickness in the Niagara Falls area. It is brownish-gray to dark gray, fine to medium grained, thin to thick bedded, saccharoidal, bituminous dolomite with stylolites, carbonaceous partings, vugs, minor occurrences of stromatolites, oolites, and tabulate coral fossils. The Oak Orchard exhibits the greatest degree of variability of the Lockport Formation members, being shaley and thin bedded in sections and massive in other sections.

#### *8.2.3.2.2 Eramosa Member*

The Eramosa Member is 16 to 18 feet thick and underlies the Oak Orchard Member. This unit is generally medium to dark gray, fine grained, thin to medium bedded, argillaceous and bituminous dolomite, with many shale partings, gypsum vugs, and some stylolites.[2] The contact between the Oak Orchard member and the Eramosa Member is characteristically sharp.

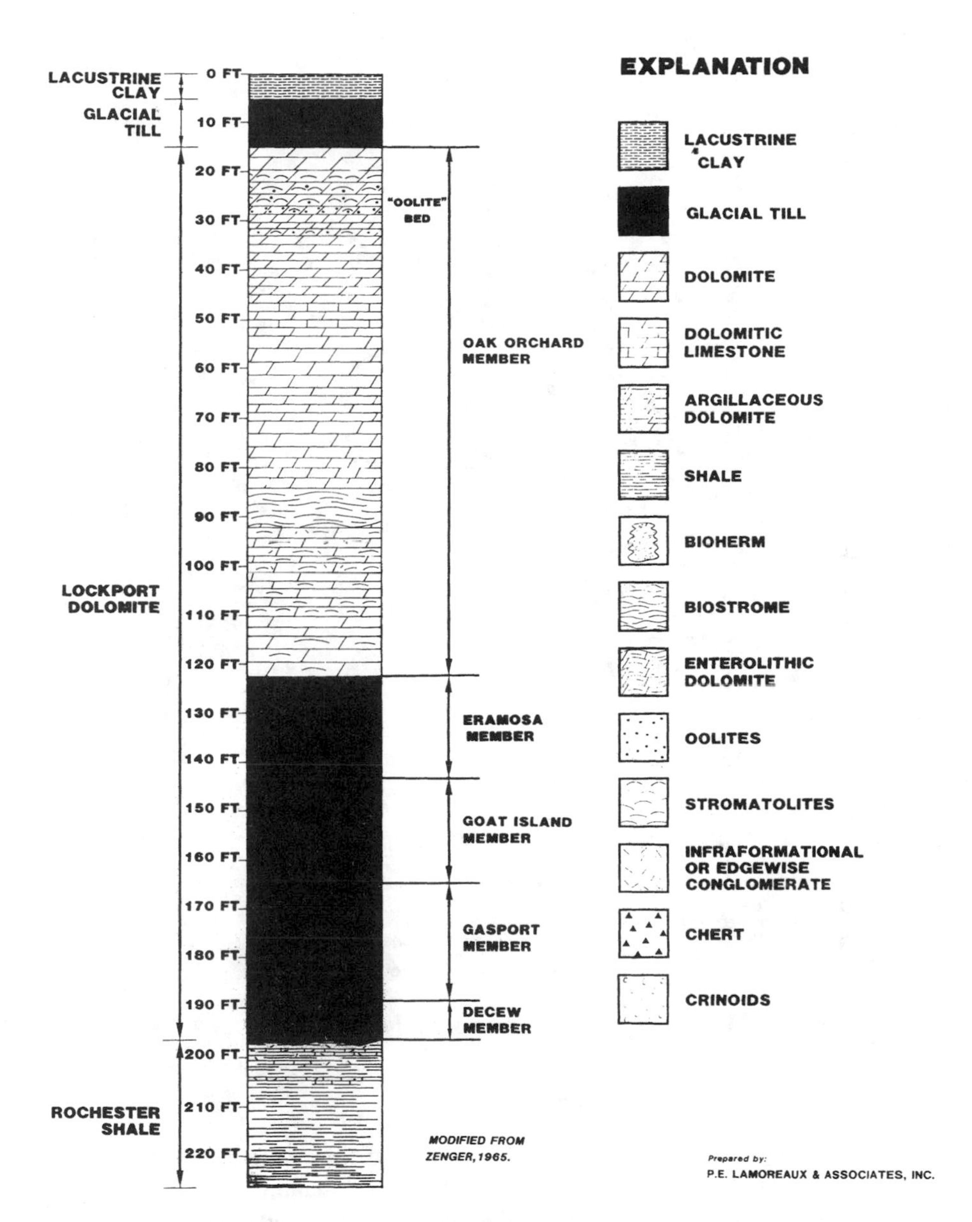


Figure 8.2.3 Stratigraphic column.

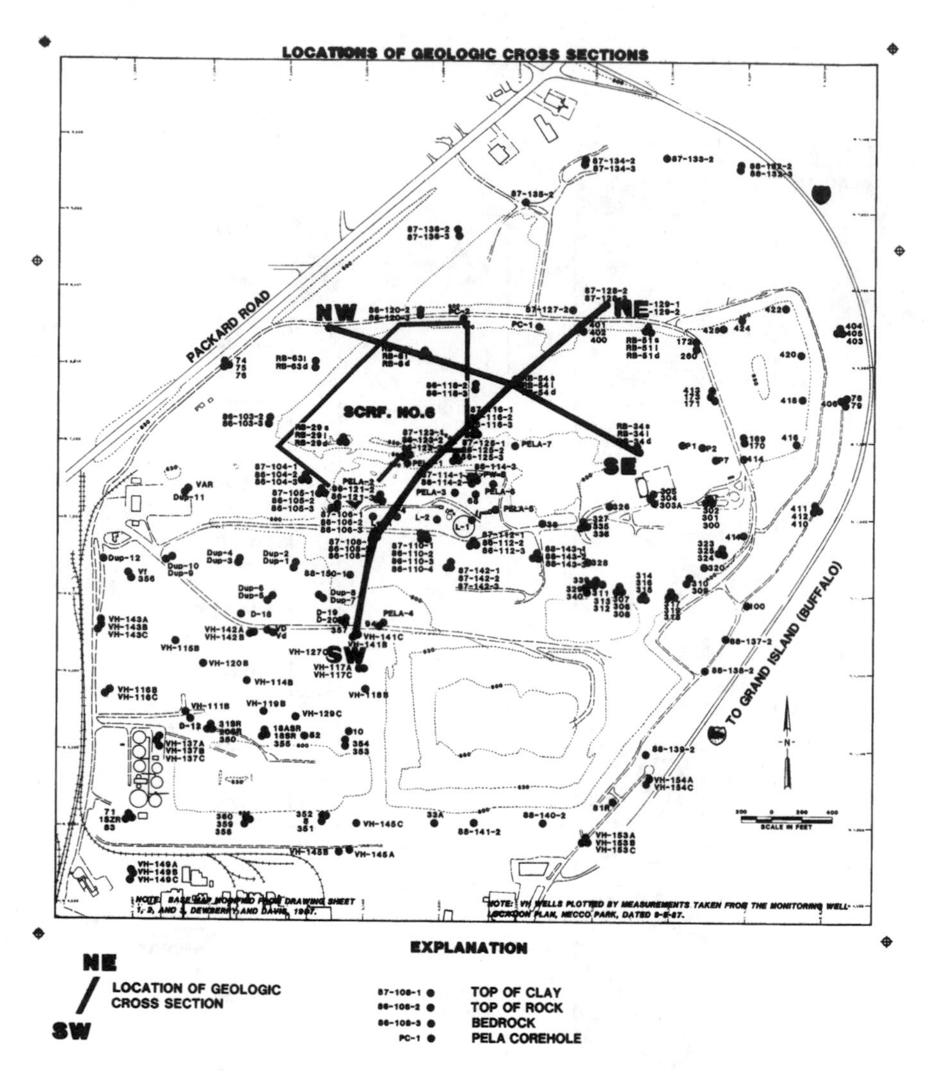


**Figure 8.2.4 Location of wells and coreholes.**

### *8.2.3.2.3 Goat Island Member*

The Goat Island Member, occurring beneath the Eramosa, is generally 19 to 25 feet thick in the Niagara Falls area. The Goat Island is light olive gray to brownish-gray, medium grained, thick bedded, saccharoidal dolomite, with abundant chert nodules near the top, stylolites, carbonaceous partings, and some vugs containing gypsum, calcite, and sphalerite.[2] The contact between the Eramosa and the Goat Island is conformable and is characterized by gradual lightening in color over a 2-foot thickness.

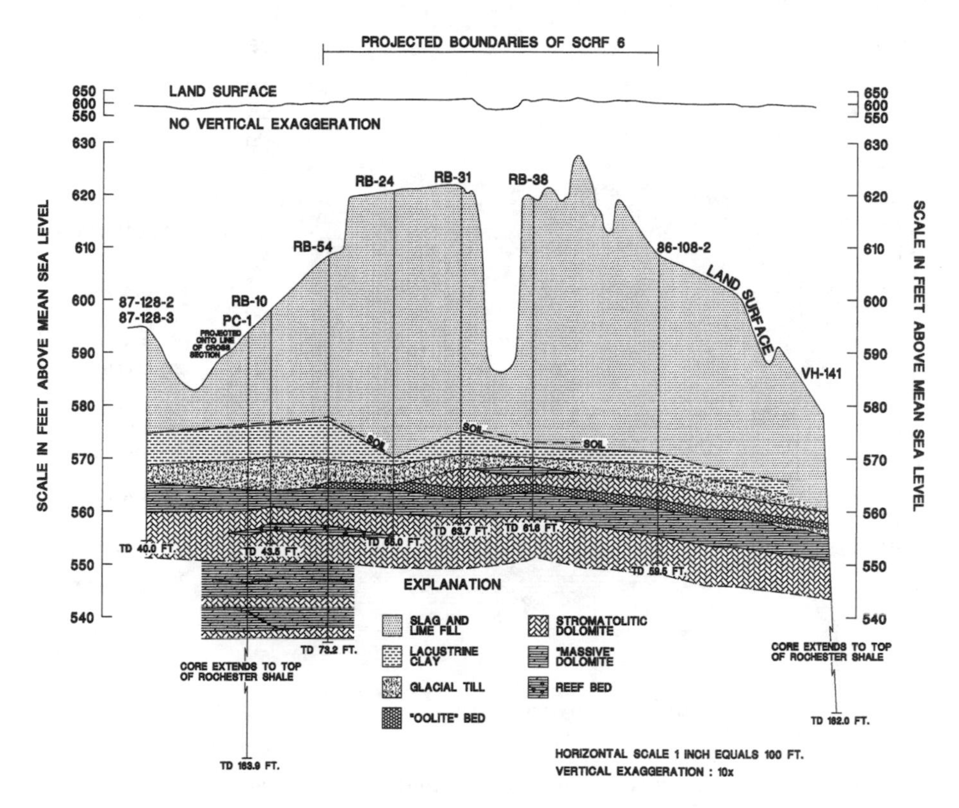


**Figure 8.2.5 Northeast-southwest geologic cross section.**

### *8.2.3.2.4 Gasport Member*

The Gasport Member occurs below the Goat Island Member and is approximately 15 to 30 feet thick. Because the Gasport occurs as a complex of different facies, this member tends to exhibit a high degree of variability between geographic localities. The Gasport is predominantly olive gray to brownish-gray, coarse grained, medium to thick bedded, fossil fragmental, crinoidal limestone or dolomite.[2] However, due to localized facies changes, this member can appear as dark gray, fine grained, argillaceous dolomite with sporadic crinoid fragments. The contact between the Goat Island and the Gasport member is conformable.[2] However, because of local facies relationships between the top of the Gasport and the bottom of the Goat Island, it is often difficult to identify. The combined thickness of the Goat Island and Gasport Members, however, is generally constant.

### *8.2.3.2.5 DeCew Member*

The DeCew Member underlies the Gasport Member and overlies the Rochester Shale. The DeCew is described as medium gray to medium dark

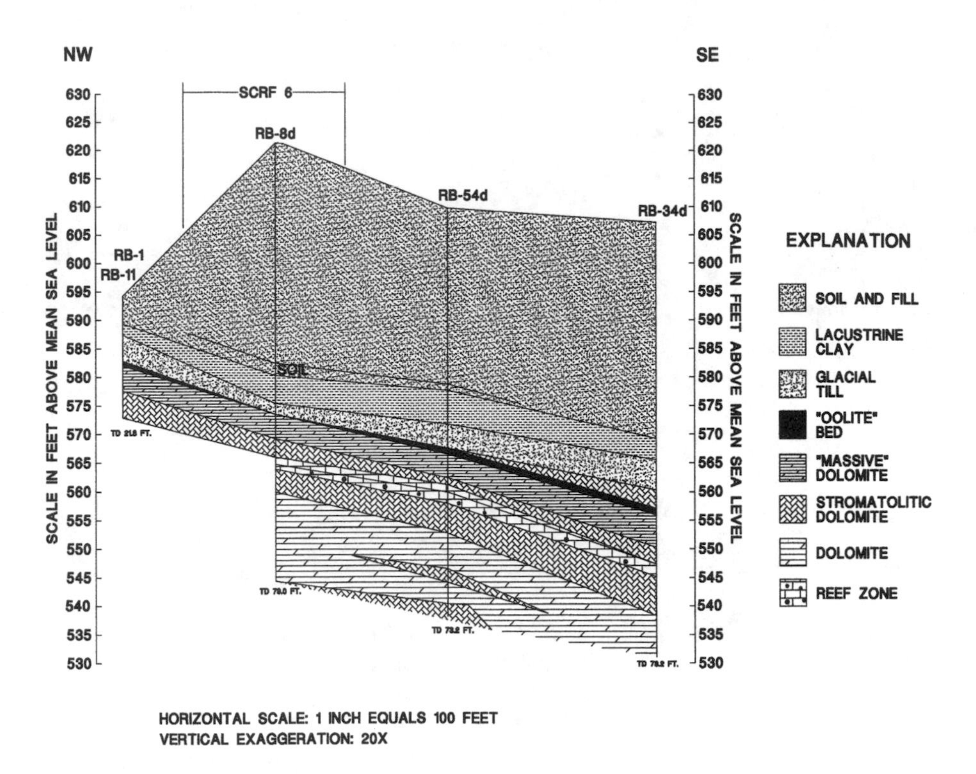


**Figure 8.2.6 Northwest-southeast geologic cross section.**

gray, fine grained, thin to thick bedded, and massive argillaceous dolomite. The thickness ranges from 8 to 10 feet in the Niagara Falls area. The contact between the DeCew and the Gasport Member is characteristically abrupt in the Niagara Falls area and is marked by a change from the massive crinoidal basal conglomerate of the Gasport Member to the fine textured, argillaceous dolomite of the DeCew.[2] Tesmer[1] separates this unit from the Lockport Formation on the grounds that a nonconformity exists between the DeCew and the Gasport indicating a break in sedimentation. However, Zenger[2] notes that the nonconformity occurs locally and that the contact between the DeCew and Gasport is conformable at other localities.

### 8.2.3.3 Rochester Shale

The Rochester Shale Formation lies below the DeCew Member and is typically 55 to 65 feet thick in the Niagara Falls area. It is considered to be fifth principal marker horizon within the study area. It is described as dark bluish- to brownish-gray, calcareous shale with occasional argillaceous limestone layers. The upper Rochester Shale tends to be more dolomitic than the lower, especially at the contact with the DeCew. This contact, although gradational at most locations, tends to be more abrupt and undulating in the Niagara Falls area. This has been attributed to localized channeling at the top of the Rochester Shale in the Niagara Falls area prior to the deposition of the DeCew Member.[1]

## 8.2.4 STRUCTURAL GEOLOGY

A south-dipping homocline, which affects the Paleozoic rocks of western and southern New York, is the dominant structural feature in the Lockport Formation, as well as in the sedimentary formations beneath it. Bedding dips are characteristically gentle. Local deviations in the dominant regional structure do occur, and may be attributed to monoclinal flexures and faulting. A large scale, tectonically related, structural pattern is believed to affect the rocks of western New York.[3]

Joints, high angle to vertical fractures related to regional stress patterns, are common in the Lockport Formation. These joints are probably most open or developed in the upper part of the Lockport Formation, where a relatively high degree of weathering has occurred.[4] Where dissolutioned, these joints may serve as conduits for vertical and horizontal movement of groundwater between bedding plane fractures. The prominent sets of vertical joints in the Niagara Falls area are oriented N65°E and N30°W.[4] Near the bedrock surface, joints tend to be open and well developed; however, they become relatively tight and poorly developed at depth.[5] The incidence or frequency of vertical fractures may vary with depth between areas. Studies conducted by the U.S. Geological Survey suggest that vertical fracture frequency may increase along regional structural lineaments.[3] These lineaments are related to the large scale structural pattern mentioned above.

Bedding plane fractures, near horizontal fractures parallel to formation bedding, tend to lie within particular stratigraphic intervals. Bedding plane fracture zones transmit the majority of the groundwater flow in the Lockport Formation.[3-5] Water-bearing bedding plane fracture zones are developed due to variations in lithology, differential weathering, solutioning, and tectonic or isostatic rebound related to stress release.

## 8.2.5 HYDROGEOLOGY

The Lockport Dolomite is the principal aquifer in the Niagara Falls area, but it is not heavily pumped because the Niagara River is the major source of water supply. Well production not affected by induced infiltration commonly ranged from 10 to 100 gallons per minute (gpm), but production as high as 950 gpm has been developed. Near the river, induced infiltration augments yields from the Lockport to industrial wells, some of which produce more than 2,000 gpm.[4]

The Lockport Dolomite has been divided into two zones on the basis of water-transmitting properties. The upper zone is 10 to 25 feet thick and has well-connected horizontal and vertical fractures. The horizontal hydraulic conductivity of the upper permeable zone is estimated to be 3ft/d ($1.76 \times 10^{-5}$ cm/sec).[4] The lower zone contains the separate water-bearing bedding planes, which generally are poorly connected by vertical joints.

The distribution of well production from the Lockport Dolomite indicates areas of high transmissivity that may be related to fractures within the bedrock. The average production of a well in the area is estimated to be 30 gpm.[4] An analysis of the distribution of wells producing more than 50 gpm shows that yields from the water-bearing zones are highest near the Niagara River, probably as a result of induced river infiltration. Johnston[4] states that the most productive of these wells are within a narrow zone that trends northeastward from about 2 miles east of Niagara Falls.

The high production along the conduits (950 gpm) and from a more recently installed well 5 miles to the northeast (370 gpm) supports Johnston's[4] hypothesis that a band of high transmissivity exists in the Lockport and is possibly caused by fracturing within the horizontal bedding planes.

Groundwater in the Niagara Falls region generally flows southwestward from recharge areas near the escarpment toward the Niagara River (Figure 8.2.7), the major discharge zone. Near the city of Niagara Falls, however, the direction of flow has been altered by man-made structures. The Niagara Power Project Reservoir is a source of additional recharge to the Lockport Dolomite, and the buried-conduit system, which carries water from the Niagara River to the power plant, is a point of groundwater discharge. Groundwater also discharges to the Falls Street Tunnel, which crosses the conduit system.

Water level data were collected from wells tapping separately top of clay and top of rock. The data were converted and tabulated as water surface elevation and plotted (Figures 8.2.8 and 8.2.9). The flow of groundwater in the top of clay is to the south-southeast. The flow of groundwater in the top of rock is also to the south-southeast.

Groundwater flow through the Lockport Formation in the area occurs through horizontal water-bearing bedding plane fracture zones. This was reported by Johnston[4] based on observations along the exposed walls of the New York Power Authority (NYPA) conduits which cut through the Lockport Formation west of the study area. Johnston[4] identified seven water-bearing zones, each consisting of either a single open bedding plane or an interval of rock layers containing several open bedding planes. Although the concept of separate and hydrologically distinct fracture zones has been an issue of dispute in the past, the U.S. Geological Survey (USGS) concurs with Johnston.[5]

A series of bedding plane fracture zones were identified during site investigations. The identification of water-bearing fracture zones was based on field observations, circulation fluid losses (expressed as percent water loss) during drilling, bedrock core examination, and hydraulic conductivity test results greater than $10^{-4}$ centimeters per second (cm/sec). Over two hundred core

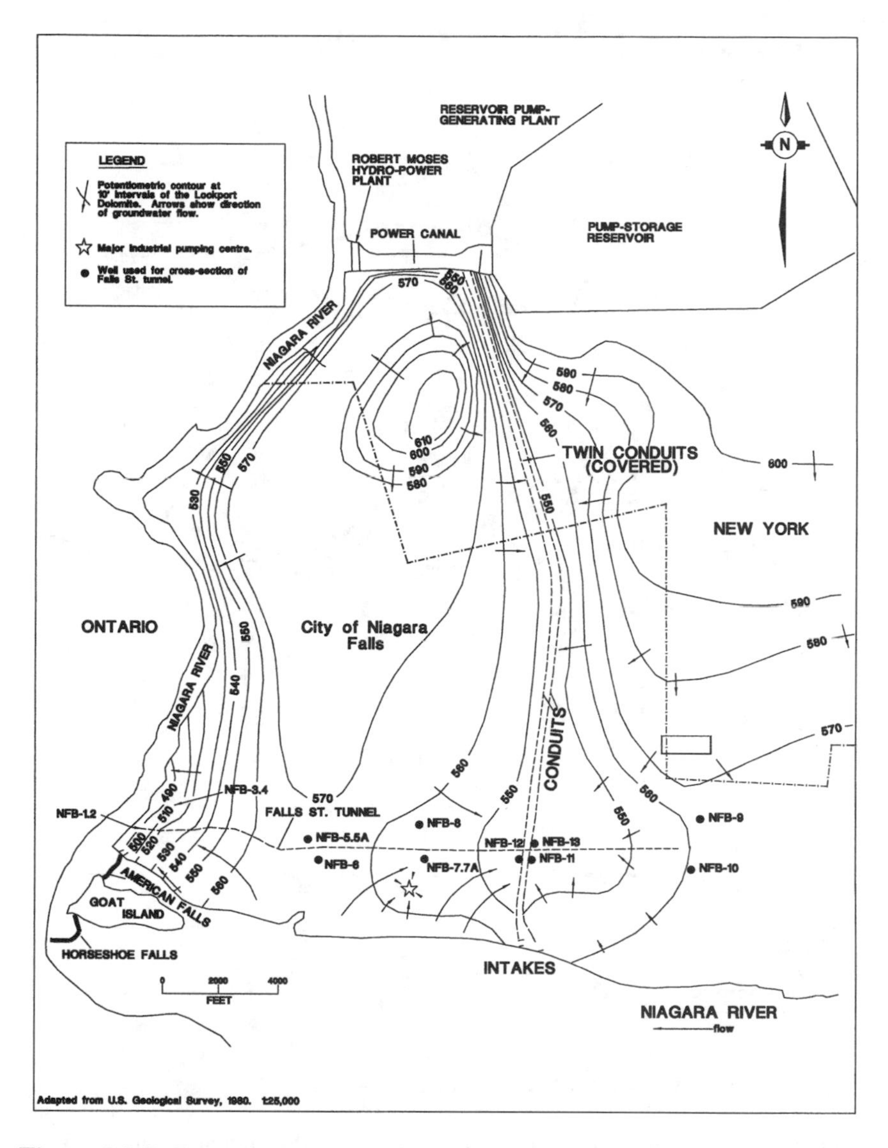


**Figure 8.2.7 Potentiometric surface of the upper Lockport (contours approximate).**

observations were used to verify the depth of a fracture zone. Usually a weathered fracture or series of fractures were observed at approximately the same depth as the noted circulation fluid losses. Moderate to high hydraulic conductivity test results, greater than $1 \times 10^{-4}$ cm/sec, usually corresponded to water-bearing fracture zones where water loss was observed. Low hydraulic conductivity values, less than or equal to $1 \times 10^{-4}$ cm/sec, usually corresponded to intervals where no circulation loss occurred. For the purpose of identification

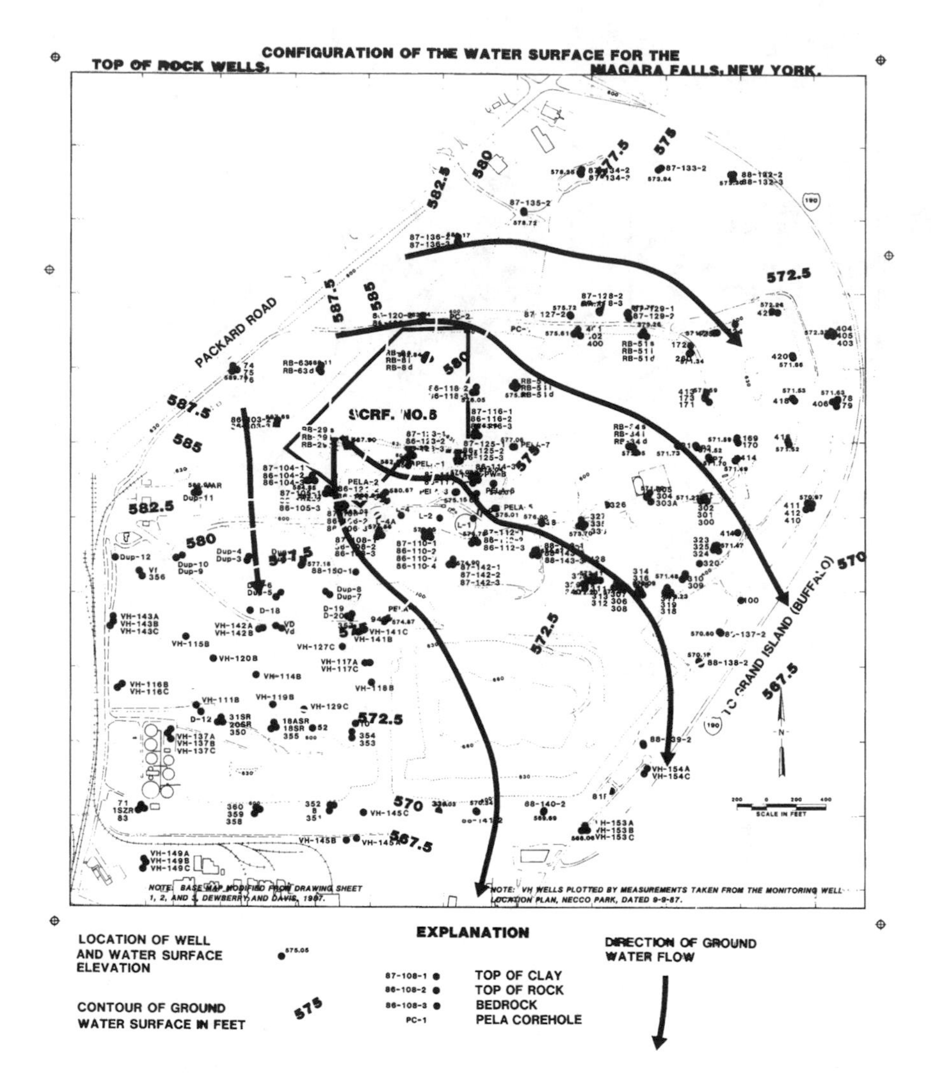


## Figure 8.2.8 Location of wells and coreholes.

and discussion, the water-bearing fracture zones in the area are designated the 7- through 1-zones (see Figure 8.2.10).

**7-Zone.** The uppermost water-bearing bedding plane fracture zone in the Lockport Formation within the area is designated the 7-zone. It generally exists approximately 4 feet below the top of rock. The 7-zone dips mainly southeast at an average angle of 0.6 of a degree. The 7-zone subcrops northwest of the area of investigation and is recharged through vertical fractures. Similar subcrop areas may exist for fracture zones 6 through 3 further northwest of the area of investigation.

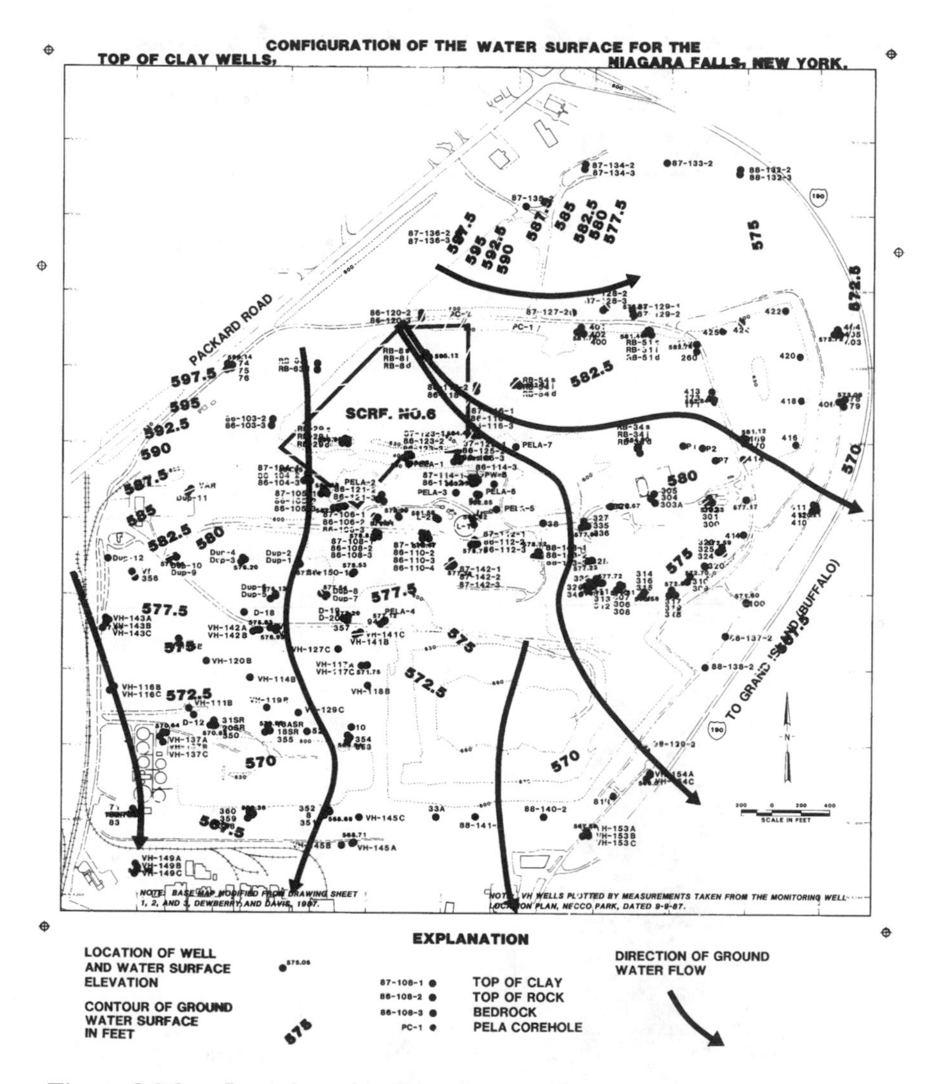


**Figure 8.2.9 Location of wells and coreholes.**

**6-Zone.** The 6-zone generally occurs approximately 10 feet below the 7-zone. This zone dips to the southeast with bedding at an angle of approximately 0.7 of a degree. This zone was not observed within the southeastern half of the area. Hydraulic conductivity results in the range of $10^{-5}$ and $10^{-6}$ cm/sec support observations during drilling and core inspection.

**5-Zone.** The 5-zone generally occurs approximately 30 feet below the 6-zone. The 5-zone is well represented in the northern half of the area but is poorly represented in the southern half. Approximate dip angle is 0.7 of a degree to the southeast. Since the 5- and 4-zones tend to be very close to one another, 5 to 10 feet, discretion was used when assigning a zone designation

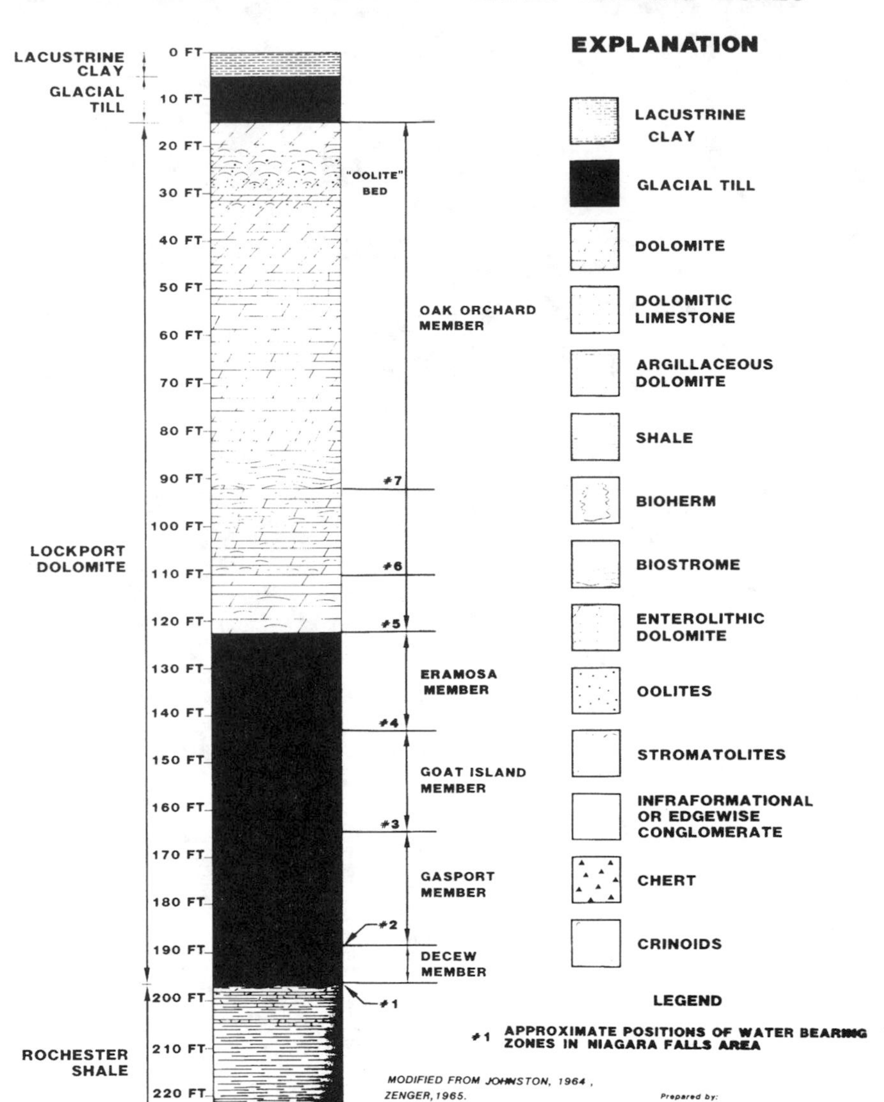


**Figure 8.2.10 Stratigraphic column.**

to either of these fractures. In locations where both of these zones are present, indications are that they may be hydraulically connected based on proximity and similar hydraulic heads.

**4-Zone.** The 4-zone usually occurs 5 to 10 feet below the 5-zone. The Eramosa-zone has not been observed in the southwestern corner of the area. It is inferred that the presence of this water-bearing zone, although widespread throughout the area, tends to be locally discontinuous. The approximate dip angle of this fracture zone is 0.4 of a degree to the southeast.

**3-Zone.** The 3-zone occurs approximately 17 feet below the 5-zone and/or 7 feet below the 4-zone. The 3-zone dips toward the southeast at approximately 0.7 of a degree. The 3-zone has not been observed in the southwest and southeast sections of the area.

**1- and 2-Zones.** A fracture zone, given the designation 1- and 2-zones, was identified through test drilling as existing approximately 60 feet below the 3-zone and 30 feet about the top of the Rochester Shale. Three water-bearing fracture zones exist within this zone in the Lockport Formation below the bottom of the Oak Orchard Member and above the top of the Rochester Shale.

## 8.2.6 AQUIFER TEST

The distribution of drawdowns produced by an aquifer test provides direct information on hydraulically significant features within the area affected by the test. The orientation of these features is determined by computing directional transmissivities between the pumped well and observation wells by the method of Papadopulos[6] as modified by Maslia and Randolph.[7] In two dimensions, a Poler plot of directional transmissivity in relation to the azimuth of the observation well* for a homogeneous, isotropic, confined aquifer forms a circle. If the aquifer is homogenous and anisotropic, the plot approximates an ellipse with major and minor axes paralleled to the principal directions of transmissivity. If the aquifer is heterogeneous, any discrete fracture is indicated by a much larger value of transmissivity along the azimuth of the well that intersects the fracture.

Production well and observation wells were installed to perform a constant rate aquifer test. The major groundwater zone penetrated during drilling of the production well and observation wells was a horizontal interval of solutioning and fracturing occurring at 3 to 5 feet below the top of the bedrock. The interval was consistent laterally and was penetrated in all the wells installed.

The purpose of the test was to determine the hydraulic characteristics of the water-bearing zones, the degree of isotrophy, the vertical and lateral extent of the interconnection of the fracture system, and the existence of any geologic and hydrogeologic boundaries.

## 8.2.7 PROCEDURE

The design of the pumping test included the selection of the location of the pumping well and 56 wells, which were used as observation wells during

* The location of the production well is defined as the origin.

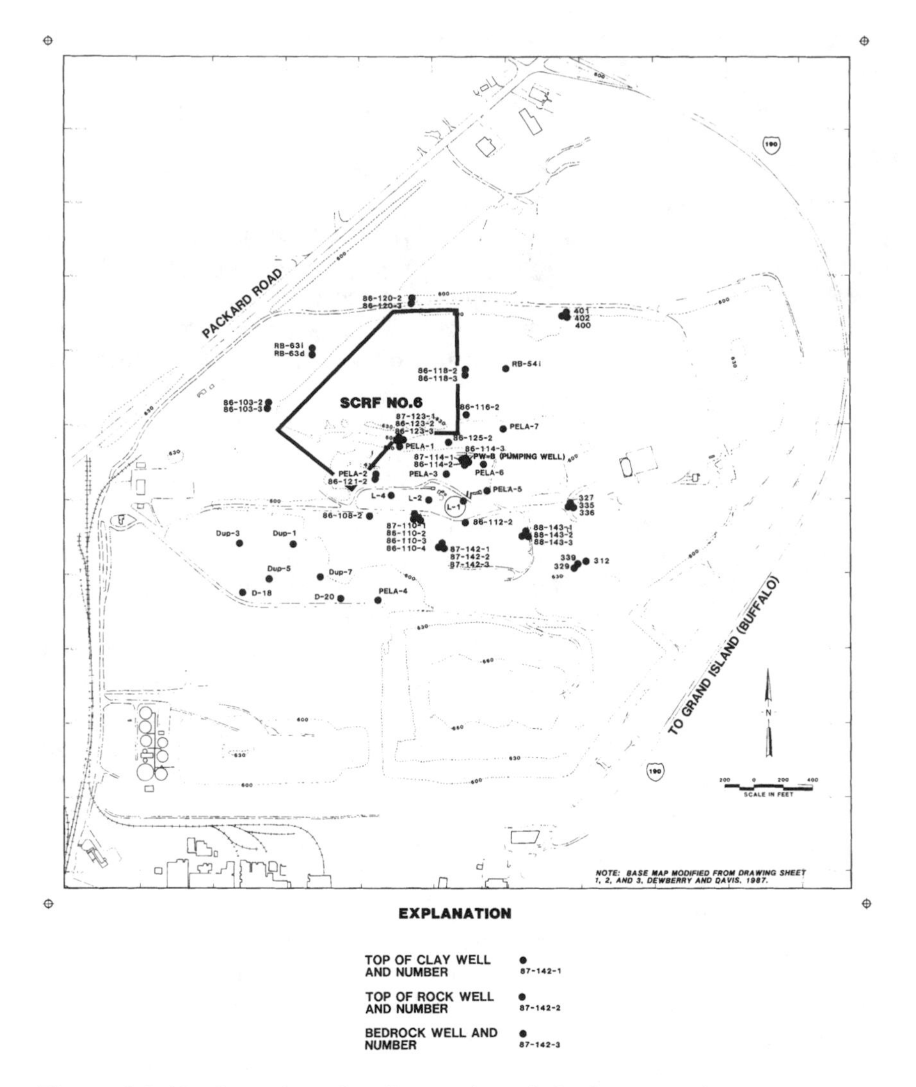


**Figure 8.2.11 Location of wells monitored during pumping test.**

the test (Figure 8.2.11). The selection of 56 observation wells was based on the review of the design and construction details of these wells and areal distance of each observation well from the pumping well (production well) to cover sufficient areal extent and obtain as much hydrogeologic information for the site as possible. The observation wells were selected for all three transmissive zones, top-of-clay, top-of-rock, and bedrock wells (see Figure 8.2.12).

Prior to performing the test, additional steps were taken to determine the natural trend of the groundwater and also the impact of ongoing activities at

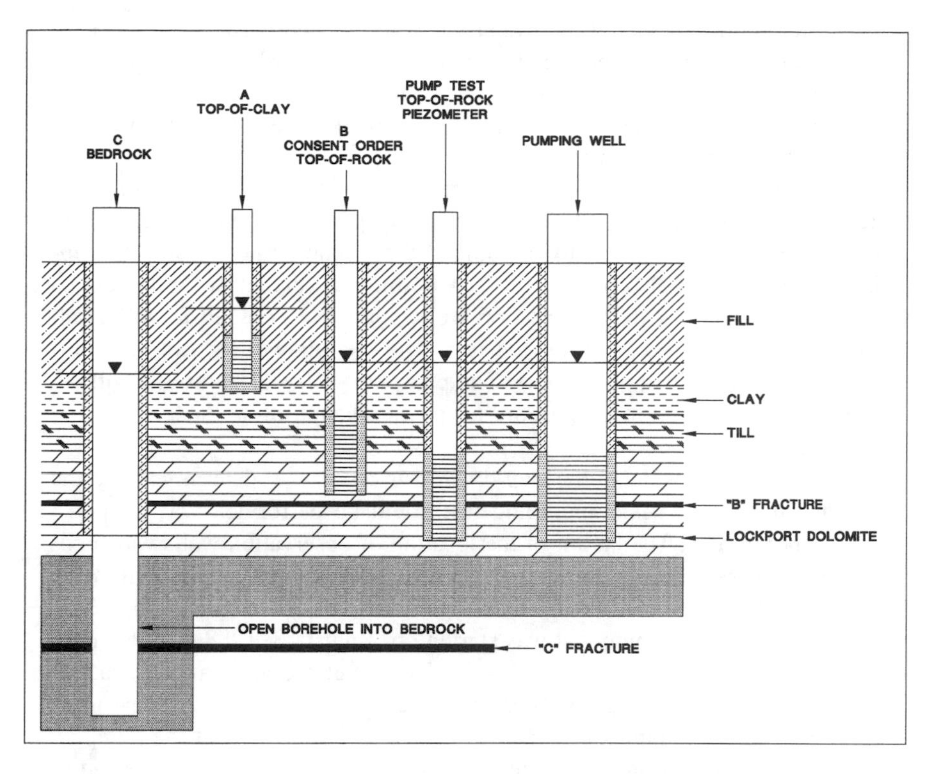


**Figure 8.2.12 Aquifer pumping test monitoring system.**

the site and at adjacent sites, such as drilling and grouting. Based on the evaluation of the data, it was decided that the test should be performed during a weekend to eliminate any impact of ongoing activities.

The selected wells were equipped with instruments that included Stevens recorders, In-Situ meters, and a Data Logger to obtain a continuous recording of the fluctuation of water level in order to determine the water-level trend, the impact of drilling and construction of wells at the site, and the grouting activities at adjacent sites. Continuous water-level data were also collected during the pumping test.

A recording barograph was used to continuously monitor barometric pressure before and during the test. Because the top-of-rock zone is confined, it was anticipated that barometric pressure changes would cause water level fluctuations that would need to be accounted for when interpreting the data.

At the beginning of the test, water level increased when barometric pressure was declining and decreased later in the test when barometric pressure was rising. It is possible that water level rises at the beginning were at least partially due to infiltration of rainfall but the timing of water level changes corresponded closely with barometric pressure change.

Before pumping, the groundwater levels in the well clusters were 5.65 to 11.86 feet higher in the top-of-clay zone than in the top-of-bedrock zone. The water level difference between the top-of-bedrock zone and bedrock zone was less than 1 foot. Therefore, the potential for vertical movement was downward during non-pumping conditions.

A step-drawdown test was performed on August 12, 1988, using four steps, pumping at rates of 8, 22, 30, and 45 gpm, to determine the rate at which the pumping well could be pumped during the 72-hour test. The step-drawdown test was also performed to determine the extent of the impact so that the locations for the observation wells could be modified to collect all critical water-level data during the test. The step-drawdown test was performed for approximately 6 hours. The pumping well was equipped with an In-Situ meter for continuous recording of the change of water level during the test.

Based on the results of the step-drawdown test, it was determined that the rate at which the 72-hour pumping test could be performed was 14 to 15 gallons per minute. The layout of the location of observation wells was modified as necessary and additional observation wells added in the monitoring program.

The 72-hour pumping test was started on August 13, 1988, and run until August 16, 1988. During the test, the well was pumped at an average rate of 14 gallons per minute and 56 observation wells were monitored to determine the impact of the pumping on water levels. After termination of the pumping on August 16, 1988, water level recovery data were collected for a period of 48 hours from all observation wells and from the pumping well.

Data from the tests were corrected for pressure fluctuations using the barometric efficiency. The water-level data were tabulated, plotted, and analyzed using standard aquifer evaluation methodologies.

Graphical plots of log time versus log drawdown and log time versus log recovery were made for each of the observation wells (Figure 8.2.13). These plots were matched against standard type curves presented in Lohman[8] for nonleaky confined aquifers using the Theis method, for leaky confined aquifers using the Hantush-Jacob method,[9] and for leaky confined aquifers where storage in the confining bed is accounted for using the Hantush modified method.[10] These three methods were chosen to cover the possible groundwater conditions in the top-of-rock zone. These methods were used to calculate transmissivity, storativity, and leakance.

The leakance factor was used to calculate permeability of the confining bed (K′). In this situation, water level measurements in the three zones would indicate that the majority of vertical flow is between the bedrock and top-of-rock zones with relatively minimal flow between the top-of-clay and top-of-rock zones. The head difference between the top-of-clay and top-of-rock zones ranges from 5 to 12 feet, while the head difference between the top-of-rock and bedrock zones ranges from 0 to 8 feet. Thus, the leakance calculation is

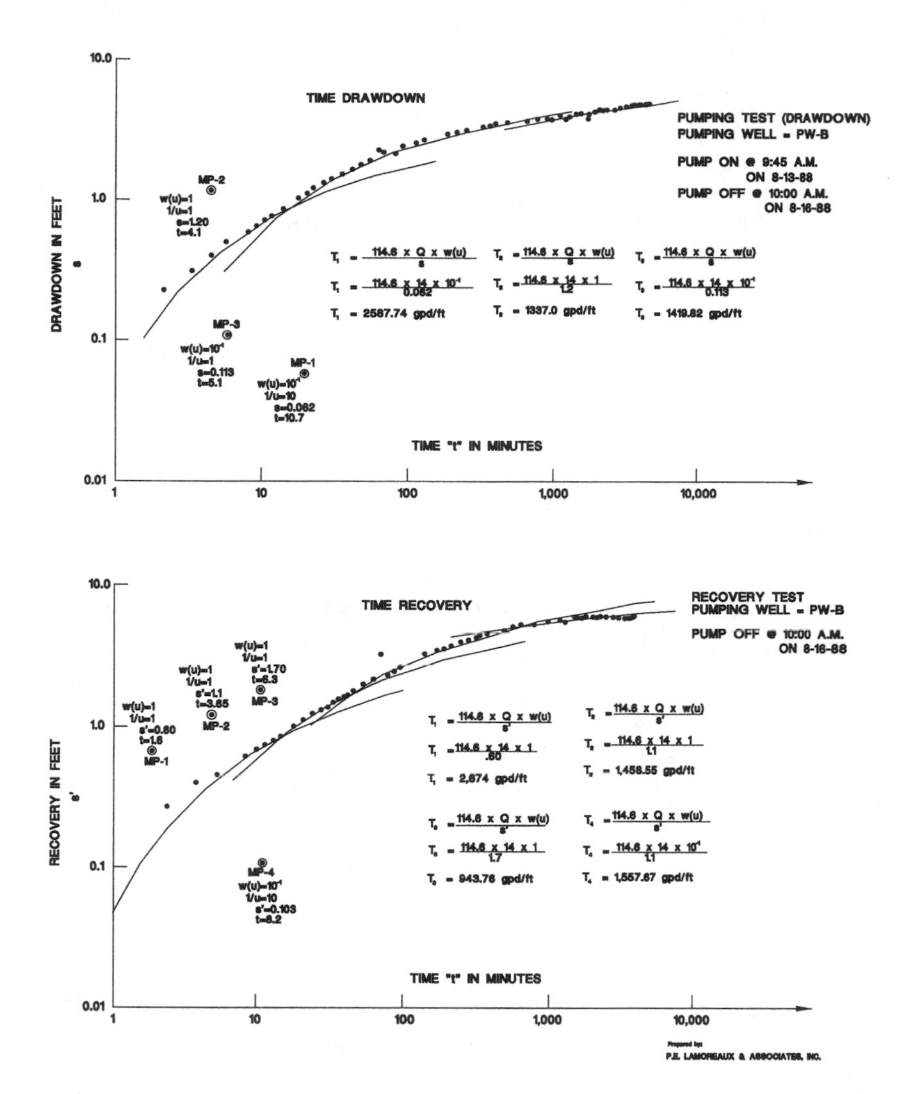


**Figure 8.2.13 Time drawdown and time recovery plots for pumping well PW-B.**

assumed to be representative of flow between the top-of-rock and bedrock zones, a thickness of 10 feet. The calculated vertical H-conductivity ranges from 0.02 to 6 gpd/ft$^2$ ($9 \times 10^{-7}$ to $3 \times 10^{-4}$ cm/sec). The summaries of the results of the aquifer test are contained in Tables 8.2.1A, 1B, and 1C.

The drawdown was computed and plotted in the form of contour maps for various time periods from 1 hour to 72 hours for the top-of-rock to determine the extent and configuration of the cone of depression, the degree of anisotropy, associated geological structural features, and solution channels

**Table 8.2.1A Calculated Aquifer Parameters, Niagara Falls**

| | T | S | v | b′ | K′ | |
|---|---|---|---|---|---|---|
| | | | | Thickness | Permeability of leakance zone | |
| Well | Transmissivity (gpd/ft) | Storativity | Leakance factor | of leakance zone (ft) | $gpd/ft^2$ | cm/sec |
| P1 | 5,900 | $5 \times 10^{-4}$ | 0.05 | 10 | 0.02 | $9 \times 10^{-7}$ |
| P2 | 2,200 | $2 \times 10^{-4}$ | 0.2 | 10 | 0.6 | $3 \times 10^{-5}$ |
| P8 | 4,500 | $6 \times 10^{-4}$ | 0.15 | 10 | 0.1 | $7 \times 10^{-6}$ |
| 170 | 4,000 | $1 \times 10^{-4}$ | 0.15 | 10 | 0.1 | $6 \times 10^{-6}$ |

**Table 8.2.1B Summary of Pumping Test Results for Top-of-Rock Wells**

| Well number | T Transmissivity (gpd/ft) | T Transmissivity ($cm^2$/sec) | K Permeability (cm/sec) | S Coefficient of storage |
|---|---|---|---|---|
| 88-143-2 | 1,725 | 2.48 | $1.63 \times 10^{-2}$ | $8.86 \times 10^{-6}$ |
| PELA 4 | 2,766 | 3.97 | $2.60 \times 10^{-2}$ | $2.56 \times 10^{-5}$ |
| DUP-7 | 4,062 | 5.84 | $3.83 \times 10^{-2}$ | $4.12 \times 10^{-5}$ |
| PELA 3 | 2,891 | 4.15 | $2.72 \times 10^{-2}$ | $8.08 \times 10^{-5}$ |
| PELA-2 | 2,139 | 3.07 | $2.01 \times 10^{-2}$ | $2.02 \times 10^{-4}$ |
| 86-114-2 | 1,420 | 2.04 | $1.34 \times 10^{-2}$ | $2.00 \times 10^{-2}$ |
| PELA 7 | 1,472 | 2.11 | $1.38 \times 10^{-2}$ | $4.73 \times 10^{-5}$ |
| 402 | 1,744 | 2.51 | $1.65 \times 10^{-2}$ | $2.44 \times 10^{-4}$ |
| 86-116-2 | 1,637 | 2.35 | $1.54 \times 10^{-2}$ | $1.80 \times 10^{-4}$ |
| PW-B | 2,057 | 2.96 | $1.9 \times 10^{-2}$ | |
| PELA-1 | 3,085 | 4.43 | $2.91 \times 10^{-2}$ | $2.96 \times 10^{-4}$ |
| 86-112-2 | 3,630 | 5.22 | $3.42 \times 10^{-2}$ | $8.31 \times 10^{-6}$ |
| 87-142-2 | 1,445 | 2.08 | $1.36 \times 10^{-2}$ | $5.91 \times 10^{-5}$ |
| 86-125-2 | 2,084 | 2.99 | $1.96 \times 10^{-2}$ | $1.70 \times 10^{-4}$ |
| 86-110-2 | 1,866 | 2.68 | $1.76 \times 10^{-2}$ | $9.80 \times 10^{-5}$ |
| PELA-5 | 2,198 | 3.16 | $2.07 \times 10^{-2}$ | $7.32 \times 10^{-5}$ |
| 86-121-2 | 3,913 | 5.62 | $3.69 \times 10^{-2}$ | $2.37 \times 10^{-5}$ |
| 86-108-2 | 2,487 | 3.57 | $2.34 \times 10^{-2}$ | $9.79 \times 10^{-5}$ |
| RB-54i | 3,145 | 4.52 | $2.97 \times 10^{-2}$ | $3.91 \times 10^{-4}$ |
| 335 | 1,163 | 1.67 | $1.10 \times 10^{-2}$ | $1.42 \times 10^{-5}$ |
| PELA-6 | 2,139 | 3.07 | $2.01 \times 10^{-2}$ | $1.62 \times 10^{-4}$ |
| 86-123-2 | 1,689 | 2.43 | $1.59 \times 10^{-2}$ | $6.58 \times 10^{-5}$ |

**Table 8.2.1C Summary of Pumping Test Results for Top-of-Bedrock Wells**

| Well number | T Transmissivity (gpd/ft) | T Transmissivity ($cm^2$/sec) | K Permeability (cm/sec) | S Coefficient of storage |
|---|---|---|---|---|
| 88-143-3 | 3,775 | 5.42 | $1.18 \times 10^{-2}$ | $2.94 \times 10^{-4}$ |
| 87-110-4 | 1,472 | 2.12 | $4.64 \times 10^{-3}$ | $1.10 \times 10^{-3}$ |
| D-18 | 4,336 | 6.23 | $1.36 \times 10^{-2}$ | $1.52 \times 10^{-3}$ |
| 336 | 7,640 | 10.98 | $2.40 \times 10^{-2}$ | $2.41 \times 10^{-4}$ |
| 86-123-3 | 5,942 | 8.54 | $1.87 \times 10^{-2}$ | $5.16 \times 10^{-4}$ |
| 86-110-3 | 4,650 | 6.68 | $1.46 \times 10^{-2}$ | $3.41 \times 10^{-4}$ |
| 86-142-3 | 5,093 | 7.32 | $1.60 \times 10^{-2}$ | $2.36 \times 10^{-4}$ |
| 312 | 8,444 | 12.13 | $2.65 \times 10^{-2}$ | $1.62 \times 10^{-4}$ |
| 87-114-3 | 3,913 | 5.62 | $1.23 \times 10^{-2}$ | $1.10 \times 10^{-1}$ |
| 86-118-3 | 6,418 | 9.22 | $2.02 \times 10^{-2}$ | $2.36 \times 10^{-3}$ |
| 400 | 1,371 | 1.97 | $4.31 \times 10^{-3}$ | $2.48 \times 10^{-4}$ |
| 86-118-3 | 5,014 | 7.21 | $1.58 \times 10^{-2}$ | $2.21 \times 10^{-3}$ |
| D-20 | 4,011 | 5.75 | $1.26 \times 10^{-2}$ | $1.72 \times 10^{-5}$ |
| 400 | 1,371 | 1.97 | $4.31 \times 10^{-3}$ | $2.48 \times 10^{-4}$ |
| 86-103-3 | 3,414 | 4.91 | $1.07 \times 10^{-2}$ | $8.80 \times 10^{-4}$ |
| RB-63-d | 9,274 | 13.33 | $2.91 \times 10^{-2}$ | $7.43 \times 10^{-4}$ |

and trends, and to graphically portray the impact on the water level in top-of-rock. A similar set of drawdown contour maps was prepared for the bedrock.

Isotropy is the condition of a medium which has hydraulic properties that are uniform in all directions. Anisotropy or non-isotropy is the condition of a medium which has hydraulic properties that are not uniform, but different in different directions.

The results of the pumping test determined that the permeability of the top-of-rock, in the area of the test, ranges from the $10^{-2}$ to $10^{-3}$ cm/sec and that of bedrock ranges from $10^{-2}$ to $10^{-3}$ cm/sec. The storage coefficient in the area of the test generally ranged from $10^{-3}$ to $10^{-6}$ for top-of-rock and $10^{-2}$ to $10^{-5}$ for bedrock. The other result obtained from the pumping test was the indication of lateral and vertical extent of the cone of depression (cone of influence). Figures 8.2.14, 8.2.15, 8.2.16, and 8.2.17 show the extent of the cones of depression in the top-of-rock wells and the bedrock wells for a period of 24 and 72 hours, respectively. Plots of the cone of depression depict the total drawdown for a period of 24 hours and 72 hours. These cones were drawn separately to show the extent of the impact of the test at different time periods throughout the test and the degree of anisotropy. Evaluation of cones of depression, which are in general circular, indicates little degree of anisotropy and non-homogeneity. The data further indicate that there is a well-developed fracture system in a particular direction and, therefore, no preferential flow system.

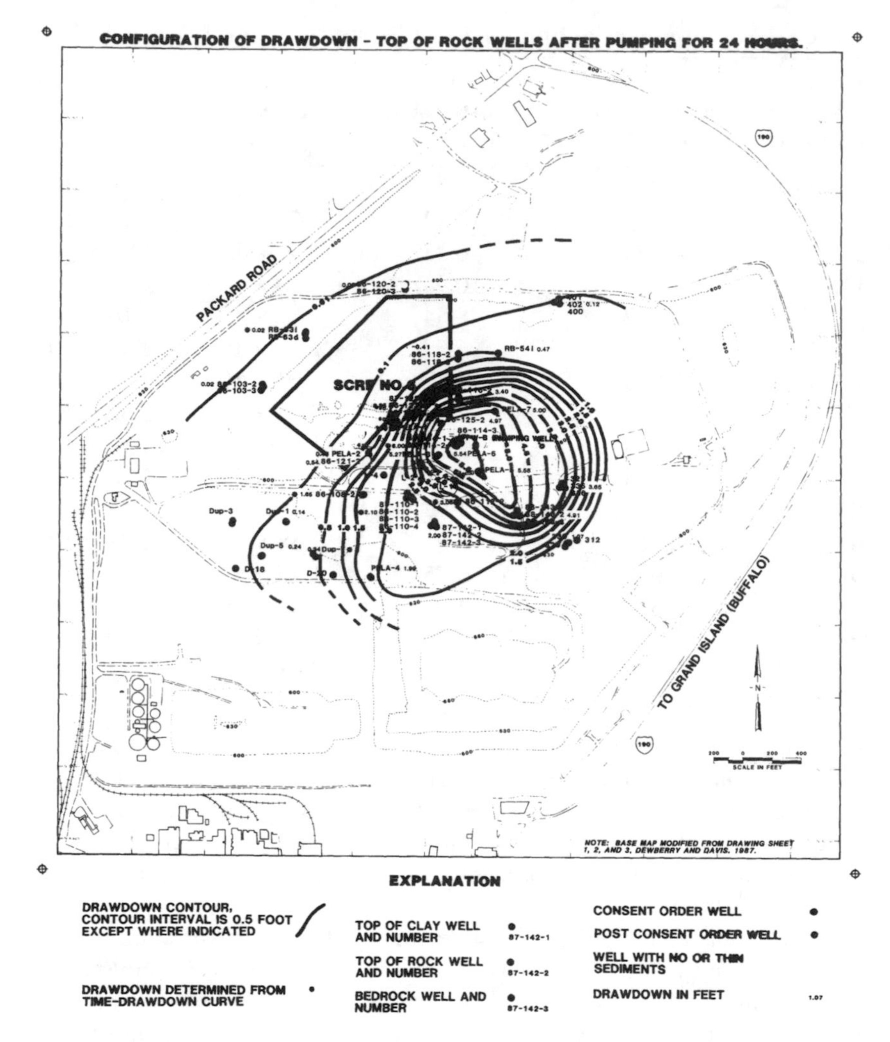


**Figure 8.2.14 Location of wells monitored during pumping test.**

Facts learned from the pumping test included: (1) water levels in monitoring wells tapping from the top of clay did not show impact from pumping, which means that there is no measurable hydraulic communication between the top-of-clay and top-of-rock; (2) water levels in some of the wells tapping the bedrock show drawdown, indicating restricted hydraulic communication between top-of-rock and bedrock; (3) the cone of depression is almost circular showing little degree of anisotropy in a discrete water-bearing zone, the top-of-rock; and (4) the configuration of the cone of depression indicates that there is no hydraulic boundary that would be indicative of faulting or fracturing associated with faulting.

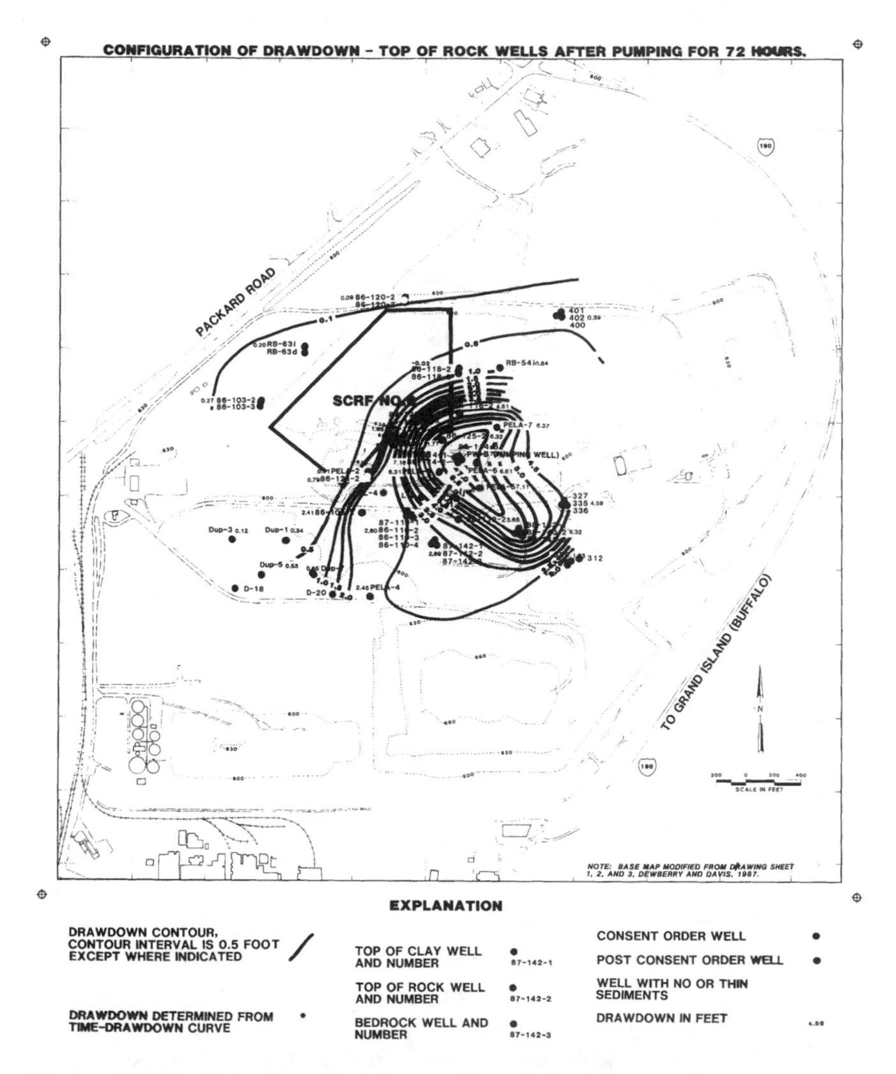


**Figure 8.2.15 Location of wells monitored during pumping test.**

## 8.2.8 CONCLUSIONS

Review of technical reports, on-site investigations and observations made during drilling and construction of the piezometers and wells, and analyses of the data collected during the step-drawdown test and aquifer test were used to characterize the top-of-rock transmissive zone in the vicinity of the site.

1. The major groundwater bearing zone determined during drilling was a porous interval of solutioning and fracturing in the dolomite occurring 3 to 5 feet below the top of bedrock. This interval was pene-

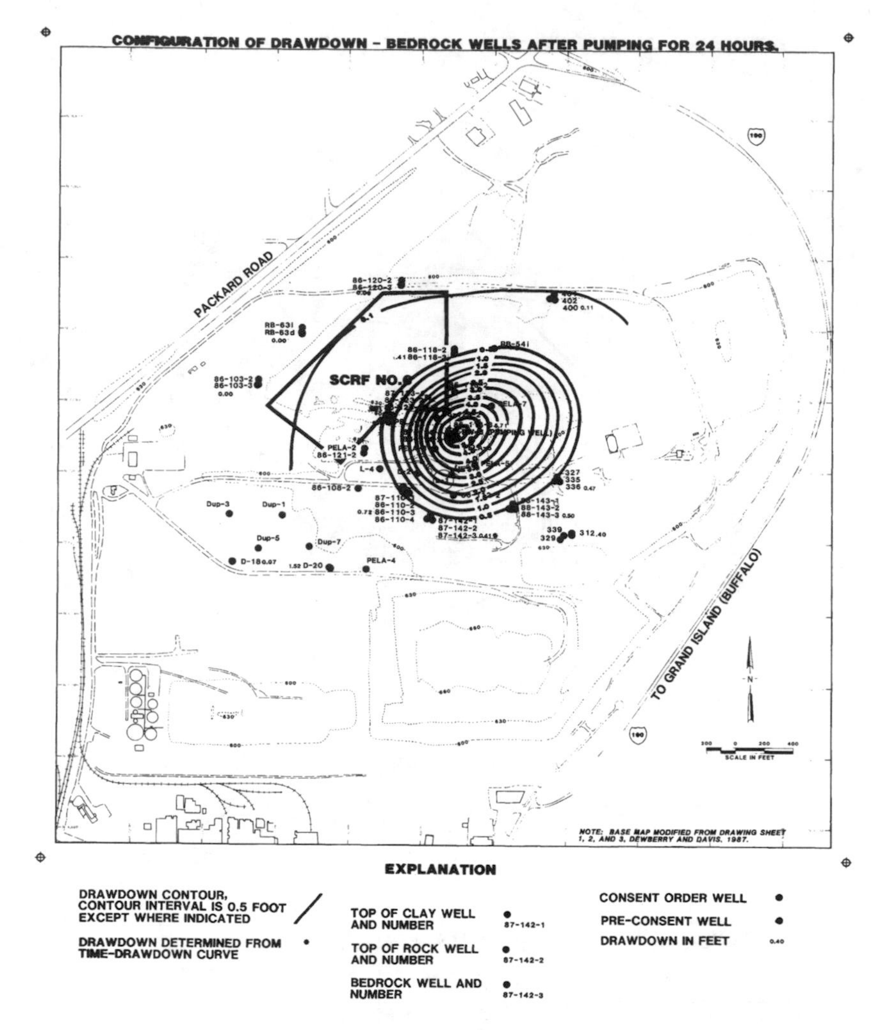


**Figure 8.2.16 Location of wells monitored during pumping test.**

trated in all the piezometers and wells installed and in nearby existing monitoring wells.

2. The site is underlain by a fractured water-bearing zone, top-of-rock, that comprises a flow system which is connected hydrologically beneath the site. It is sufficiently homogeneous so that it can be systematically monitored and, if necessary, effectively remediated.
3. The general direction of groundwater flow in the top of clay is south-southwest. The flow in both the top-of-rock and bedrock is south-southeast.

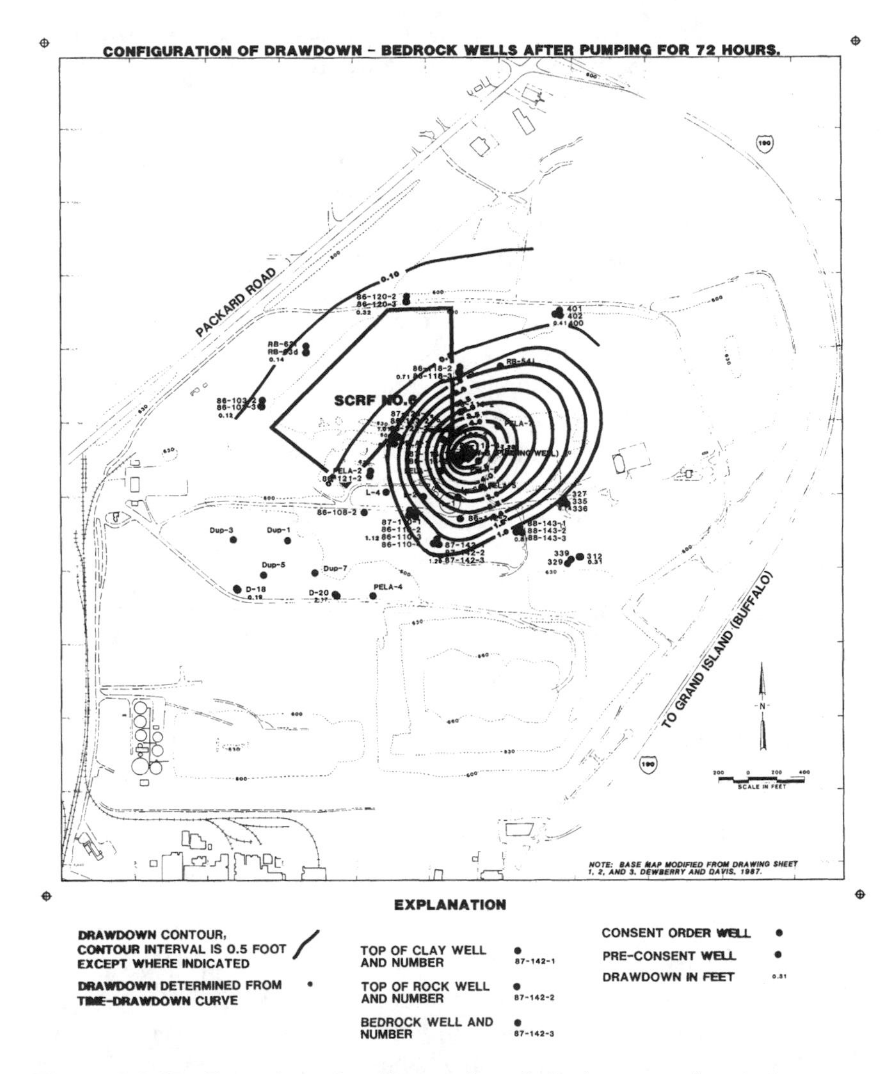


**Figure 8.2.17 Location of wells monitored during pumping test.**

4. Under non-pumping conditions, a south-southeast trending water level depression is present in the top-of-rock zone, indicating localized groundwater interflow between the top-of-rock and bedrock zones.
5. Barometric pressure changes cause water level fluctuations in the top-of-clay, top-of-rock, and bedrock zones, indicating confinement in all three zones.
6. The barometric efficiency of the top-of-rock zone appears to be on the order of 25 to 50%.

7. The hydraulic conductivity of the top-of-rock ranges from $10^{-2}$ to $10^{-3}$ cm/sec and the bedrock ranges from $10^{-2}$ to $10^{-3}$ cm/sec. The storage coefficient ranges between $10^{-3}$ to $10^{-6}$ for top-of-rock and $10^{-2}$ to $10^{-5}$ for bedrock. The vertical hydraulic conductivity between the top-of-rock and bedrock zone is 0.2 gpd/ft$^2$ ($5.28 \times 10^{-4}$ cm/sec).
8. Pumping at a rate of 14 gpm from the top-of-rock zone caused draw down in all directions at a distance in excess of 500 feet away from the test pumping well. Elongation of the cone of depression indicates that this zone is somewhat anisotropic.
9. Analysis of the aquifer test data indicates localized linear (non-radial) flow conditions in the top-of-rock zone coinciding with the orientation of prominent jointing in the top of the Lockport Dolomite in the Niagara Falls area. The groundwater conditions in this zone are somewhat anisotropic, but influence of pumping was observed in all directions away from the pumping well. Vertical fracturing contributes to the groundwater conditions as a secondary control to the flow.
10. The groundwater monitoring plan will detect any potential release from the site. The detection monitoring wells are to be properly located along the boundary of the site. If any release of contamination occurs, then contamination would be detected. The remedial plan for the site could be prepared and remedial action could be taken during this period.
11. Based on understanding of the design of the site, there should be no appreciable change in the groundwater flow if the facility is constructed as planned. In the unlikely event of release occurring from the site, the facility could be remediated by implementing a remedial plan.

The results of geologic and hydrologic investigations characterize this site to be suitable for a hazardous waste landfill provided the landfill is properly designed and engineered including installation of double liner, effective leachate collection, and adequate pre- and post-monitoring systems.

This study illustrates that proper knowledge of geology, structural controls, and hydrogeology is the key and best way to understand groundwater flow regimes and watershed to select, as best as possible, locations for landfills. The understanding of subsurface geologic conditions further aids to mitigate and/or minimize potential problems including capturing of leachate by leachate collection systems, post-monitoring, and protecting groundwater resources.

## REFERENCES

1. Tesmer, I.H., *Colossal Cataract — The Geologic History of Niagara Falls*, State University of New York Press, Albany, 1981.

2. Zenger, D.H., *Stratigraphy of the Lockport Formation (Middle Silurian) in New York State*, New York State Museum and Sciences Geological Survey, Bulletin 404, 1962.
3. Yager, R.M., and Kappel, W.K., *Detection and Characterization of Fractures and Their Relation to Groundwater Movement in the Lockport Dolomite, Niagara County, New York*, U.S. Geological Survey Water Resources Division, 1987.
4. Johnston, R.H., *Groundwater in the Niagara Falls Area, New York, with Emphasis on the Water-Bearing Characteristics of the Bedrock*, New York State Conservation Department, Bulletin GW-53, 1964.
5. Miller, T.S., and Kappel, W.M., *Effect of Niagara Power Project on Ground-Water Flow in the Upper Part of the Lockport Dolomite, Niagara Falls Area, New York*, U.S. Geological Survey Water Resources Investigations Report 86-4130, 1987.
6. Papadopulos, I.S., Non-steady flow to a well in an infinite anisotropic aquifer, *Proceedings of the Dubrovnik Symposium on the Hydrology of fractured rock*, International Association of Scientific Hydrology, 1965, pp. 21–31.
7. Maslia, M.L., and Randolph, R.B., *Methods and Computer Program Documentation for Determining Anisotropic Transmissivity Tensor Comments of Two Dimensional Groundwater Flow*, U.S. Geological Survey Open-File Report 86-227, 1986, 64 pp.
8. Lohman, S.W., *Ground-Water Hydraulics*, U.S. Geological Survey Professional Paper 70B, 1972.
9. Hantush, M.S., and Jacob, C.E., Nonsteady radial flow in an infinite leaky aquifer, *American Geophysical Union Transactions*, 36, 1, 1955, pp. 95–100.
10. Hantush, M.S., Modification of the theory of leaky aquifers, *Journal Geophysical Research*, 65, 11, 1960, pp. 3713–3725.
11. Cooper, H.H., Jr., Type curves for nonsteady radial flow in an infinite leaky artesian aquifer, in *Short-Cuts and Special Problems in Aquifer Testing*, Bentall, R., Ed., U.S. Geological Survey Water-Supply Paper 1545-C, 1963, pp. C48–C55.
12. Jenkins, D.N., and Prentice, J.K., Theory for aquifer test analysis in fractured rock under linear (nonradial) flow conditions, *Ground Water*, 20, 1, 1982, pp. 12–21.

# 8.3 CATASTROPHIC SUBSIDENCE: AN ENVIRONMENTAL HAZARD, SHELBY COUNTY ALABAMA

## 8.3.1 INTRODUCTION

The sudden formation of sinkholes or "catastrophic subsidence" in recent years has focused attention on a little-understood geologic hazard. Few people realize that thousands of sinkholes have formed in the United States since 1950. Costly damage, some accompanied by injuries and loss of life, has resulted from sudden collapses beneath highways, railroads, bridges, buildings, dams, reservoirs, pipelines, vehicles, and drilling operations. Perhaps one of the most spectacular was the "Golly Hole" collapse on December 2, 1972, in Shelby County, Alabama; another was the surface collapse of part of a city block in Winter Park, Florida, in 1981.

Sinkholes can be separated into categories defined as "induced" and "natural." Induced sinkholes are those caused or accelerated by human activities, whereas natural ones occur in nature. Recognition of induced sinkholes or catastrophic subsidence, the subject of this study, and their investigation have been confined mainly to this century. Almost all investigations dealing with triggering mechanisms or processes have been made since 1950.

The purpose of this section is to present mechanisms triggering the development of induced sinkholes resulting from water level declines, to identify predictive capabilities relating to sinkhole occurrence, and to describe techniques used in a case history to relocate a gas pipeline in a highly vulnerable karst setting.

## 8.3.2 GENERAL HYDROGEOLOGIC SETTING

The karst terrain chosen to illustrate catastrophic sinkhole development is Dry Valley, Shelby County, Alabama (Figure 8.3.1). It is a youthful basin that contains a perennial or near-perennial stream. Water is stored in underlying carbonate rocks and moves through interconnected openings along bedding

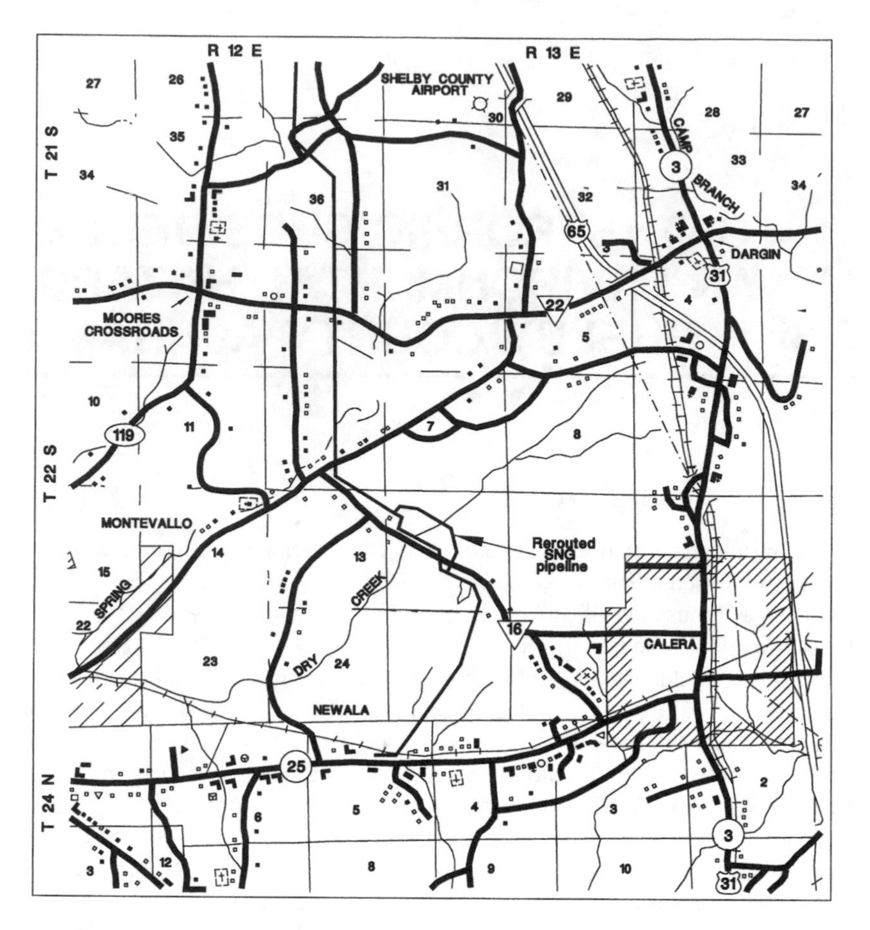


**Figure 8.3.1 Location of study area.**

planes, joints, fractures, and faults, some of which are enlarged by solutioning. Recharge from precipitation, in response to gravity, moves downward into this system of openings or toward the stream channel where it discharges and becomes streamflow. A schematic cross section illustrating the conditions described is shown in Figure 8.3.2.

Water in rocks underlying the basin occurs under water table and artesian conditions; however, this study is concerned with water table conditions only. The configuration of the water table conforms to that of the topography but is also influenced by precipitation, geologic structure, and water withdrawal. Bedrock openings underlying lower parts of the basin are water-filled and those underlying upland areas north of County Highway 16 (see Figure 8.3.1) are air-filled.

A mantle of unconsolidated deposits resulting from the solution of the underlying rocks consists chiefly of residual clay (residuum). This clay com-

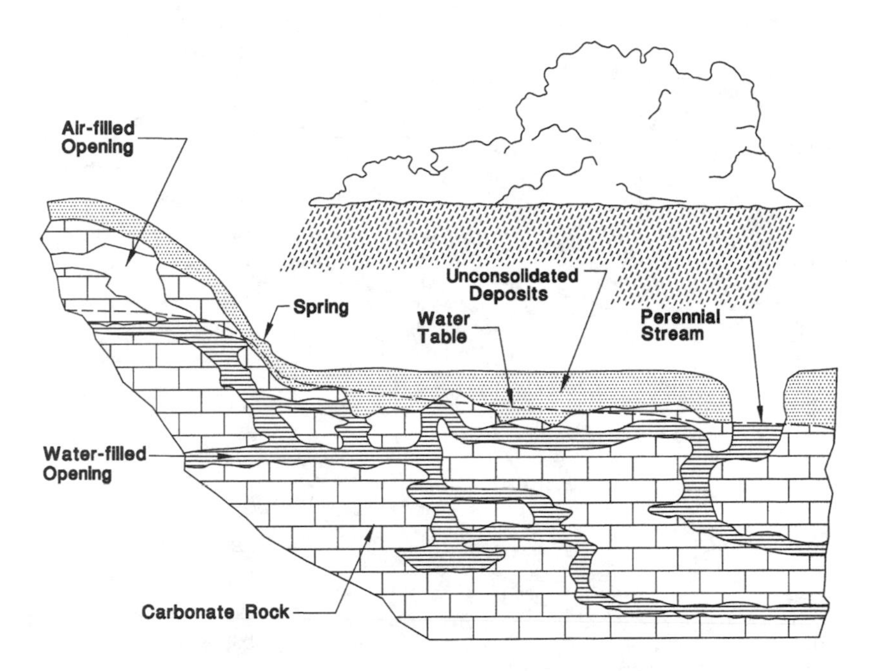


**Figure 8.3.2 Schematic cross section characterizing geologic and hydrologic conditions in a youthful karst terrain.**

monly contains chert debris and covers most of the bedrock surface. Alluvial or other unconsolidated deposits often overlie the clay adjacent to streams. The contact between residuum and underlying bedrock is highly irregular because of differential solution of the bedrock. Unconsolidated deposits commonly fill openings in bedrock to depths of 30 feet or more.

## 8.3.3 GEOLOGY OF THE DRY VALLEY AREA

Dry Valley is within the Cahaba Valley District of the Valley and Ridge physiographic province which is characterized by northeast-southwest trending valleys and ridges. The Cahaba Valley was formed by differential erosion of folded and faulted rock formations composed primarily of chert, limestone, and dolomite (Figure 8.3.3).

Rock formations in the Dry Valley area outcrop in northeast-southwest trending parallel bands. The rocks dip to the southeast at 20–60° and range in age from Cambrian to Mississippian. From northwest to southeast, the rock formations include the Copper Ridge Dolomite of Cambrian age; the Chepultepec Dolomite, Longview Limestone, Newala Limestone, Lenoir Limestone, and Athens Shale of Ordovician age; the Chattanooga Shale of Devonian age; and the Fort Payne Chert and Floyd Shale of Mississippian age (Figure 8.3.3).

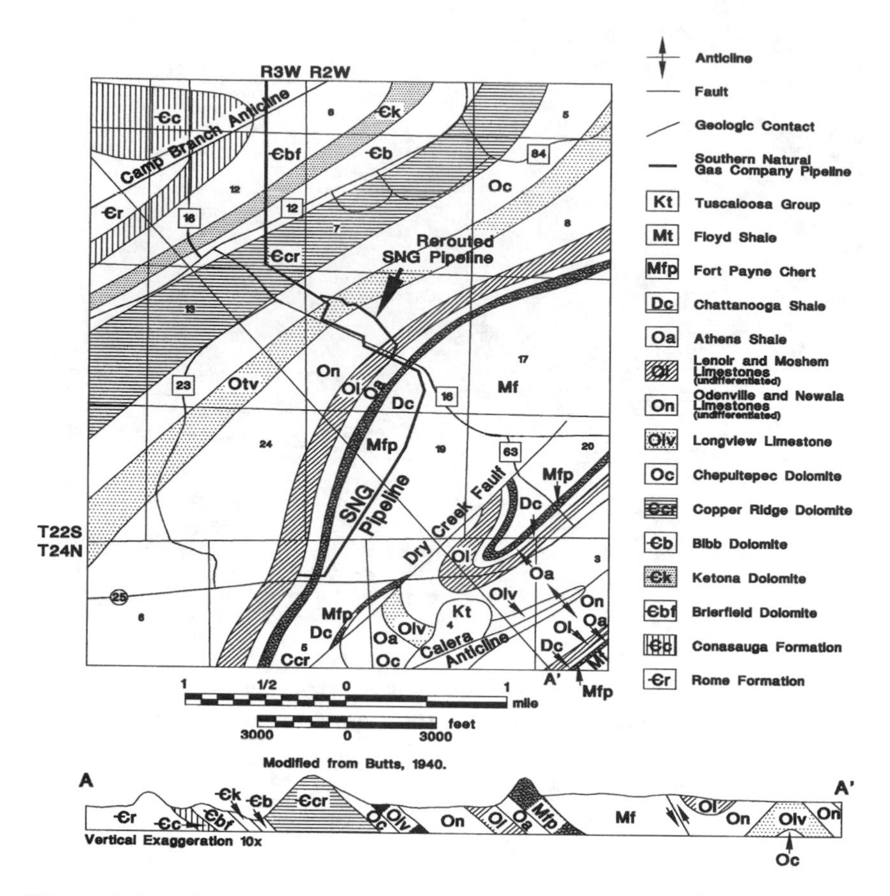


**Figure 8.3.3 Geologic map and cross section of the Dry Valley area (modified from Butts, 1940).[1]**

The Copper Ridge and Chepultepec dolomites form the western boundary of Dry Valley and support a stream-dissected ridge that is locally more than 100 feet above Dry Creek. The valley of Dry Creek is underlain by the Longview, Newala, and Lenoir Limestones. The Newala is mined from recessed quarries and underground mines in the valley as a source of raw material for the manufacture of cement. The Athens Shale and Fort Payne Chert outcrop in a sinuous, narrow ridge that forms the eastern boundary of the valley.

A mantle of unconsolidated material consisting of residual clay covers the bedrock in the area, obscuring surface exposures of geologic contacts and faults. This unconsolidated material, or residuum, has resulted from the solution of underlying carbonate rocks. It commonly contains varying amounts of insoluble chert debris. Some of this unconsolidated material fills solutionally enlarged fractures and solution openings in the bedrock underlying the valley floor. Because of differential solution, pinnacles of bedrock extend in places upward into the residuum, and boulders of "floating" rock occur within the residuum. These can easily be mistaken for the bedrock surface.

## 8.3.4 WATER LEVEL DECLINE AND CATASTROPHIC SUBSIDENCE

Sinkholes resulting from water level declines are not unique to the Dry Valley area. Foose,[2] in a study in Pennsylvania, first identified sinkhole activity associated with pumping and a decline in the water table. He determined that these sinkholes were confined to areas in which a drastic lowering of the water table had occurred and that the sinkhole occurrence ceased when the water table recovered. He stated that the shape of collapses indicated a lowering of the water table and withdrawal of support. Robinson et al.[3] added that sinkhole occurrence in a cone of depression was related to the increased velocity of groundwater. Spigner[4] attributed intense sinkhole development near Jamestown, South Carolina, to a water level decline resulting from pumpage and provided descriptions indicating loss of support and downward movement of unconsolidated deposits due to piping. Sinclair[5] attributed similar activity in Florida to loss of support and water level fluctuations.

Cited reports have described only in part the geologic and hydrologic impacts from a decline of the water table that cause the downward migration of unconsolidated material. It is important to understand, that in almost all instances, only the unconsolidated overburden becomes unstable and flows downward causing a collapse or failure, whereas the bedrock remains stable (Figure 8.3.4). The most common cause of induced subsidence is the decline of the water table as a result of groundwater withdrawal from wells or from underground mines and quarries. This occurred in Dry Valley. Similar problems are documented in published reports for other areas of Alabama (Powell and LaMoreaux,[7] Newton and Hyde,[8] and Newton et al.[9]), Pennsylvania, Florida, South Africa, Europe, and elsewhere in the karstic areas of the world (Newton[10]).

The following processes or activities are generally recognized as causing or accelerating subsidence following a decline of the water table.

1. The loss of buoyant support exerted by groundwater to unconsolidated materials overlying bedrock. Based on comparative specific gravities, for instance, this support to an unsaturated clay overlying a bedrock opening would amount to about 40% of its weight.
2. An increase in the velocity of groundwater movement resulting from an increased hydraulic gradient toward a discharge point. This water velocity results in the flushing of sediments filling openings in the cavity system, which, in turn, results in the downward movement of overburden into bedrock openings that forms a sinkhole.

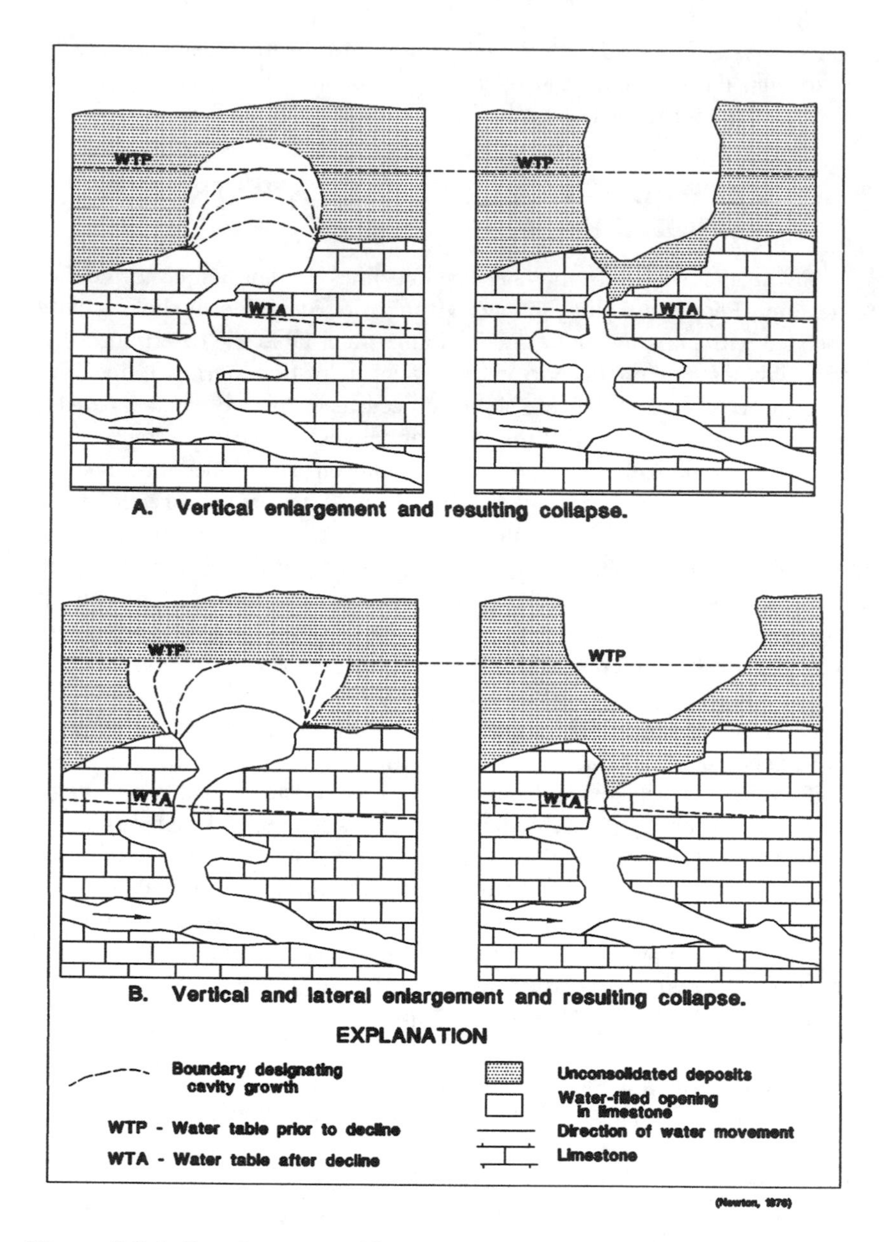


**Figure 8.3.4 Development of induced sinkholes (modified from Newton, 1976).[6]**

3. The weakening of unconsolidated bridging materials and downward erosion of these materials caused by alternate repeated addition and subtraction of buoyant support and alternate wetting, drying, and lubrication brought about by water level fluctuations.

4. Induced recharge to previously water-filled bedrock cavities by infiltrating surface water passing through and eroding overlying unconsolidated material downward. This process, most active during periods of heavy or prolonged rainfall, is the same process described by many authors as "piping" or "subsurface mechanical erosion."
5. Grading, ditching, or other human-related disturbances that result in thinning of overburden or concentrations of drainage at the surface or in the subsurface. These activities induce more water to move more rapidly along preferential flow patterns through soils or overburden and into bedrock. Triggering mechanisms are piping, saturation, and loading. Other examples include leaking pools, pipes, gutters, irrigation, and broken lined canals or ditches. Collapses resulting from leakage from underground pipes are well documented in the literature. Such a collapse in a gold mining district in South Africa resulted in the loss of a three-story building and the lives of 29 men.
6. Heavy construction, traffic, or explosives that disturb the soil or overburden and trigger its downward movement into solution openings in bedrock.
7. Removal of vegetation or the planting of large deep-rooted trees that increases recharge by creating avenues for more rapid movement of water from the land surface through soils and overburden to bedrock.
8. Drilling, augering, or coring where surface water gains access to uncased or unsealed holes. These activities cause erosion of overburden into underlying openings in bedrock. This occurrence has resulted in collapses at and near drill rigs or the holes created.
9. Impounding of water results in saturation of overburden and loss of cohesiveness of unconsolidated deposits overlying bedrock openings. This, accompanied by loading caused by the weight of impounded water, results in the collapse of unconsolidated material into a bedrock opening. Similar collapses beneath impoundments are also caused by piping. This occurs where the water table has declined below the top of bedrock and where openings at the surface are interconnected with those in bedrock. Collapses resulting from saturation and loading have been described by Aley et al.[11] and those resulting from saturation and piping have been described by Warren.[12] Collapses resulting in draining of impoundments in cones of depression are not uncommon.

## 8.3.5 HYDROLOGY OF DRY VALLEY

### 8.3.5.1 Present Conditions

Hydrologic conditions in Dry Valley are characterized by a water table that has been lowered by extensive groundwater withdrawals by mines and wells. The approximate decline has been illustrated by Warren.[13] Natural surface water drainage patterns in the Dry Creek area have been extensively modified by road construction and mining operations. Dry Creek is now intermittent north of Shelby County Highway 16 all year because of the small drainage area and rapid downward infiltration of water from the main channel to the lowered water table in bedrock. A tributary of Dry Creek originating from Simpson Spring and an impoundment on the Floyd Shale east of the rerouted pipeline (Figure 8.3.3) has appreciable flow over the shale during the "dry season." However, this flow disappears into induced sinkholes near the contact between the Athens Shale and Lenoir Limestone. This water and that from the upper reach of Dry Creek Valley move southward in the subsurface to a mine where it is pumped back into a downstream reach of Dry Creek. It then flows southwestward to Spring Creek (see Figure 8.3.1).

Discharge measurements made by the U.S. Geological Survey in October 1973 (low-flow period) indicated the runoff from Simpson Spring and the impoundment at 0.07 cfs (cubic feet per second). A discharge measurement made downstream by the U.S. Geological Survey at Dry Creek on Shelby County Highway 23 (see Figure 8.3.1) in October 1973 indicated the discharge was about 32 cfs. At the time of this measurement, Dry Creek north of Highway 16 was dry.[13] Therefore, all natural runoff originating in the Dry Valley drainage basin does not flow through this part of the basin as surface water. Stream flow in the area immediately north and south of Highway 16 discharges directly into sinkholes. For example, runoff from the Simpson Spring tributary of approximately 5 cfs on March 4, 1977, was discharging into a recent collapse near County Highway 16 in the SE 1/4 NW 1/4, sec. 18, T. 22 S., R. 2 W. In April 1977, a runoff of about 6 cfs was observed to be flowing into a second collapse farther upstream (Figure 8.3.5).

Surface water discharging into sinkholes in the area enters the solution cavity system in the underlying carbonate rocks and, from this part of the karst system, is pumped by dewatering wells into a downstream reach of Dry Creek that is south of Highway 16. Groundwater withdrawals in October 1973 amounted to about 14,000 gpm (gallons per minute), or about 32 cfs.[13] This is approximately the same discharge measured in Dry Creek at Shelby County Highway 23 during the same period.

Water table decline in the area is the result of extensive groundwater pumping from wells, recessed quarries, and underground mines. Water levels at the center of the cone of depression are more than 400 feet below land surface. The center of the cone of depression corresponds closely with the location of the deepest underground mine in the area (Figures 8.3.6 and 7).

A schematic cross section of a cone of depression (Figure 8.3.8) superimposed on a youthful terrain (Figure 8.3.2) illustrates the downward migration of unconsolidated deposits, the creation of cavities in overburden, and "catastrophic sinkhole" development. The creation of a cone of depression in an area of large water withdrawal results in a loss of buoyant support to overburden and an increased hydraulic gradient toward the point of discharge. Both

Figure 8.3.5 Stream discharging into sinkhole on April 3, 1977.

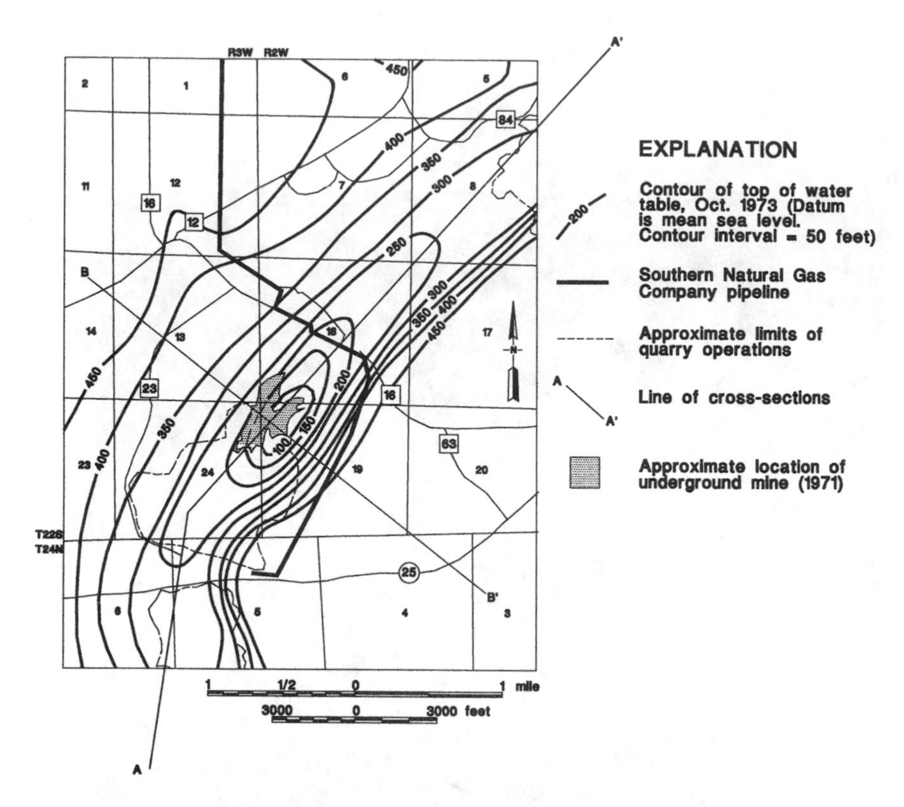


Figure 8.3.6 Water table map of the Dry Valley area. (Modified from Warren, 1976.)

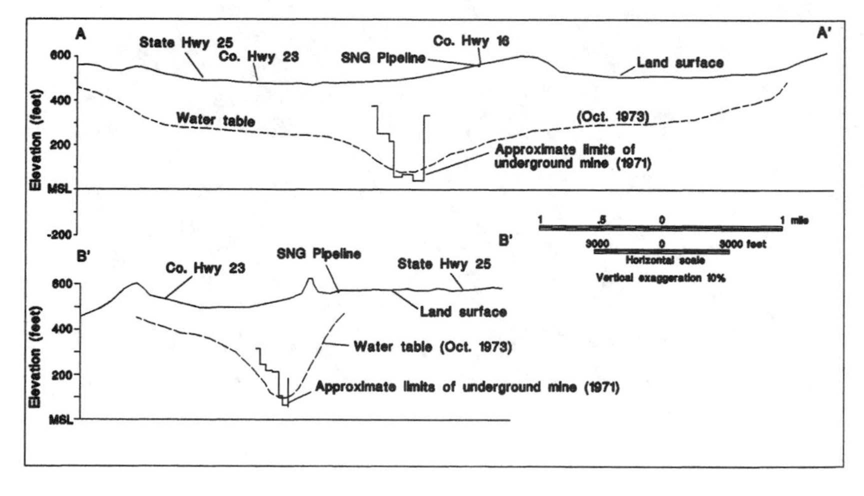


Figure 8.3.7 Hydrologic cross sections of the Dry Valley area.

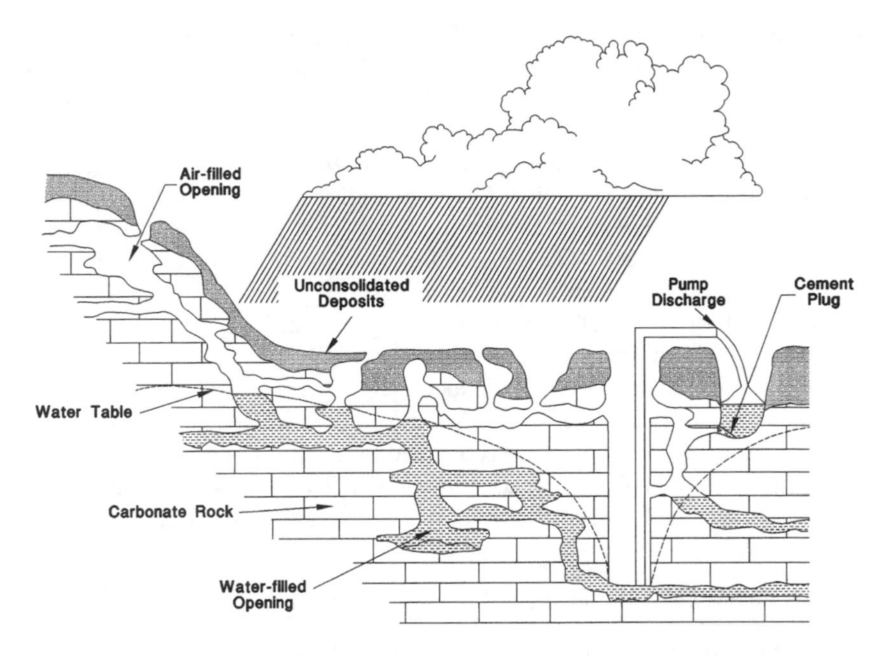


**Figure 8.3.8 Schematic cross section showing changes in geologic and hydrologic conditions resulting from groundwater withdrawal in a youthful karst terrain.**

can cause sinkhole development. The increased gradient results in increased velocity of groundwater movement. Erosion caused by this movement of water through a system of openings such as joints, fractures, faults, or solution cavities partly filled with clay or other unconsolidated sediments results in the creation of cavities that enlarge toward the surface and eventually collapse (Figures 8.3.4 and 8.3.8).

Variations in pumpage and recharge result in water level fluctuations far greater in magnitude than those occurring under natural conditions. The repeated movement of water through openings in bedrock against overlying unconsolidated deposits causes repeated addition and subtraction of buoyant support to them and repeated saturation and drying. This triggers the downward migration of the deposits that creates or enlarges cavities in the overburden (Figure 8.3.4).

The inducement of recharge through openings in unconsolidated deposits interconnected with openings in bedrock also results in the creation of cavities in the unconsolidated deposits. The material immediately overlying the bedrock openings is eroded to lower elevations. The water table, previously located above the top of bedrock (Figure 8.3.2), is no longer in a position to dissipate the mechanical energy of downward moving recharge. Repeated rains result in the progressive enlargement of this type of cavity. A corresponding thinning of the cavity roof due to its enlargement toward the surface eventually results in collapse. The position of the water table below unconsolidated deposits and openings on the top of bedrock favorable to induced recharge is illustrated in Figure 8.3.8. The creation and eventual collapse of cavities in the deposits by induced recharge is described as "piping."

Where the cone of depression is maintained by constant pumpage, all mechanisms described are active. Any one or all mechanisms may be responsible for the development of a collapse at a specific site. For example, in an area near the outer margin of the cone, the creation of a cavity and its collapse can result from all mechanisms. It can originate from a loss of support, can be enlarged by water level fluctuations, can be enlarged by increased velocity of water movement against sediment that originally filled the openings, and can be enlarged and collapsed by induced recharge.

Many induced sinkholes in Dry Valley are near the center of the cone of depression and their locations are controlled by the decline in the water table. The number of sinkholes and related features such as "pipes" or fractures has increased since 1967. The number, to date, is estimated to exceed two thousand.

## 8.3.6 USE OF REMOTE SENSING METHODS

A broad definition of remote sensing includes all methods of collecting information about an object without being in physical contact with it. If a more restrictive definition is used, it could include only those methods that employ electromagnetic energy, including light, heat, and radio waves, as a means of detecting and measuring target characteristics.[14] The major types of remote sensing used in carbonate hydrology are aerial photography, satellite imagery, thermography, and radar. Remote sensing techniques or surveys used on the land surface include sonar, down-hole geophysical logging and television cameras, and seismic resistivity, gravity, magnetic, radar.[15]

### 8.3.6.1 Satellite Imagery

The determination of the optimum satellite imagery or remote sensing band or band ratios (i.e., range of detected wave lengths) and the type of other remote sensing techniques to be employed depends on the objectives of the study and what features are sought to be enhanced. This optimum band or band ratio selection can be done by statistical methods (which generally require use of a computer) or by manual techniques such as the coincident spectral plot methods described by Shourong.[16] This method was used to map regional and site-specific geologic features and lineaments. These results, in turn, were integrated with findings obtained from aerial photography. They were then used to interpret regional and local geology and geologic structure that provided an understanding of subsurface karstification trends and preferential flow paths (Figure 8.3.9). Similar investigations have been made in Alabama by Powell et al.[17] and elsewhere.

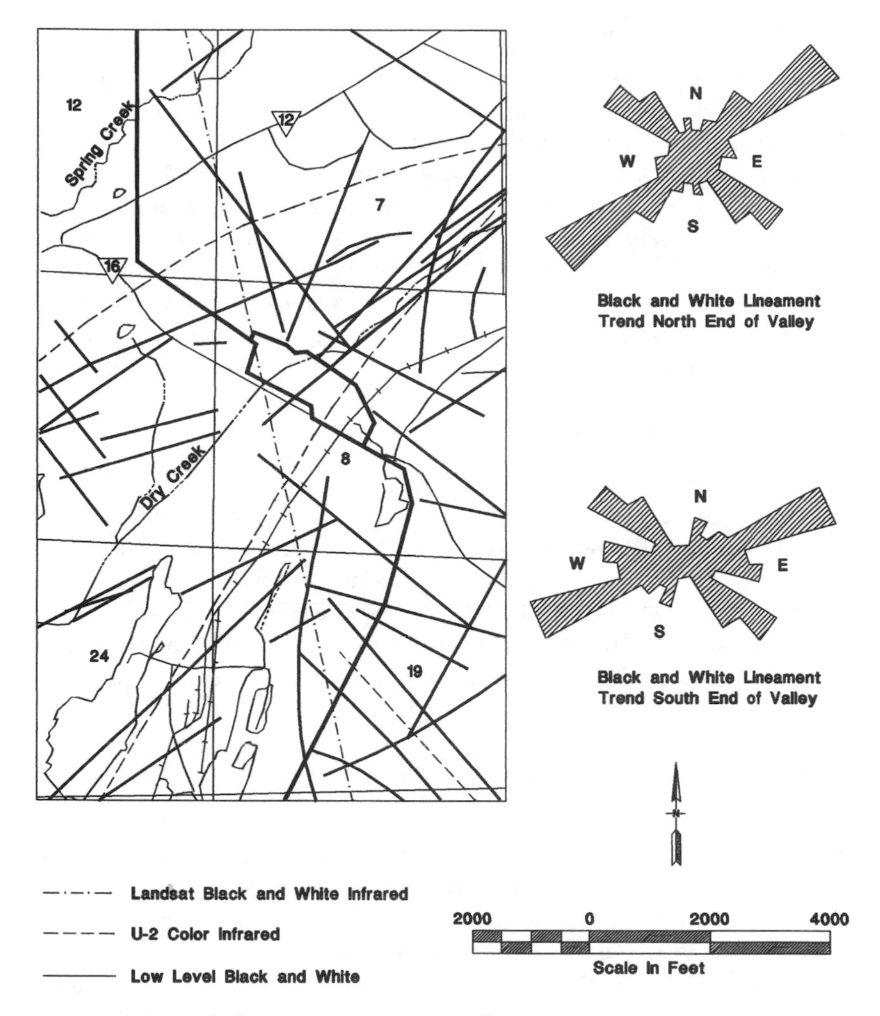


**Figure 8.3.9 Lineaments interpreted from aerial photographs.**

## 8.3.6.2 Aerial Photography

Sequential high- and low-altitude color infrared and black-and-white aerial photographs were obtained and analyzed to determine drainage patterns and geomorphic trends and to define lineaments. Other information on soil, vegetation, and rock outcrops was defined. The photographs were also used to locate and monitor induced sinkholes and to define trends in their development. This information was used to interpret the regional and local geologic structure and to identify potential preferential groundwater flow patterns. Findings and data were then integrated with those from the satellite imagery studies (Figure 8.3.9). Similar work utilizing photography has been accomplished in Alabama by Newton et al.,[9] Pennsylvania by Lattman and Parizek,[18] and elsewhere.

#### 8.3.6.3 Seismic Survey

Seismic refraction profiling was selected to provide a rapid and accurate definition of the bedrock surface trends and depths and to provide the thickness of unconsolidated material overlying bedrock (Figure 8.3.10).

A total of 130 seismic profiles were completed from May 17 to July 21, 1977, along the existing pipeline route and along possible alternate routes. After the final pipeline route was selected, an additional 47 seismic profiles were completed between August 22 and August 30, 1977, to obtain greater detail. An additional 10 profiles were completed along Shelby County Highway 16 in response to a request from the County Highway Department to identify areas of potential future subsidence for highway planning purposes.

### 8.3.7 TEST DRILLING

To verify and provide increased definition of the bedrock surface, 29 holes were augered in the study area on June 7 and 8, 1977, and an additional 48 holes were augered between September 12 and 15, 1977. A Central Mine Equipment Model 55 rig equipped with 4-inch auger flights was used to determine the depth to bedrock. All holes were augered to refusal. Samples of cuttings were described and geologic logs were prepared and correlated with the geophysical data to define the top of bedrock and overburden thickness along Alternate Route 1 and County Highway 16 (Figure 8.3.10).

### 8.3.8 INVENTORY AND MONITORING OF SUBSIDENCE

Catastrophic sinkholes within the study area were mapped using aerial photography and field investigation by LaMoreaux and Associates.[19] The area was field inspected on a monthly basis to detect subsidence occurrence and growth. Over 100 new subsidence features were identified and described. During dry weather and relatively stable hydrologic conditions, the land surface was stable. After prolonged dry periods followed by rainfall, a substantial increase in subsidence occurred. For example, during a monitor run on February 13, 1980, there were 13 active subsidence features. Rains preceding this inspection followed a dry period from October to December 1979.

Low-level black-and-white infrared and color photography was used to identify points of probable subsidence effectively. Criteria used included changes in drainage patterns, ponding of water, vegetation stress, lineament and geologic structural features, and construction activity. One such feature was predicted and observed as it collapsed.

The monthly monitoring findings were plotted on a base map showing Shelby County Highway 16, the existing Southern Natural Gas pipeline, and Alternate Route 1 across Dry Creek Valley. These monthly maps were used to determine the pattern, size, and trend of subsidence features with respect to geologic and dynamic hydrologic conditions that, in turn, provided a basis for the choice of the location for an alternate pipeline route.

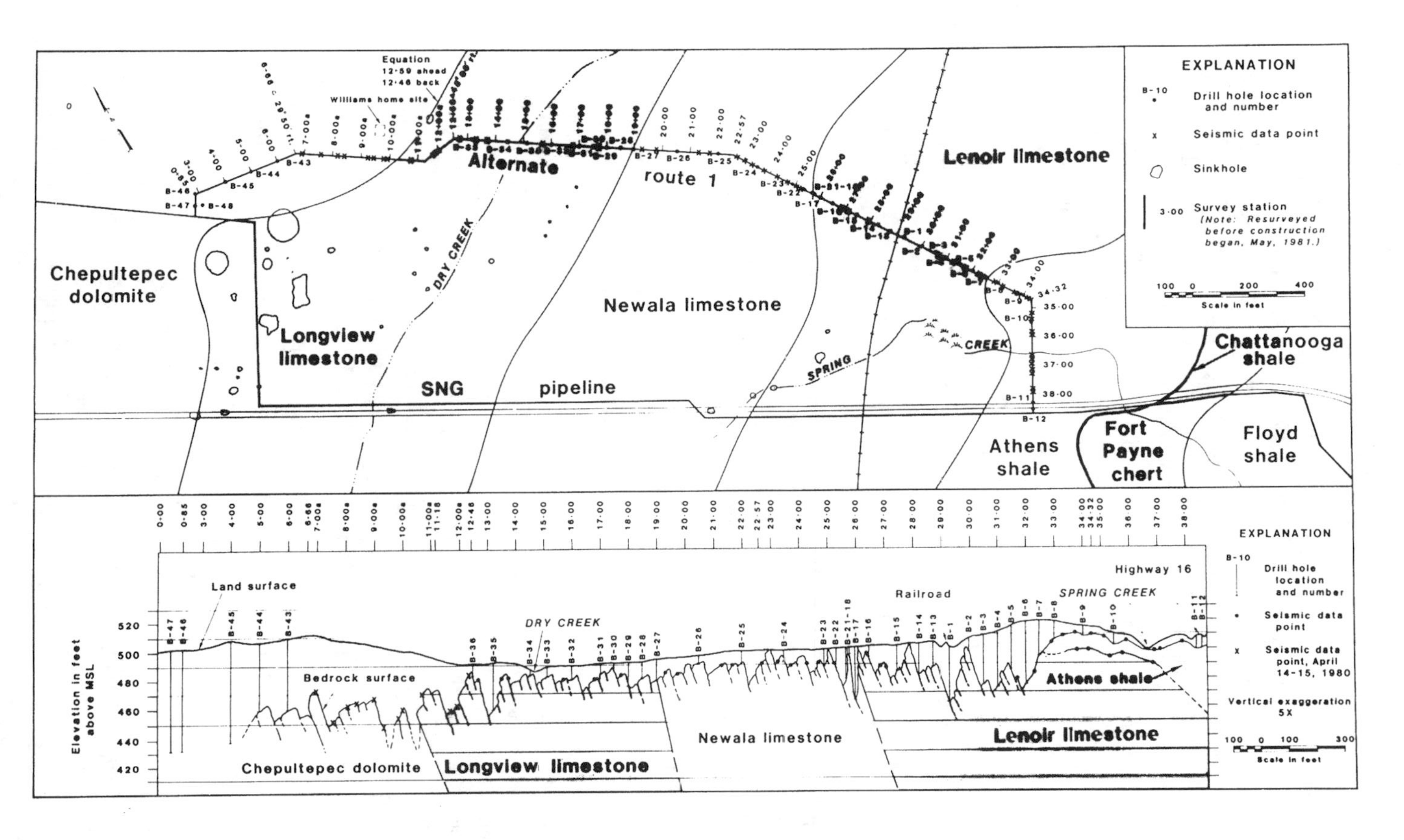


Figure 8.3.10 Geologic map and cross section, Shelby County, Highway 16.

## 8.3.9 PREDICTION OF INDUCED SINKHOLES

Induced sinkholes are predictable in the context that they will occur within the area impacted by activities such as dewatering. Also predictable, in some instances, are their alignment with other sinkholes and their shape, size, and depth. Predictive capabilities would be most significant in the type of terrain described in this article, and would be dependent on amount of geologic and hydrologic data available.

The most predictable induced sinkhole development is that resulting from water level declines due to dewatering by subsurface mines, recessed quarries, and wells. This occurs where the water level, previously above the top of bedrock during all or most of the year (Figure 8.3.2), is maintained below it by pumping (Figure 8.3.8). All mechanisms that trigger sinkhole development in unconsolidated deposits are activated by the decline.

Conversely, the unconsolidated deposits are not impacted and sinkholes will not occur where the zone in which the water level fluctuates is located below the top of bedrock prior to dewatering. Determining the position of the water table in relation to the top of bedrock aids in predicting whether sinkholes will or will not occur at a given site.

Where and when some sinkholes will occur in a dewatered area is also predictable to a limited degree. Many occur where concentrations of surface water are greatest such as streambeds, natural drains, or poorly drained areas. Large numbers occur where natural drainage has been altered and where natural recharge has been increased as a result of activities such as ditching and timber removal. Most of the sinkhole activity occurs during or immediately after rains, especially deluges, when hydrologic stresses to overburden are greatest.

Prediction of size, shape, and depth is based on a knowledge of the character of the bedrock; the extent, orientation, and size of the solution features; the thickness and stability of the overburden; and the mechanisms triggering overburden erosion. Induced recharge in an area prone to flooding, for instance, would assure maximum subsurface erosion of overburden.

## 8.3.10 SOUTHERN NATURAL GAS PIPELINE: A CASE HISTORY

Active subsidence (catastrophic collapse) in Dry Valley presented a danger to highways, railroad, buried telephone cables, personal property, farm animals, and oil and gas pipelines, including the Southern Natural Gas 10-inch Bessemer to Calera pipeline. Many collapses had occurred along the pipeline right-of-way. Some collapses directly underneath the pipeline had exposed it (Figure 8.3.11).

Geologic, geophysical, and hydrologic surveys along the pipeline identified the areas where sinkholes could occur. Extensive collapse sinkholes resulted from a combination of factors, including groundwater withdrawal,

**Figure 8.3.11 Collapse exposing pipeline.**

modification of surface drainage, construction activities, and heavy and prolonged rain. Catastrophic collapse in the area will continue indefinitely until these conditions change. Therefore, an alternative pipeline route had to be chosen to anchor the pipeline to bedrock with anchor points not greater than 20 feet apart because of the strength of the pipe. Determining the geographic distribution, frequency, and probability of catastrophic sinkhole occurrence was accomplished by the following work:

1. Preparation of a detailed map of the geology and structure along the pipeline in the critical areas of subsidence (Figure 8.3.10).

2. Mapping of exposures of bedrock limestone in quarries, road cuts, and sinkholes to determine dip and strike of bedding and joints and fault trends to relate to preferential solution zones and groundwater flow patterns.
3. Acquisition and analysis of satellite imagery and high- and low-altitude aerial photography (black-and-white, black-and-white infrared, color infrared, color). Resulting regional and local geological structural trends, lineaments, and sinkhole and drainage alignments were studied to project preferential groundwater flow patterns and solution zones in bedrock limestone (Figure 8.3.9).
4. Use of seismic geophysical studies and test drilling to define the top of bedrock and overburden thickness along the alternate pipeline route (Figure 8.3.10).
5. Determination of geology along the new pipeline route before construction, which was verified during its construction to ensure that the pipeline was securely connected to bedrock.
6. Monitoring of sinkhole-subsidence occurrence on a monthly basis over a period of 38 months. Each month, photography from an overflight was analyzed and located subsidence features were checked in the field and documented.

Based on the various studies, two alternate pipeline routes were delineated that would reduce the danger from catastrophic subsidence beneath the pipeline. Alternate Route 1, the final route chosen, was the best and most direct route across Dry Valley. It followed shallower bedrock, had less overburden thickness, fewer sinkholes, and undisturbed drainage, and crossed an area underlain by less limestone and a larger area underlain by Athens Shale (Figure 8.3.10).

Construction for the new pipeline route involved opening the ditch line twice. The ditch was first dug to remove all bedrock float and pinnacles. The ditch was backfilled and dressed at the end of each day to prevent rainfall and surface runoff from entering. During excavation, it was noted that bedrock was shallower and pinnacles were more frequent beneath Dry Valley than previously identified. Construction was redesigned to obtain maximum bedrock support to the pipeline.

The ditch was then reopened to lay pipe. Fractured bedrock zones were grouted, where necessary. Supports of steel piling driven to bedrock with a crossbar were erected in areas of deep unconsolidated overburden and large solution features. A steel casing was placed beneath a railroad and Shelby County Highway 16 to protect the pipeline from excessive weight. These supports were tied directly to bedrock.

The new pipeline was cleaned, tested, and tied into the old pipeline. Valves were placed on both the old and the new pipelines so that the old line could be reactivated if necessary. The original pipeline across Dry Valley was purged with nitrogen.

Tree roots and logs were removed from the overburden and a clay lining was placed in high subsidence risk areas adjacent to the pipeline to prevent downward migration of water into the ditch. After the pipe was laid, the ditch was backfilled and a clay crown was spread over it. The right-of-way was then graded and restored to approximate original land surface. It was then properly terraced to control surface drainage and seeded with grass and all fences previously crossing it were replaced. Natural drainage was left unobstructed.

Subsequently, the right-of-way has been monitored through a period of rains during which catastrophic subsidence might be expected. No subsidence has been recorded to date along the right-of-way and the area is now completely reclaimed and vegetation recovered.

## REFERENCES

1. Butts, C., Description of the Montevallo and Columbiana Quadrangles (Alabama), U.S. Geological Survey Atlas Folio 226, 1940.
2. Foose, R.M., Ground water behavior in the Hershey Valley, Pennsylvania, *Geol. Soc. Amer. Bull.*, 64, 1953, pp. 623–645.
3. Robinson, W.H., Ivey, J.B., and Billingsley, G.A., Water supply of the Birmingham area, Alabama, *U.S. Geological Survey Circular*, 254, 1953, 53 pp.
4. Spigner, B.C., Land Surface Collapse and Ground-Water Problems in the Jamestown Area, Berkley County, South Carolina: South Carolina Water Resources Commission Open-File Report no. 78-1, 1978, 99 pp.
5. Sinclair, W.C., Sinkhole Development Resulting from Ground-Water Withdrawal in the Tampa Area, Florida, U.S. Geological Survey Water Resources Investigations 81-50, 1982, 19 pp.
6. Newton, J.G., Early Detection and Correction of Sinkhole Problems in Alabama, with a Preliminary Evaluation of Remote Sensing Applications, Alabama Highway Department, Bureau Research and Development, Research Report no. HPR-76, 1976, 83 pp.
7. Powell, W.J., and LaMoreaux, P.E., A problem of subsidence in a limestone terrane at Columbiana, Alabama, *Alabama Geological Survey Circular*, 56, 1960, 30 pp.
8. Newton, J.G., and Hyde, L.W., Sinkhole problem in and near Roberts Industrial Subdivision, Birmingham, Alabama — a reconnaissance, *Alabama Geological Survey Circular*, 68, 1971, 42 pp.
9. Newton, J.G., Copeland, C.W., and Scarbrough, L.W., Sinkhole problem along proposed route of Interstate 459 near Greenwood, Alabama, *Alabama Geological Survey Circular*, 83, 1973, 53 pp.
10. Newton, J.G., Natural and induced sinkhole development — eastern United States, *International Association of Hydrological Sciences Proceedings*, Third International Symposium on Land Subsidence, Venice, Italy, 1984.
11. Aley, T.J., Williams, J.H., and Masselo, J.W., Groundwater contamination and sinkhole collapse induced by leaky impoundments in soluble rock terrance, *Missouri Geological Survey and Water Resources Engineering Geology Series*, 5, 1972, 32 pp.

12. Warren, W.M., Retention basin failures in carbonate terranes, *Water Res. Bull.*, 10, 1, 1974, pp. 22–31.
13. Warren, W.M., Sinkhole occurrences in western Shelby County, Alabama, *Alabama Geological Survey Circular*, 101, 1976, 45 pp.
14. Sabins, F.F., Jr., *Remote Sensing*, W.H. Freeman, San Francisco, 1978, 426 pp.
15. LaMoreaux, P.E., Wilson, B.M., and Memon, B.A., Eds., *Guide to the Hydrology of Carbonate Rocks*, UNESCO, 1984, 354 pp.
16. Shourong, Shu, The coincident spectral plot method for selecting the remote sensing bands of carbonate rocks, *Carsologica Sinica*, 1, 2, 1982, pp. 158–166.
17. Powell, W.J., Copeland, C.W., and Drahovzal, J.A., Delineation of linear features and application to reservoir engineering using Apollo 9 multispectral photography, *Alabama Geological Survey Information Series*, 41, 1970, 37 pp.
18. Lattman, L.H., and Parizek, R.R., Relationship between fracture traces and the occurrence of ground water in carbonate rocks, *J. Hydrol.*, 2, 1964, pp. 73–91.
19. LaMoreaux, P.E. and Associates, Inc., Unpublished reports to the Southern Natural Gas Company, Tuscaloosa, Alabama, 1982.

## SELECTED READING

Remote-sensing techniques and the detection of karst, in *Bulletin of the Association of Engineering Geologists,* XVI, 3, 1979, pp. 383–392.

Environmental aspects of the development of Figeh Spring, Damascus, Syria, in *Proceedings of the Third Multidisciplinary Conference on Sinkholes and the Engineering and Environmental Impacts of Sinkholes and Karst,* St. Petersburg Beach, Florida, Oct. 2–4, 1989; Beck, B.F., Ed., A.A. Balkema, Rotterdam, The Netherlands, pp. 17–23.

Environmental planning for karst areas, in *Proceedings of the International Symposium and Field Seminar on Hydrogeologic Processes in Karst Terranes* (abs.), UKAM, Antalya, Turkey, 1990.

LaMoreaux, P.E., Assaad, F.A., McCarley, A.E., Eds., *Annotated Bibliography of Carbonate Rocks,* Vol. V, Karst Commission IAH, Verlag, Heinz, Heise, Hannover, Germany, Vol. 14, 1993, 425 pp.

LaMoreaux, P.E. and Newton, J.G., Catastrophic subsidence: An environmental hazard, Shelby County, Alabama, in *Environmental Geology and Water Sciences,* Vol. 8, No. 1/2, Springer-Verlag, New York, 1986, pp. 25–40.

LaMoreaux, P.E., Wilson, B.M., and Memon, B.A., Eds., Guide to the Hydrology of Carbonate Rocks, United Nations Educational, Scientific and Cultural Organization 7, place de Fontenoy, 75700 Paris, 1984, 345 pp.

# 8.4 ENVIRONMENTAL HYDROGEOLOGY OF FIGEH SPRING, DAMASCUS, SYRIA

## 8.4.1 INTRODUCTION

A hydrogeological assessment and a stress pumping test at Figeh Spring were performed to determine groundwater flow paths; recharge, storage, and discharge characteristics; and the maximum reliable yield. The project was designed to augment low-season flows from the spring and to supplement the water supply for the City of Damascus, Syria. The evaluation was directed toward installation of permanent pumping facilities at the spring (Figure 8.4.1). Work included a detailed analysis of published literature on the geology and hydrology of the area, as well as a review of data from the files of Figeh.

Before additional modification to the spring could be made, it was necessary to understand, in detail, the recharge, storage, and discharge, as well as preferential, groundwater flow patterns in the karst limestone system. Evaluation included studies of satellite imagery, sequential aerial photography, geomorphology, stratigraphy, and geologic structure (folding, faulting, and jointing). The studies were performed to determine the relationships among geologic control, karstification, preferential groundwater flow patterns, and storage characteristics of the aquifer system.

Interpretation of remote-sensed data, including satellite imagery and high- and low-level air photography, was verified by field studies. Geologic and geomorphic parameters controlling discharge and preferential flow were described. A well and spring inventory, stream flow measurements, and sequential sampling and analyses of surface water and groundwater provided water quality parameters, which were correlated with natural geologic phenomena and stress pumping from wells and springs. The final phase of work included a series of synchronized pumping tests at Figeh Spring, Side Spring, Ain Harouch, and PELA test wells (Figure 8.4.2).

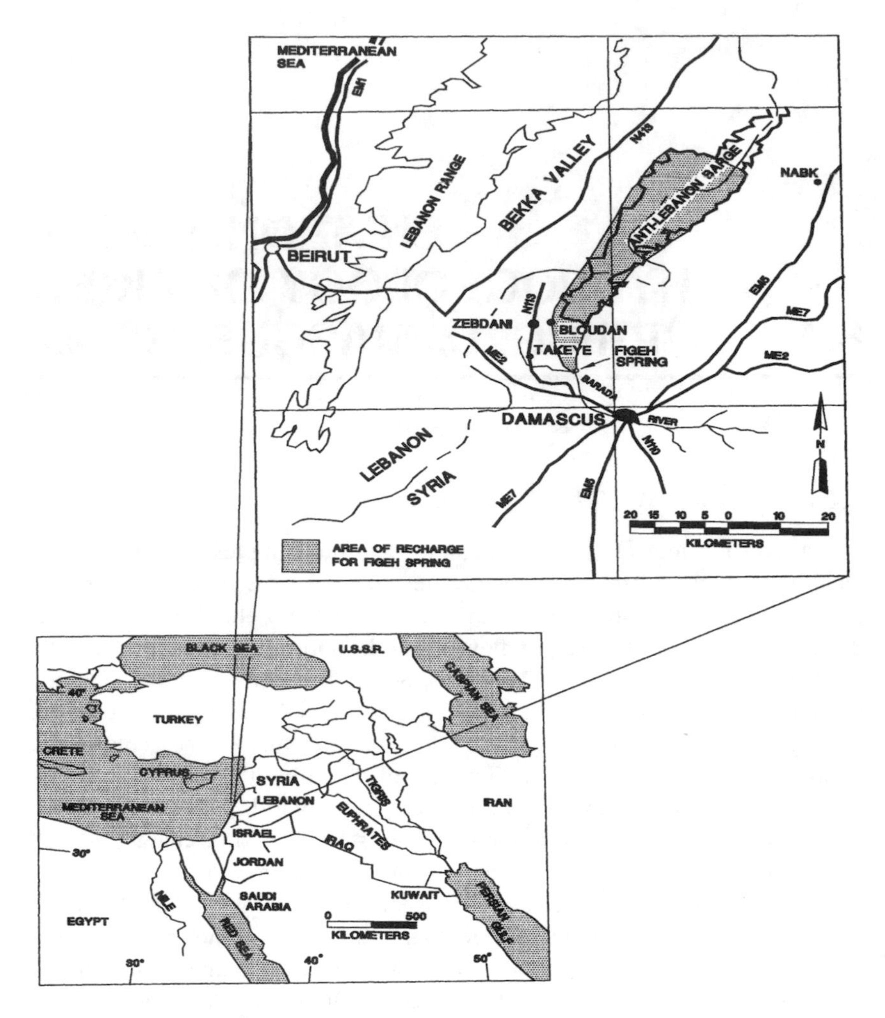


**Figure 8.4.1 Regional location map.**

## 8.4.2 GEOMORPHOLOGY

The geomorphology of the recharge area is the result of Jurassic to Recent deposition, tectonics, and vulcanism. Sporadic uplift along with compressive folding and faulting at shallow depths has resulted in a variety of surface forms and geologic structures in the Anti-Lebanon range in the area of Figeh Spring. Dominant ridges of the area are anticlines, questas, or hogbacks which have resulted from folding and/or faulting. Major wadis (valleys) follow synclinal structures, occur as strike valleys parallel to hogbacks, or are the result of Pleistocene erosion along normal faults of significant displacement. Immediately north of the Barada River near Deir Qanoun, even minor anticlines form

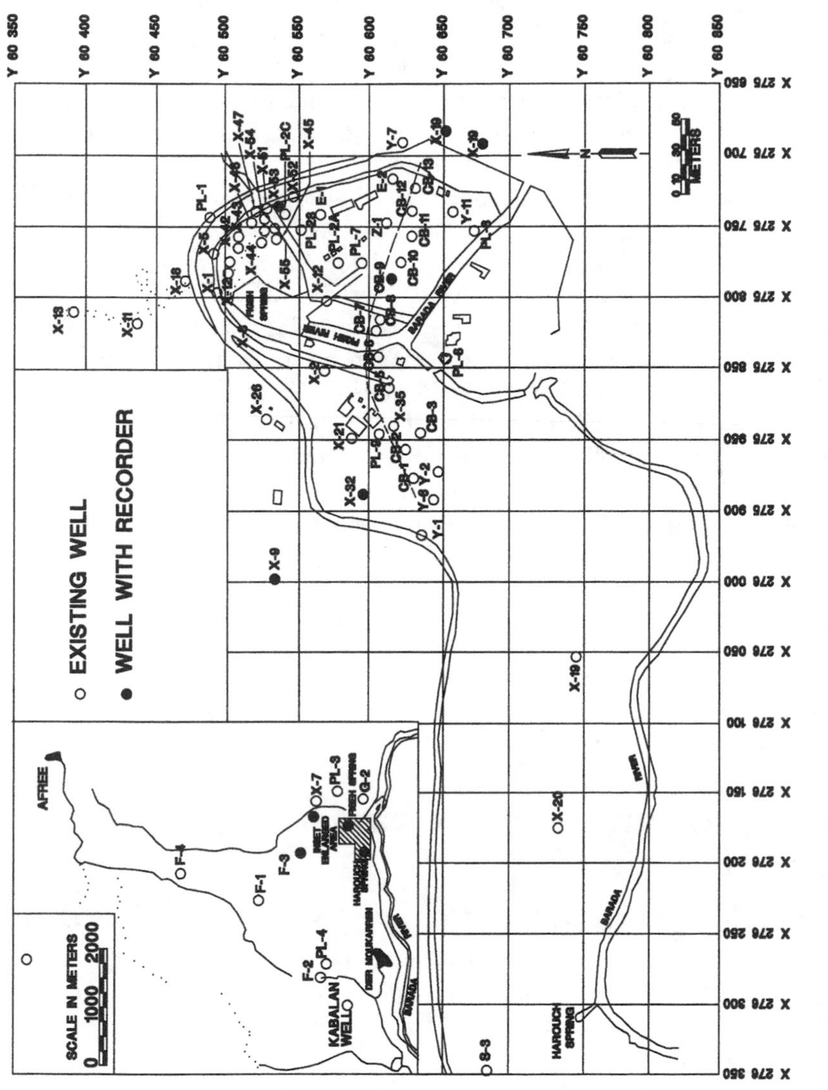


Figure 8.4.2 Ain Figeh location map — Ain Figeh wells and springs.

ridges and small wadis follow synclinal axes. Structural features in the recharge area can easily be recognized on aerial photographs and in the field.

Synclinal valleys often contain the trunk of consequent, trellis drainage. Wadis of the trellis system most frequently follow transverse fractures, which are either perpendicular to or cross fold-axes at angles of 30 to 40 degrees. The fractures can be easily observed on a drainage map of the Zebdani quadrangle and depicted on rose diagrams of the Arrsal quadrangle. In the Zebdani quadrangle, east-west trending normal faults become the loci of major wadis with consequent tributaries which are controlled by minor fractures that have prevailing north to northwest trends. Reverse faults appear to have remained tight during landscape development.

Paleokarst features had their origin along bedding planes in the Cenomanian and Turonian limestone during Paleocene and Eocene at a time of major structural movement. During Pliocene and Pleistocene, major changes in groundwater base level occurred and karstification was accentuated.

Paleokarst features have been preserved since Pleistocene and are abundant. However, of major significance to the Figeh karst system is the presence of extensive solution cavity systems and caves in the Turonian limestone and dolomite, as well as collapse features in Senonian strata which have been buried by Pleistocene alluvial deposits of the Barada River. The most significant karst feature of the area is the cave which serves as the conduit for discharge of groundwater at Ain Figeh. The cave was formed by groundwater discharge that rises subparallel to the Khadra anticlinal axis in Turonian dolomite. The cave floor is underlain by as much as 70 m of breccia which indicates that the cave has migrated upward through the dolomite by a process of roof collapse. Continuation of cave evolution, through roof collapse, will provide changes in position of the discharge point of Ain Figeh with time. For this reason any construction or development of the spring must be carefully planned and carried out with extreme caution.

## 8.4.3 GEOLOGY: STRATIGRAPHIC SEQUENCE

### 8.4.3.1 Cretaceous System

Essentially all exposed rock units in the area of recharge for Figeh Springs belong to the Cretaceous System and include important aquifers and aquicludes (Figure 8.4.3 and Figure 8.4.4). The most complete and continuous section of these Cretaceous rocks, more than 1800 m thick, is exposed between Bloudan and Wadi Hubeidi (Figures 8.4.5 and 8.4.6).

The Cretaceous system was subdivided and described by Dubertret.[1] Later work under Ponikarov[2] provided extensive information about the Cretaceous system, but resulted in only minor modification to the stratigraphy. The present study included interpretation of aerial photographs and field mapping to verify the distribution of lithostratigraphic units of the Cretaceous System (Figure 8.4.5).

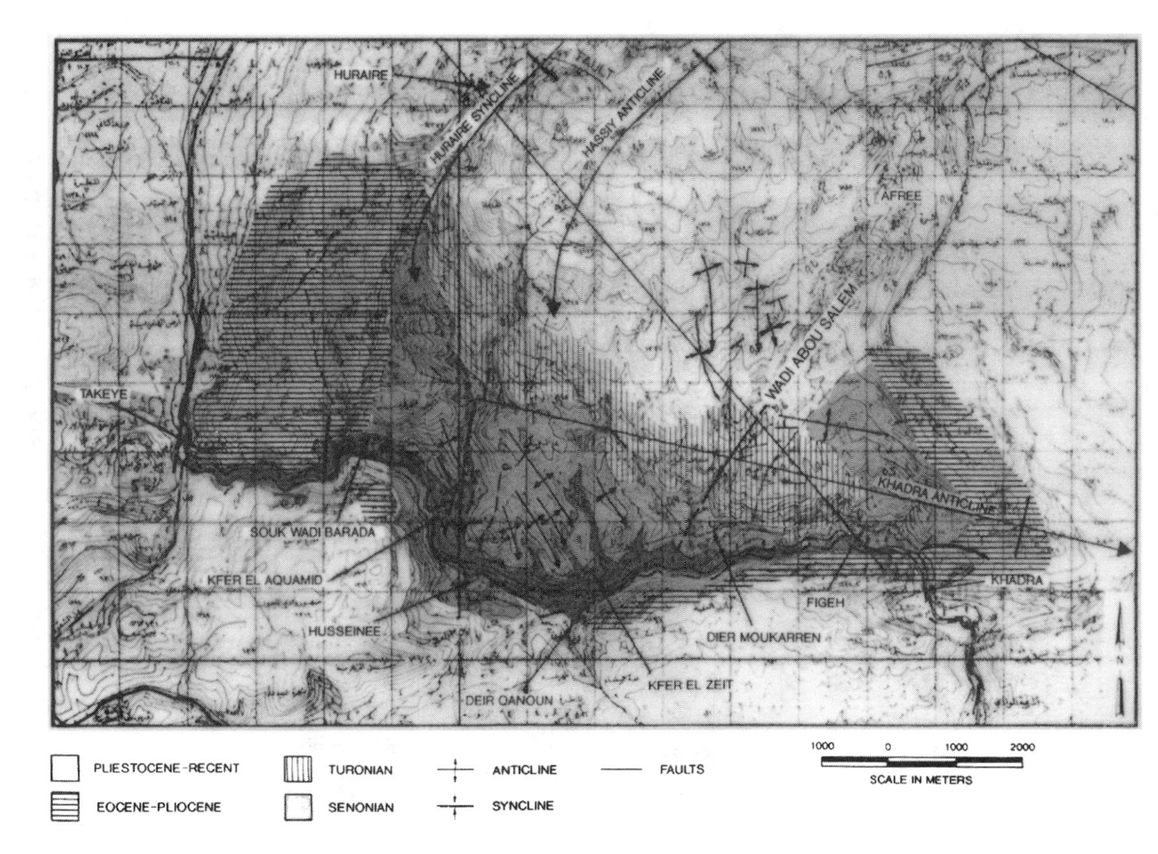


**Figure 8.4.3 Significant geologic features along the Barada River.**

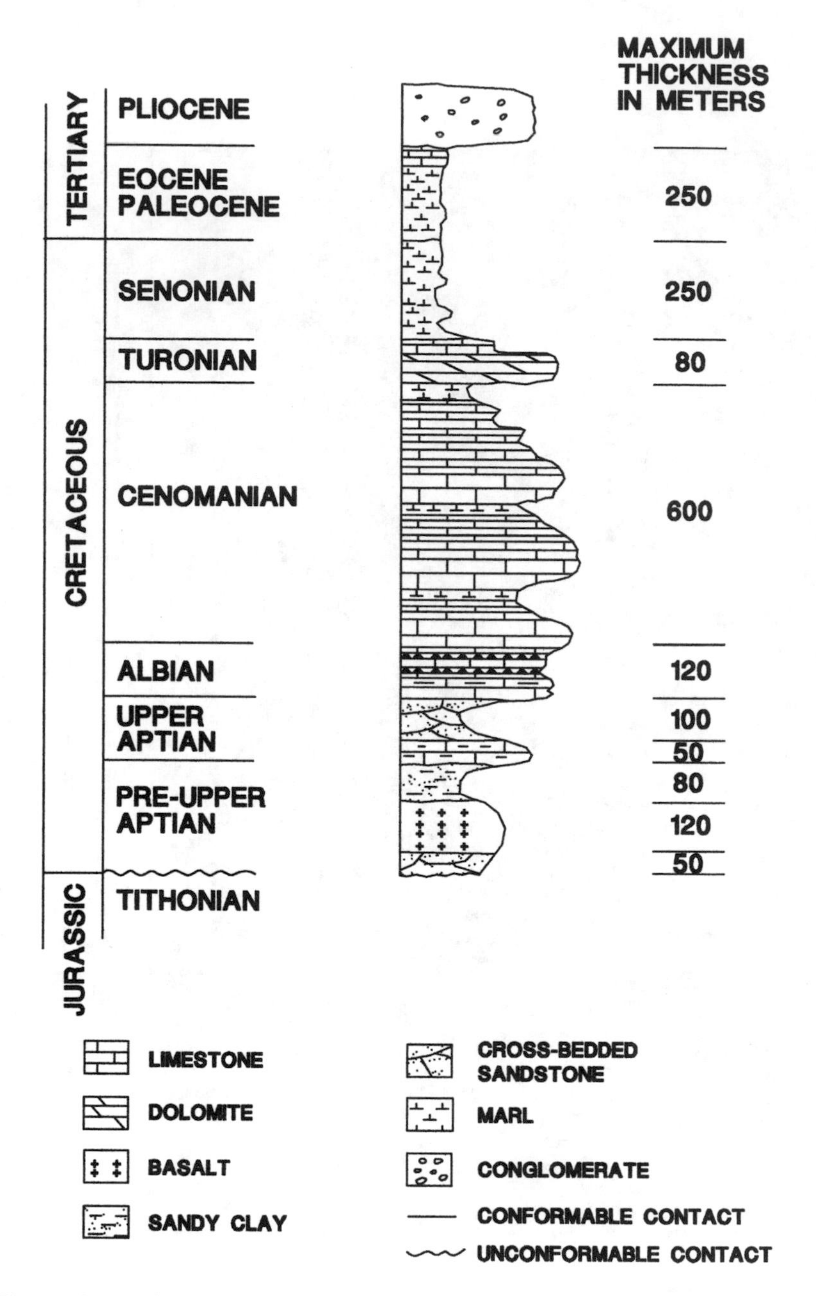


Figure 8.4.4 Geologic column of the recharge area for Figeh Spring.

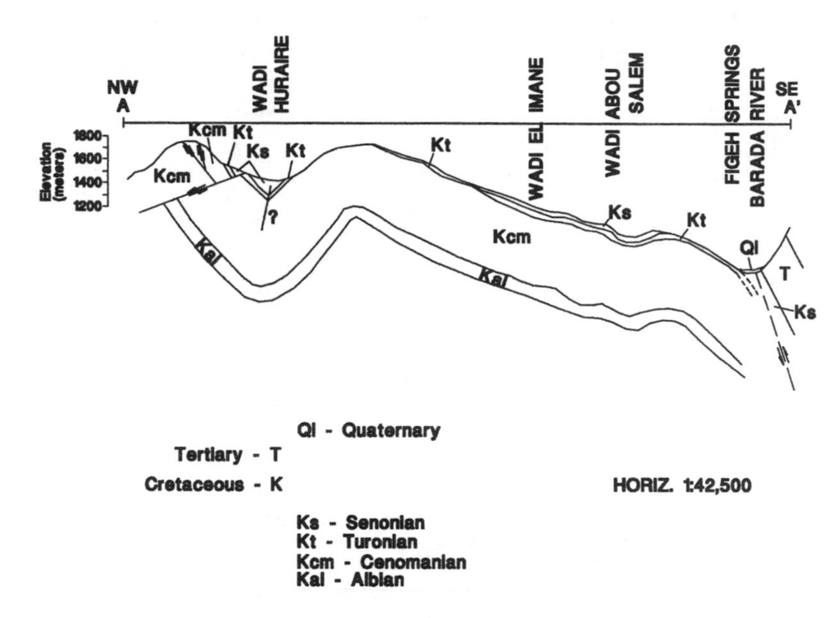


Figure 8.4.5 Geologic cross section A-A′.

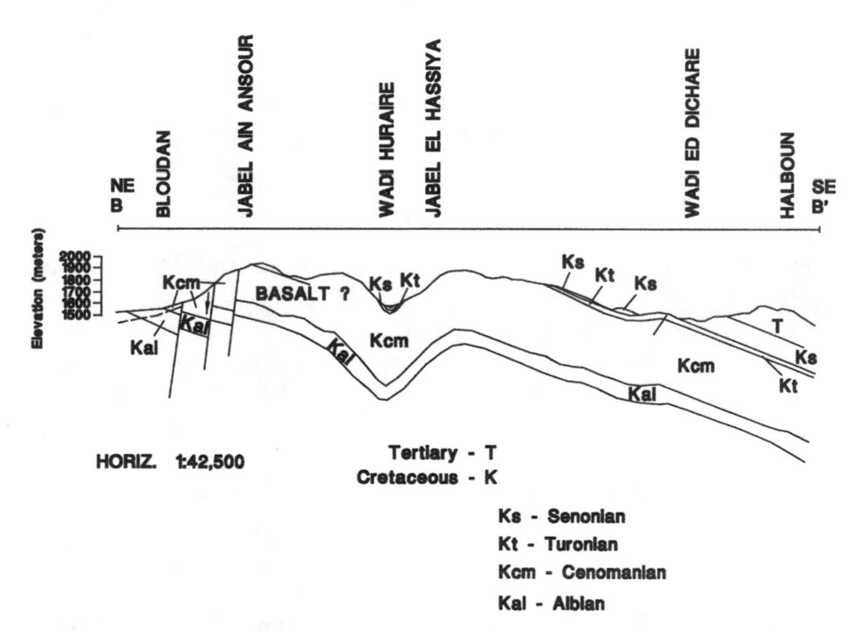


Figure 8.4.6 Geologic cross section B-B′.

Dubertret[1] distinguished the Lower and Upper Cretaceous series and divided them into stages as follows:

Lower Cretaceous Series
- Pre-Upper Aptian
- Upper Aptian Stage
- Albian Stage

Upper Cretaceous Series
- Cenomanian Stage
- Turonian Stage
- Senonian Subseries
  - Maastrichtian Stage
  - Campanian Stage
  - Santonian Stage
  - Coniacian Stage

#### *8.4.3.1.1 Lower Cretaceous Series*

##### *Pre-Upper Aptian*

Pre-upper Aptian rocks unconformably overlie rocks of Tithonian (Jurassic) age. They consist of a basal, cross-bedded, quartz sandstone with ferruginous cement, with a basal conglomerate being present locally. The sandstone is coarse- to medium-grained and attains a total thickness of about 50 m and basalt flows locally overlie the basal sandstone. Where present, the basalt may be as thick as 120 m. The upper part of the Pre-upper Aptian unit is composed of argillaceous sandstone, clays, lignites, and sandy clays. Maximum thickness of Pre-upper Aptian rocks in the Figeh recharge area is about 200 m; however, the thickness of the unit is variable, depending upon the thickness of the basalt. These rocks are not resistant to weathering and are exposed on lower slopes or in valleys. Springs may be associated with the lower sandstone. Clay beds near the top of the Pre-upper Aptian sequence act as the lower confining unit for the overlying Cretaceous aquifers and the Figeh karst system.

##### *Upper-Aptian Stage*

The upper part of the Upper Aptian strata consists of ferruginous, quartz sandstones that have a total thickness of up to 100 m. The sandstones are similar to those of the Pre-upper Aptian strata.

Rocks of the Upper Aptian stage conformably overlie Pre-upper Aptian strata and are divided into two distinct lithologic units. A basal unit consists of about 50 m of well-cemented, yellowish-white, compact limestone and marl. The limestones are light gray, fossiliferous, lumpy grainstones and contain about 50% clay. The limestone forms cliffs which are easily identified ("la muraille de Blance" of Dubertret). Solution pits, karren, and karst features are present. These limestone beds are intensely fractured; springs may occur near the base of the unit.

*Albian*

Limestones, argillaceous limestones, and marls conformably overlie Upper Aptian sandstones. Limestones have a characteristic yellowish-green color and lumpy structure. They are fossiliferous and oolitic and are interbedded with argillaceous limestones. The argillaceous limestones are thin-bedded and platey and may contain marls. The Albian strata have a total thickness of about 120 m in the recharge area. They are compact and indurated and have primary and secondary porosity. In outcrop, they are a source of recharge for water and do not appear to be aquicludes. Albian strata have not been tested by drilling or by pumping tests in the Figeh area.

### *8.4.3.1.2 Upper Cretaceous Series*

The Upper Cretaceous series includes Cenomanian, Turonian, and Senonian strata, with an aggregate thickness of about 1000 m.

*Cenomanian Stage*

Rocks of the Cenomanian age conformably overlie Albian strata and are the most extensively exposed strata in the recharge basin. Rhythmically alternating sequence of rocks characterize Cenomanian strata, and lithologies include massive compact limestone at the base; gray, fossiliferous limestone; platey, argillaceous limestone and yellowish lumpy marl; and marls at the top. Alternating sequences typically contain two or more of the above lithologic units and may include layers of dolomite limestone.

Maximum thickness of Cenomanian strata is about 600 m. The unit thins to about 400 m near the crest of the Hassiya Anticline. Closely spaced fractures are present in Cenomanian strata throughout the area of recharge. Karst features (solution pits, caves, etc.) are limited to exposures of massively bedded limestones and dolomitic limestone, which rarely exceed a thickness of 20 m. Infiltration and movement of groundwater is controlled by the density of fractures and solution openings and the relatively thin-bedded nature of the rocks. The direction of groundwater flow is in the direction of dip of the strata and along the orientation of fractures.

*Turonian Stage*

Gray, massive-bedded, unfossiliferous, dolomites, and dolomitic limestones, as much as 80 m thick, are present along the flanks of the El Hassiya Anticline and the Dome of Figeh. The unit is absent in all other parts of the recharge area. The rocks are fine- to medium-grained and massively bedded. The Turonian dolomites and dolomitic limestones contain karst features in the vicinity of Figeh Spring including the cave systems at the outlet of the spring. This is the only unit exposed in the recharge area of sufficient thickness, competence, and aerial extent to support extensive cavern systems in the subsurface.

Disagreement exists about the age of the massively bedded dolomitic limestone. Dubertret[1] and Ponikarov[2] describe the unit as upper-Cenomanian. However, others[3-5] place the unit in the Turonian stage. This report has complied with this more recent classification.

Overlying the dolomites is a light gray to white, thin-bedded, shaley limestone containing some marl. These beds are aquicludes and impede the flow of groundwater.

#### *Senonian Strata Subseries*

Rocks of the Senonian subseries include Coniacian, Santonian, Campanian, and Maastrichtian stages and constitute the lower portion of a thick sequence of confining beds which overlie the principal Cenomanian and Turonian aquifers of the Figeh aquifer system. The Senonian consists predominantly of marl and has a total thickness of 225 to 250 m. Senonian strata are conformably overlain by about 200 m of lithologically similar marls of Paleocene and Eocene age.

Geologic mapping of Senonian strata is facilitated by the presence of dolomitic sandstone in rocks of the Coniacian Stage, by gray-brown flint beds in rocks of the Campanian Stage, and by massive chalk beds in the Maasthrichtian strata.

Figeh Spring exists near the downstream exposure of the Cenomanian-Turonian aquifer system. The contact between these systems is penetrated by cores for test holes in the Figeh Spring area and has been observed in the Spring during scuba diving exercises. It occurs about 755 m msl or 30 m below land surface in Figeh cave system. Senonian strata in this locality are locally fractured and karstified and act as a semi-confining series of beds that restrict lateral and vertical movement of groundwater in the Figeh system.

### *8.4.3.1.3 Geologic Structure*

The recharge area for Figeh Spring includes a northeast trending part of the Anti-Lebanon Range that is bounded on the northwest by the Bekka Valley rift-system and by the Barada River to the south; the southeastern boundary is marked by the limits of large Tertiary basins (Figure 8.4.7).

The dominant structure of this portion of the Anti-Lebanon Range is that of a southeastward dipping homocline. The homocline originates along faults of the Bekka Valley or its branches, such as Serghaya Fault system. The homoclinal form of the Anti-Lebanon Range is disrupted by anticlines and

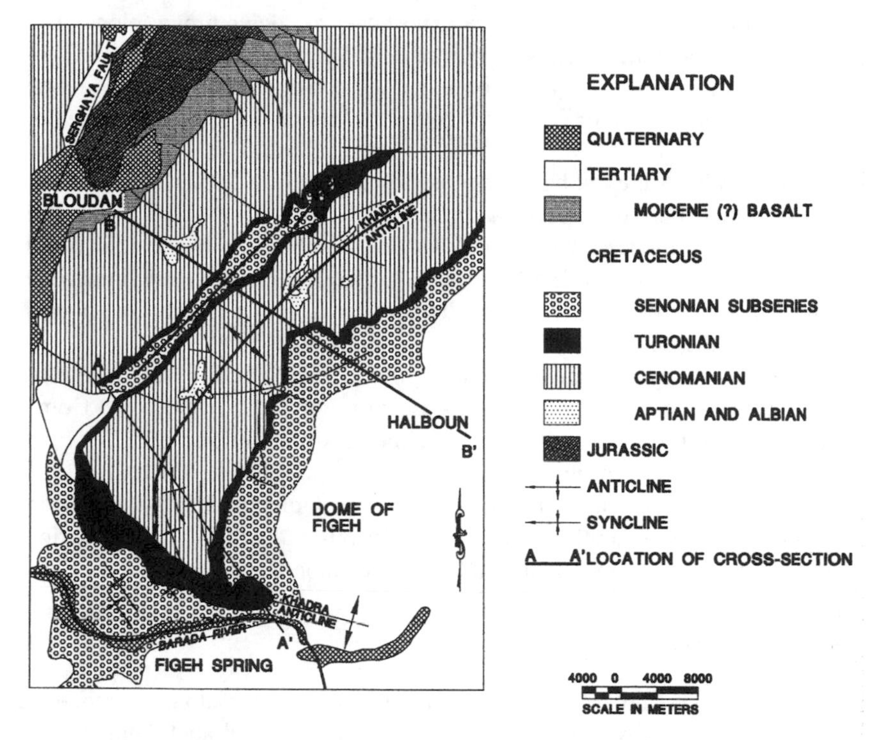


**Figure 8.4.7 Geologic structural features of the Figeh Spring system.**

synclines that trend subparallel to the regional strike of the homocline (northeast). The most complex folding of the recharge area occurs in the area immediately north of the Barada River and Figeh Spring and includes about one-third of the recharge area (approximately 200 km$^2$ of the total 700 km$^2$ in the recharge area).

Dominant folds of the area include the Hurarie Syncline, the Hassiya Anticline, the Khadra Anticline, and the Dome of Figeh (Figure 8.4.7). The Hurarie Syncline and Hassiya Anticline are northeast-trending, subparallel folds, which have been intermittently active since late Cretaceous (Cenomanian). The Hurarie Syncline and Hassiya Anticline are dominant structural features for a distance of about 20 kilometers north of the Barada River. The syncline is nearly symmetrical with a dip of 45 to 60 degrees on each limb. However, the anticline is asymmetrical with a dip of 45 to 60 degrees on the northwest limb. The southeast limb has a dip of 15 to 20 degrees (Figures 8.4.5 and 8.4.6). Both folds plunge gently toward the southwest, except near their southern limits where plunge steepens and cross-folding causes the axes to trend almost due south.

The Hurarie Syncline is terminated by faulting at its southern end near the Village of Hurarie. Cenomanian, Turonian, and Senonian strata, exposed in the syncline, terminate abruptly at the fault. South of the fault, approximately

300 m of Pliocene conglomerate are exposed in the downthrown block. The Pliocene conglomerate extends approximately 5 km to the Barada River.

Minor folds of low amplitude occur on the flanks of the Hurarie Syncline and the Hassiya Anticline. The minor folds are seldom coaxial with the major folds.

The Khadra Anticline has a total exposed length of 12 km. The axis of the fold plunges 11 degrees in a direction of north 78 degrees west. The anticline is asymmetrical with a near vertical southern limb, the fold being well exposed north of Ain Khadra (Figure 8.4.7). The Khadra Anticline is one of the most recent folds (Pliocene-Pleistocene) in the recharge area and is transverse to the dominant folds of the Anti-Lebanon Range.

Intersection of the northwestern trending Khadra Anticline with the northeastern trending structures of the Anti-Lebanon Range has created the Dome of Figeh, has refolded the Hassiya Anticlines and the Hurarie Syncline, and has created southeastern plunging minor folds exposed west of the Dome.

The path of the Barada River is controlled by faults parallel to the Khadra Anticlinal axis. At the southeast end of the fold axis faults controlled deposition in the Neogene basin, marginal to the Anti-Lebanon Range.

#### *8.4.3.1.4 Faults and Fractures*

The northwest boundary of the recharge basin occurs along a branch of the major rift system which forms the Bekka Valley. The branch fault system is well exposed near Serghaya and will be referred to as the Serghaya fault zone. The zone itself is 1 to 3 km wide; however, intense fracturing of adjacent Jurassic and Cretaceous rocks has occurred. The Serghaya fault zone behaves as a hinge fault with normal, left-lateral movement. Stratigraphic throw increases to the northeast. Maximum throw in the Zebdani quadrangle is over 1,000 m.

The structural configuration strongly suggests the presence of major bounding faults hidden beneath sediments of Paleogene and Neogene Age along the southeastern and southern margins of the recharge area. The rectangular path of the Barada River provides strong indication of fault control along the southern border of the recharge area. The structural control of the Barada has been verified south of Zebdani, near Takeye, and near Souk Wadi Barada. The east-west segment of the Barada Valley suggests control by faults which also serve as the boundary of the large Neogene basin south of Figeh Spring.

The southeastern border of the recharge area is coincident with a major lineament (Jarajir lineament) which is subparallel to the Serghaya and Bekka rift faults. The Jarajir lineament extends more than 100 km northeast from Halboun. The Jarajir lineament is presumed to be the surface expression of a major fault system at depth.

The Jarajir lineament serves as the boundary between the Cretaceous rocks, exposed in the recharge area for Figeh Spring, and the large Tertiary-

Quaternary basin (13 by 50 km) in which Assal el Ward is located. Northeast of the Assal el Ward basin the Jarajir lineament appears to bifurcate. One branch follows the central axis of the Tertiary-Quaternary basin in the vicinity of Jarajir and Quara. The other branch follows the western margin of the basin.

Within the recharge area, faults and joints are numerous, and no exposure of rock is devoid of fractures. The major faults in the southern part of the recharge area strike northwest, almost perpendicular to fold axes, and consist of both normal and reverse faults, some of which have a right-lateral component of movement. North, northeast, and east-west trending normal faults may have significant displacement but are not as numerous as the northwest-trending faults.

Orientation of fractures, joints, and faults measured in the field (in the northern part of the Zebdani Quadrangle) between Bloudan and Figeh are illustrated on a rose diagram in Figure 8.4.8. In order of decreasing importance, the dominant fracture orientations are nearly perpendicular to fold-axes (23.7% of the fractures), cross the fold-axes at angles of about 40 degrees (18.6% of fractures), or are almost parallel to fold-axes (13.6% of fractures). The fracture systems are open and, in conjunction with the southeastward stratigraphic dip, can readily transport groundwater toward Figeh Spring.

In the central part of the recharge area (northern part of the Assal El Ward Quadrangle), dominant fractures are oriented more nearly in a north-south direction. Groundwater flow in this area is controlled by the southeastern dip of the strata; the nearly north-south orientation of dominant, open fractures; and subsurface faults associated with the Jarajir lineament, all of which provide components of flow toward Figeh Spring.

North of the recharge area (in the NABK Quadrangle), stratigraphic dip is toward the southeast. The trend of the most abundant fractures is northwestward (Figure 8.4.8). Groundwater in the area moves along the bedding planes and fractures in a southeast direction toward Jarajir, Nebek, and Quara. The area cannot be considered as a source of recharge to Figeh Spring.

The northern two-thirds of the recharge area has a more simple homoclinal form than that near Figeh Spring. Strata undulate, but north to northeast strikes prevail; and dip of the strata becomes progressively more gentle in the northern part of the recharge area. Minor anticlines and synclines occur but have small amplitudes and appear to have little effect in diverting the flow of groundwater. The dominant trend of fractures in the central portion of the recharge area is approximately north-south; whereas the orientation of the most abundant fractures rotates progressively toward the northwest in the northern part of the recharge area.

Straight segments of wadis have characteristically developed along faults or joints. Fracture patterns can be accurately interpreted from aerial photographs or from drainage maps. Fracture-controlled surface morphology is characteristic of the northern two-thirds of the recharge area. Groundwater infiltration and groundwater flow in the central part of the recharge basin is controlled by the south to southeastern stratigraphic dip, southward-dipping,

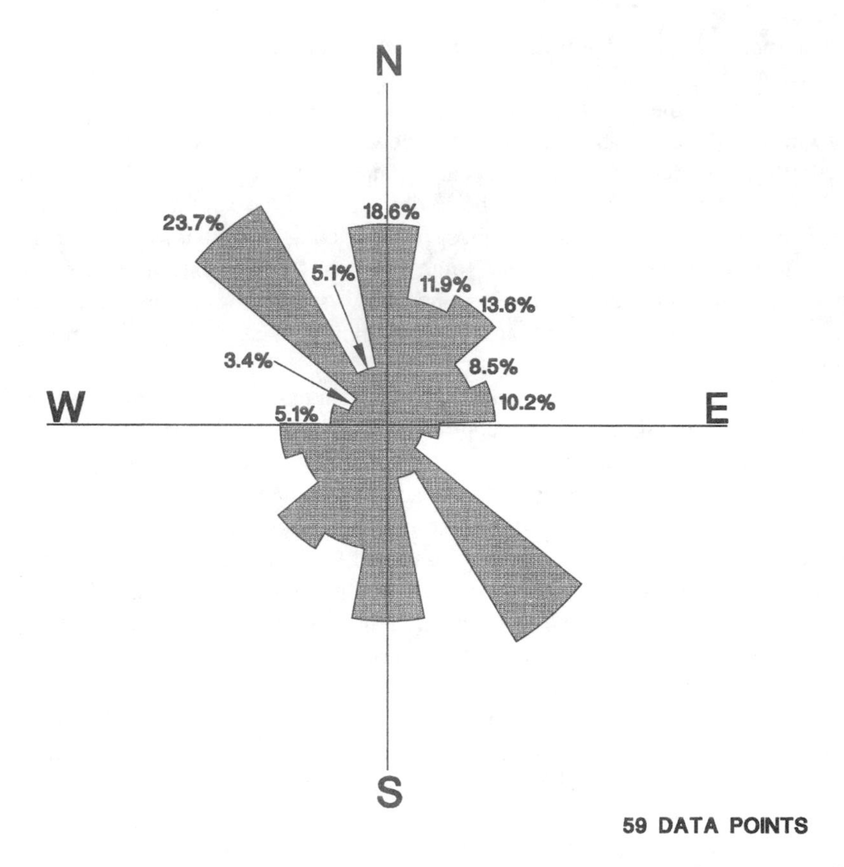


**Figure 8.4.8 Rose Diagram of faults and joints between Bloudan and Figeh.**

normal faults in valleys, and north-south trending, minor fracture systems. The Cenomanian and Turonian strata are intensely fractured, thus causing these beds to be porous and permeable. The significant directional component of flow is toward Figeh Spring.

## 8.4.4 HYDROGEOLOGY OF THE FIGEH AREA: GEOLOGIC STRUCTURAL SETTING AND KARST DEVELOPMENT

For karstification to develop and for solution action to progress to form caverns, it is necessary that:

1. water rich in carbon dioxide be available to recharge the system,
2. sufficient permeability (in the form of fractures or bedding plane or both) be available for water to move in the rocks, and
3. water be able to discharge from the system.

A fourth criterion, which expedites the process, is a steady source of recharge water, such as from snow melt, or an overlying blanket of sediments. Under watertable conditions, the zone of higher permeability tends to develop in the zone of greatest circulation and solution, which is commonly at or just below the water table. Topography and position of carbonates below ground are important; carbonates that allow at least moderate circulation and are entrenched by perennial streams tend to develop solution openings and increase circulation. Some circulation of water and, to a certain extent, solution cavities may locally occur at a depth of several hundred meters below the level of the major stream of a region provided that a good discharge system and well-established gradient are available and the water is chemically aggressive. This is true in the case of the Barada River Valley which flowed at a lower elevation during the late Pleistocene.

The relation of the recharge to the discharge area in a karst region determines, to a large extent, the patterns of the lateral solution channels or openings. The size and frequence of these channels will depend on many factors involving the conditions in the recharge area, the volume of water that enters the recharge area, the solubility of the karst rocks, and the rate at which the base level is lowered as perennial streams entrench.

Where the discharge area is along a more or less straight line, such as the Barada River Valley, the lateral solution channels tend to be more or less parallel to each other and at right angles to the line of discharge. Discharge is therefore in a line of springs like Figeh, Ain Harouch, and Ain Kadra into the Barada. The direction of movement of the water between the recharge areas and discharge areas is affected by faulting, folding, jointing, and other geologic structures. Cave passages may occur at more than one level.

Water-level behavior in space and time is a primary consideration for interpreting karst hydrology. The position of the water table is important because:

1. The water table defines generally the zones of greatest circulation and solution.
2. The configuration of the water table aids in identifying the general direction of flow, the hydraulic gradient, and areas of recharge and discharge.
3. Information about the water table provides general information about permeability of the aquifer system.
4. The position of the water table locally indicates the extent to which caverns are filled with air or water.

Carbonate aquifers often have seasonal variation in water levels of 20 to 80 meters, and the effects of recharge in the recharge area from isolated rains are nearly instantaneous. Large seasonal variation in water level beneath the carbonate uplands may result in the movement of groundwater to one basin in dry weather and to another basin in wet weather. A controlling factor in the range of seasonal groundwater levels is the great infiltration capacity of karst terranes. Large volumes of water from heavy storms infiltrate into the air-filled caverns of karst lands, reaching the zone of saturation quickly and causing the water table to rise rapidly. From a single storm or snowmelt event, the rise in the water table in relatively impermeable parts of the saturated carbonate rocks can be as much as 10 or more meters, whereas the rise of the water table in permeable zones, of the same carbonate rocks, can be less than a meter.

Big springs, such as Ain Figeh, the third largest spring in the world, rather than small springs and diffuse seepage, are the general rule in karst regions, and big springs often emerge from underground streams or caves. Most groundwater flow, however, occurs in large solution openings near the top of the saturated zone, which carry much of the groundwater to the springs. Zones of solution openings may tend to branch upgradient, along fractures, which in arterial fashion represent the more permeable, upper parts of the saturated zone. The water table is depressed along the arterial system so that groundwater discharge meets the surface stream almost at grade.

In karst areas the distribution of permeability beneath streams causes them to lose or gain water, depending on the position of the water table with reference to stream level. A surface stream may lose water where bedrock in the losing stretch is very permeable and the water table is low, and it will gain water where the water table is above stream level. For example, the Barada River gains water where it crosses an area immediately south of the Wadi Hurarie Syncline and loses water in the reach west of Deir Moukarren.

An understanding of the geologic history and paleokarstification is critical in the Figeh karst system in the vicinity of Ain Figeh where the fractured limestone and dolomites dip toward the Barada River Valley. Karstification of these rocks began during the Pleistocene along fractures and bedding of the more pure limestone. These beds were subsequently folded and fractured during tectonic uplift in the area. During the Pleistocene, the Barada River near Ain Khadra became impounded, probably by a landslide, and the base level of the river was raised. Erosion subsequently has taken place, dropping the base level to the present level of the Barada River. At Figeh alluvial deposits extend as much as 45 m below present land surface (Figure 8.4.9). These beds of sand, gravel, and boulders fill a buried Paleokarst feature at Ain Figeh and comprise an important segment of the Karst aquifer storage system.

Excellent plan views and cross sections of three major solution features in the immediate vicinity of Figeh Spring (Gallery Cave, Cheikh Cave, and the Karst Cavity) have been mapped and surveyed in detail.[5] Each cave is subparallel to the bedding of Turonian rocks; however, each contains small openings associated with fractures. The lower elevation of each cave is 835 meters above mean sea level and corresponds to elevations of late Pleistocene to recent river terraces. The presently active karst system of Figeh, Gallery Cave, and Cheikh Cave were originally paleokarst features of late Pleistocene to Recent age.

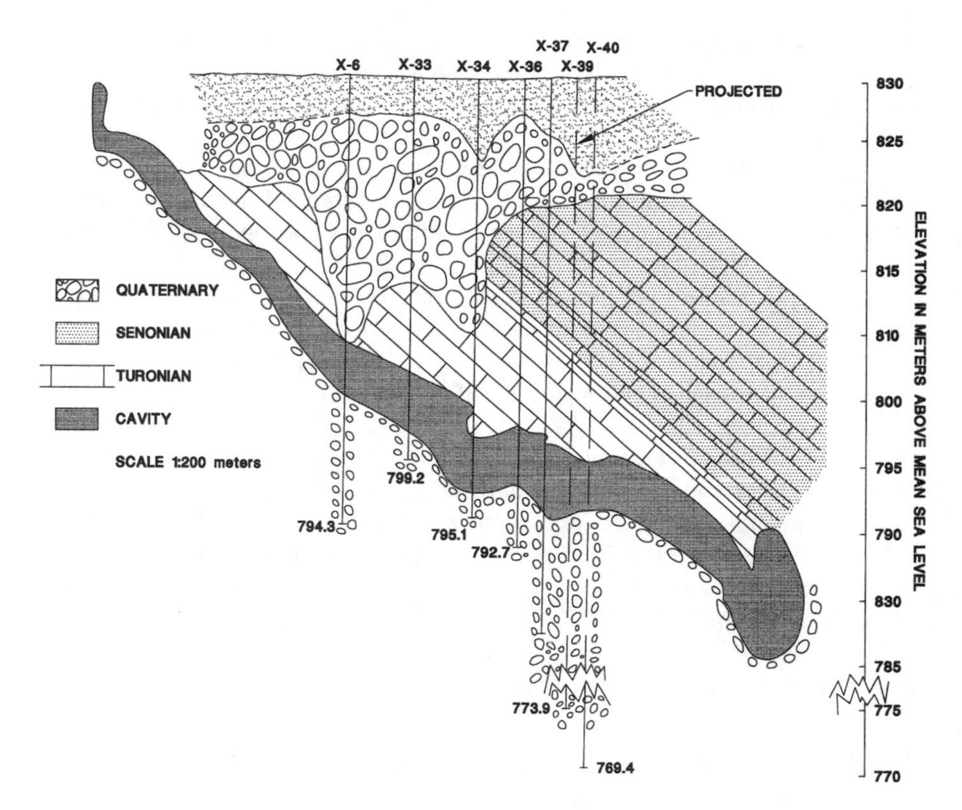


**Figure 8.4.9 Geologic cross section Ain Figeh.**

Gallery Cave and Cheikh Cave differ from the Figeh Karst Cavity, as they occur at higher elevations than the Figeh Karst Cavity, and because the floors of these two caves are underlain by solid bedrock, whereas the lower part of the Figeh Karst Cavity is filled with breccia.

The developmental history of the Figeh Karst Cavity is more complex than for Gallery and Cheikh Caves as two different, but interrelated, karst features are associated with Figeh Karst Cavity. The primary feature is the Karst Cavity; however, above the Karst Cavity is a depression in the top of the bedrock. The depression is filled with breccia overlain by soil and/or alluvial deposits, as demonstrated in the logs for test wells X-6, X-33, X-34, and X-8 (Figure 8.4.9).

The Karst Cavity is overlain by Turonian strata. Although the overlying beds are fractured, they have not been significantly displaced from their stratigraphic position except where overlying depressions occur. However, drill holes, that penetrated the bottom of Karst Cavity, tapped an underlying karst breccia. For example, well X-40 penetrated 27 m of karst breccia beneath the cavity.

A theory of origin for the Figeh Karst Cavity incorporates the upward movement of groundwater along fractures connecting Cenomanian and Turonian strata. Groundwater in Turonian strata is under artesian pressure, and has eroded, by dissolution and mechanical action, a cave system subparallel to bedding of Turonian strata. Figure 8.4.14 (see later) indicates a progressive evolution of the cave system in which the physical cavity migrates upward through the Turonian rocks due to intermittent collapse of the cave roof and back filling the cave floor with karst breccia. The stratigraphic interval at which the cave system originated is unknown as drilling has not penetrated the base of the breccia in the karst.

Filled depressions are present in the immediate vicinity of Figeh Spring that are paleokarst features and are connected to and associated with development of the main Figeh cavity. Wells X-6, X-33, and X-34 (Figure 8.4.9) penetrated the margin of one depression filled with karst breccia and connected to a cavity penetrated by well X-6. The deepest part of this depression known at present has been penetrated in test well X-8, where the breccia extended to a depth of 30 m, and is overlain by alluvial sediment representing an ancient, higher stage of the Barada River. A similar and apparently large alluvial-filled depression was penetrated in well E-3. Karst breccia and alluvial fill is also present in the surrounding drill holes X-12, E-1, X-31, and X-10. These depressions are not expressed at the surface and were probably created and filled during Pleistocene time. The karst breccia in the depressions serve as zones of groundwater movement and storage in the present Figeh system.

## 8.4.5 RECHARGE-STORAGE DISCHARGE OF GROUNDWATER

Photogeologic and surface geologic methods were used to identify boundaries of the recharge area (Figure 8.4.7). For modeling and quantitative analysis, it was necessary to have a thorough understanding of recharge, storage, and discharge in the groundwater system. Criteria for identifying the recharge area were as follows:

1. The contact of the Turonian rocks with overlying confining beds or with faults of significant displacement was used to identify the southern and southeastern limit of the recharge area.
2. The western and northwestern limit was defined by the surface-water drainage divide in the Anti-Lebanon mountains or by stratigraphic contact with underlying confining beds, where structural control allows groundwater to cross surface-water divides.

The recharge area for Figeh has boundaries influenced by location of surface-water divides, geologic structures (fractures and folds), and stratigraphic relationships. Results from these studies were confirmed by field work which included geologic structural analysis.

The recharge-storage zone within which groundwater may flow toward Figeh Spring is complex and varies depending on the porosity, permeability, structure, and season of the year.

Groundwater in the system is typically unconfined at higher elevations and becomes confined or semi-confined near the points of discharge along the Barada River, such as at Figeh Spring and at Ain Harouch. The confining beds consist of upper Cretaceous to Eocene marls that overlie Turonian strata.

### 8.4.5.1 Groundwater Movement in the Recharge Area

Groundwater movement in the portion of the Anti-Lebanon Range that serves as the recharge area for Figeh Spring is structurally and stratigraphically controlled. The homoclinal form of the Anti-Lebanon Range allows general groundwater movement in a southeastern direction by percolation along bedding planes in the thinly bedded, Cenomanian limestones. Secondary porosity is provided by open fractures, while allowing vertical movement of groundwater. The orientation of dominant fracture systems and the dip of strata (in the central and northern parts of the recharge area) provide a directional component of groundwater flow toward Figeh Spring. The southwest-plunging Hurarie Syncline, Hassiya Anticline, and associated fracture systems control the direction of groundwater movement in the southern one-third of the recharge area.

Groundwater in the Hurarie Syncline does not discharge at Figeh Spring. A portion of this water is confined beneath Senonian strata and should be available by drilling wells northeast of the faulted terminus of the syncline near the village of Hurarie. A portion of groundwater moves toward the south and southeast to discharge as springs along the Barada River west of Figeh Spring. These springs include Ain Habil, springs near Souk Wadi Barada, and Kfer el Aquamid, as well as Ain Harouch. Some groundwater underflows or discharges to the Barada River.

Groundwater on the eastern flank of the Hassiya Anticline flows in a southerly direction and becomes confined beneath Senonian strata. The groundwater moves around the eastern side of the Dome of Figeh and rises to the surface by upward flow parallel with the axis of the Khadra Anticline to discharge at Ain Figeh. Faults beneath the Barada Valley appear to restrict the amount of water that underflows the Barada River.

Groundwater and surface water in the northern part of the Anti-Lebanon Range flow either toward the Bekka Valley or toward the Tertiary basin near Jarajir and Quara.

Movement of groundwater along the border between the recharge area and the Assal el Ward basin is less certain owing to limited information. The Assal el Ward basin is a large Tertiary-Quaternary, synclinal basin. Faults associated with the Jarajir lineament may restrict groundwater movement from the recharge area into the basin. Alternatively, groundwater may flow across the fault zone, be confined beneath Senonian strata in the basin, and be directed toward Figeh Spring because of the southwestern plunge of the syncline.

### 8.4.5.2 Recharge–Discharge Relationships

Climatological factors effecting discharge include:

1. Rainfall, evaporation, and transpiration;
2. Discharge into or under the Barada River and discharge or pumping from wells and springs;
3. Seepage along the southeastern boundary of the basin due to the stratigraphic dip and the trend of fracture;
4. Leakage of groundwater through fractures crossing the northwestern boundary of the recharge area; and
5. Losses along the southern border of the system as spring flow.

Over historical times, beginning with the Romans, Figeh Spring experienced a variety of different attempts to capture water. Evidence of this early development remain in the opening to the cavity of Figeh Spring and the aqueducts leading south toward the city of Damascus. The discharge from Figeh Spring and the water obtained from the Pilot Development Project is mainly from the karst system. Water discharging from Side Spring has two sources — the Figeh Karst System and the alluvial deposit in an old stream channel that is recharged from the Barada River. Ain Harouch and Deir Moukarren represent discharge points from the karst system. Shallow large-diameter wells developed water from alluvial deposits in an old meander of the Barada River and moved it toward the sump at Side Spring.

### 8.4.5.3 Climate — Evaporation and Transpiration

The climate of most of the recharge area of Figeh is classified as “Mediterranean” by the widely recognized Koppen climatic classification system. The Mediterranean climate is one of dry, hot summers and wet, cool winters. Part of the recharge area rises over 2500 m above mean sea level, almost certainly causing average monthly winter temperatures to be below freezing in the highest elevations.

The normal vegetative response to the Mediterranean climate ranges from grassland to forest; however, in the Figeh recharge area the response is xerophytic vegetation. Thornbush and cacti of several varieties proliferate as a sparse ground cover. The area is characterized by rubbly bedrock, thin top soil, and man-induced soil and vegetation changes caused by overgrazing, agriculture, and gathering of firewood. Rapid infiltration of moisture, combined with the long, dry summers, results in low soil moisture retention.

Therefore, natural vegetation plays a minimal role in the amount of moisture available for recharge in the recharge area.

There are three operative weather stations in the recharge area, all between elevations of 1,540 and 1,560 m above mean sea level. The Bloudan and Hurarie stations appear to be well maintained and the data recorded with minimal transferral error. Data for the Afre station are not reliable and were not used. Table 8.4.1 shows the mean monthly precipitation for Bloudan and Hurarie for hydrologic years 1971 through 1978. Figure 8.4.10 shows the mean monthly temperature and humidity for Bloudan and Hurarie, respectively, for hydrologic years 1971 through 1978.

Moisture generally decreases eastward across the recharge area for any given elevation. This is a consequence of increasing distance from the principal source of water vapor, the Mediterranean Sea, and the increasing effectiveness of a rainshadow effect as the moisture is carried eastward across the Anti-Lebanons by winter storms. This decrease in precipitation from west to east across the recharge area appears to be approximately 10%. The precipitation decreases dramatically in the Syrian Desert to the east of the Anti-Lebanons. The higher elevations should receive more precipitation than lower elevations from a combined effect of orographic causes and cyclonic storms. In addition, evaporation should be less at higher elevations in response to lower temperatures and consequently higher relative humidities. This should result in the availability of more net moisture for recharge at higher elevations than at lower elevations. There appears to be little north-south variation in precipitation in the recharge area at any given level.

Humidity generally increases with a decrease in temperature in the area. Figure 8.4.10 shows the relationship of relative humidity and temperature for the hydrologic years 1971 through 1978.

The hydrologic year utilized by Figeh is from September 1 to August 31. Precipitation occurs predominantly during the late fall, winter, and early spring. Average annual precipitation at Bloudan and Hurarie for September 1971, through August 1978, by hydrologic year, was 611.10 mm and 632.08 mm, respectively, with the wettest month of January having 140.09 and 155.46 mm, respectively. Summer months of June, July, and August are very dry, averaging less than 1.0 mm. Figures 8.4.11 and 8.4.12 generalize the data from Bloudan and Hurarie, respectively, into climographs to illustrate the months of moisture surplus and deficit for a hydrologic year. The stations show a moisture surplus (for potential recharge) from November through February and a deficit for the remainder of the year.

Significant is the fact that over 88% of the Figeh recharge area of approximately 735 km is higher than 1,500 m above sea level and, therefore, higher in elevation than the weather stations of Bloudan and Hurarie and Afre (Table 8.4.2). Seventy-six percent of the total area is higher than 1,900 m above mean sea level.

**Table 8.4.1 Mean Monthly Precipitation for Bloudan (A) and Hurarie (B), September 1971–August 1978, Hydrologic Years**

| Month | S | O | N | D | J | F | M | A | M | J | J | A | Annual average |
|---|---|---|---|---|---|---|---|---|---|---|---|---|---|
| **A. Bloudan (1560 m above msl)** | | | | | | | | | | | | | |
| Rainfall (mm) | 0.20 | 13.43 | 68.80 | 90.42 | 94.09 | 71.13 | 78.27 | 47.41 | 10.54 | 0.96 | — | — | 475.25 |
| Monthly percent | 0.04 | 2.83 | 14.48 | 19.03 | 19.80 | 14.97 | 16.47 | 9.98 | 2.22 | 0.20 | — | — | 100.00 |
| Snowfall equivalent (mm)[a] | — | — | 4.57 | 17.64 | 46.00 | 28.64 | 36.00 | 3.00 | — | — | — | — | 135.85 |
| Total precipitation | 0.20 | 13.43 | 73.37 | 108.06 | 140.09 | 99.77 | 114.27 | 50.41 | 10.54 | 0.96 | — | — | 611.10 |
| Percent of precipitation as snowfall | — | — | 6.23 | 16.32 | 32.84 | 28.71 | 31.50 | 5.95 | — | — | — | — | 22.23 |
| **B. Hurarie (1540 m above msl)** | | | | | | | | | | | | | |
| Rainfall (mm) | 1.24 | 8.70 | 73.67 | 89.24 | 88.96 | 82.11 | 69.74 | 45.73 | 8.11 | 0.61 | — | — | 468.11 |
| Monthly percent | 0.26 | 1.86 | 15.74 | 19.06 | 19.00 | 17.54 | 14.90 | 9.77 | 1.73 | 0.13 | — | — | 100.00 |
| Snowfall equivalent | — | — | 11.57 | 9.21 | 66.50 | 37.00 | 37.90 | 1.79 | — | — | — | — | 163.97 |
| Total precipitation | 1.24 | 8.70 | 85.24 | 98.45 | 155.46 | 119.11 | 107.64 | 47.52 | 8.11 | 0.61 | — | — | 632.08 |
| Percent of precipitation as snowfall | — | — | 13.57 | 9.36 | 42.78 | 31.06 | 35.21 | 3.77 | — | — | — | — | 25.94 |

[a] Converted from centimeters of snow to millimeters of meltwater (based on standard conversion rate: 10 to 1).

Tables in this chapter taken from *Env. Geol. Water Sci.*, 13(2), 77–127.

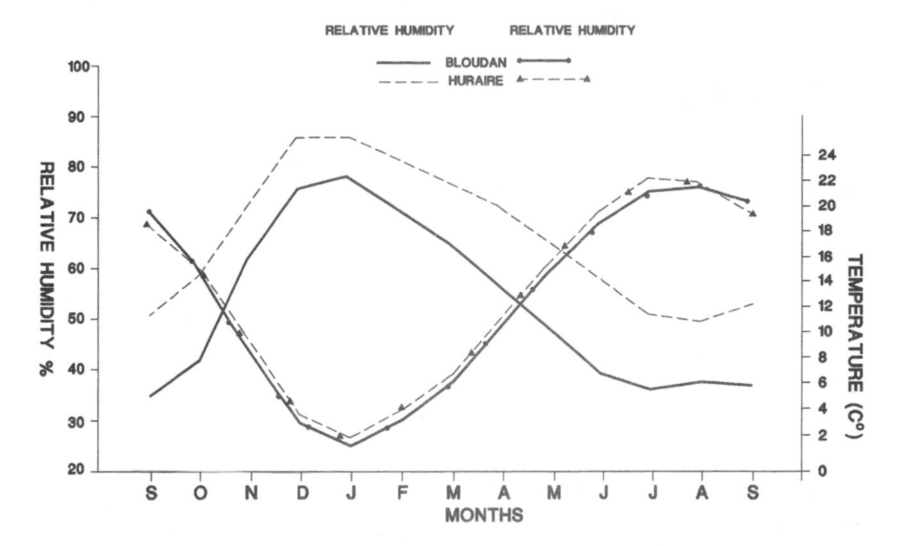


**Figure 8.4.10 Mean monthly temperature and humidity, Bloudan and Hurarie, 1971–1978.**

To estimate the availability of moisture for recharge of the Figeh System, several models were attempted using data from the weather stations at Bloudan and Hurarie. The five principal variables utilized in the evolution of the final model were:

1. Seasonal variations in precipitation;
2. Elevational differences in temperature;
3. Evaporation;
4. Snowfall and moisture storage as ice; and
5. Infiltration.

The mean precipitation for the hydrologic years September 1971 through August 1978 for Bloudan and Hurarie was used in calculating potential moisture available for recharge in the study area. A 0.6 infiltration rate was determined, based on lysimeter data from the Bloudan and Hurarie weather stations. This rate is representative of only the lower portion of the recharge area. In other words, based on the assumptions that precipitation will be higher and evaporation rates lower at higher elevations, the infiltration rate should increase with increasing elevations.

Over the seven-year period from hydrologic years 1971 through 1978, annual snowfall, converted to melt water, accounted for 22.23% of the total precipitation at Bloudan and 26% at Hurarie, although average annual snowfall amounts were approximately the same at both stations. Generally, the variability in amounts of annual snowfall was slightly greater at Hurarie than

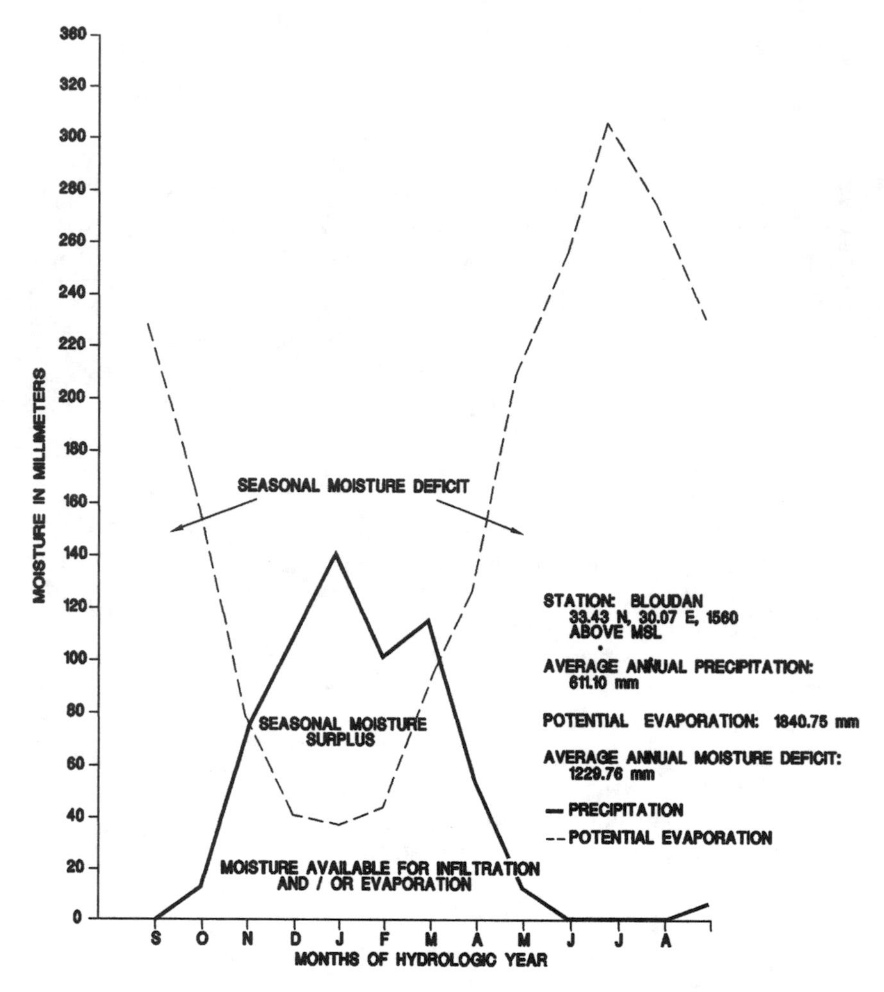


**Figure 8.4.11 Climograph for Bloudan by Hydrologic Years, 1971–1978.**

Bloudan (Table 8.4.3). Generally, 1 cm of fresh snow is equal to 1 mm of water when melted.

To estimate the snowfall in the recharge area, some rough calculations were made using only the data for Bloudan and Hurarie (Table 8.4.4):

1. Using an adiabatic rate of 0.8°C/100 m (dry adiabatic cooling rate of 1°C/100 m less the dew point lapse rate of 0.2°C/100 m), the estimated average monthly elevation of dew point (the temperature at which the air becomes saturated or at which relative humidity becomes 100%) was calculated for the recharge area (Table 8.4.4).
2. Consequently, the elevation at which the average monthly air temperature should theoretically reach 0° was calculated, representing

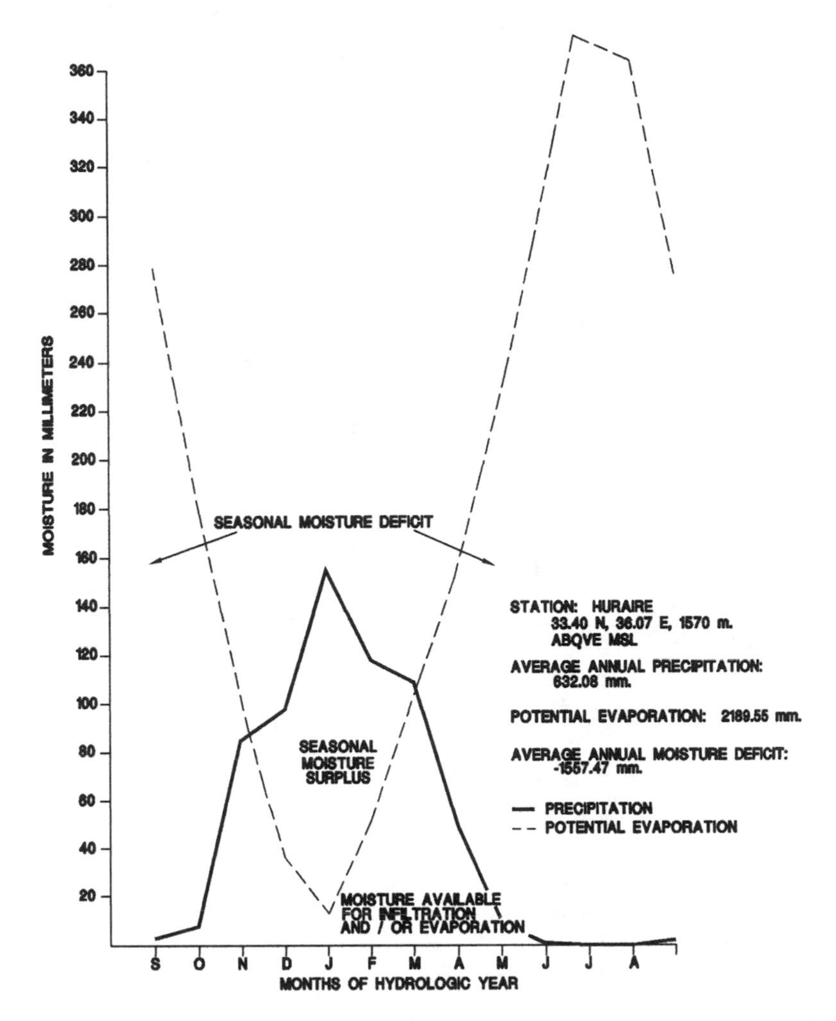


**Figure 8.4.12 Climograph for Hurarie by Hydrologic Years, 1971–1978.**

an approximately monthly snowline, or, in other words, where precipitation should occur mostly as snow or ice.

3. The relationships between the elevations of the recharge area and the elevations of the estimated mean monthly 0° centigrade were analyzed to determine how much of the recharge area was above the snowline for each month (See Figure 8.4.13). Table 8.4.4 illustrates the significance of these relationships by elevation. Precipitation occurring above the snowline (0° elevation) is assumed to be in the form of snow.

Over the recharge area an estimated average of 74% of the precipitation from December to March may have occurred as snowfall during the hydrologic

**Table 8.4.2 Elevation of Recharge Area by Quadrangle (km²)**

| Elevations (m) | 1000 | 1000–1300 | 1300–1600 | 1600–1900 | 1900–2200 | 2200–2500 | 2500+ | Total | Percent |
|---|---|---|---|---|---|---|---|---|---|
| Zebdani | 3.15 | 37.93 | 42.78 | 73.18 | 64.80 | 2.05 | — | 223.89 | 30.44 |
| Damas Nord | — | — | 0.30 | 18.18 | 1.20 | — | — | 19.68 | 2.68 |
| Rijak | — | — | — | 0.33 | 27.43 | 17.68 | — | 45.44 | 6.18 |
| Assal El Ward | — | — | — | — | 167.70 | 84.15 | — | 251.85 | 32.24 |
| Aarsal | — | — | — | — | 25.08 | 95.68 | 2.75 | 123.51 | 16.79 |
| Nabk | — | — | — | — | 46.03 | 25.10 | — | 71.13 | 9.67 |
| Total | 3.15 | 37.93 | 43.08 | 91.69 | 332.24 | 224.66 | 2.75 | 735.50 | — |
| Percent | 0.42 | 5.16 | 5.85 | 12.47 | 45.17 | 30.55 | 0.37 | — | 100.00 |

**Table 8.4.3 Estimated Mean Snowfall Equivalent (mm), September 1971–August 1975**

| Month | S | O | N | D | J | F | M | A | M | J | J | A |
|---|---|---|---|---|---|---|---|---|---|---|---|---|
| **Based on Data for Bloudan** | | | | | | | | | | | | |
| *Elevation (m)* | | | | | | | | | | | | |
| Above 2500 | — | — | 73.4 | 108.1 | 140.1 | 99.8 | 114.3 | 50.4 | — | — | — | — |
| 2201–2500 | — | — | 73.4 | 108.1 | 140.1 | 99.8 | 114.3 | 38.4 | — | — | — | — |
| 1901–2200 | — | — | 50.4 | 108.1 | 140.1 | 99.8 | 88.3 | 26.6 | — | — | — | — |
| | | | | | | | | | | | | 0°C Elevation[a] (snow line) |
| 1601–1900 | — | — | 27.5 | 62.9 | 140.1 | 64.2 | 62.2 | 14.8 | — | — | — | — |
| 1301–1600 | — | — | 4.6 | 17.6 | 46.0 | 28.6 | 36.0 | 3.0 | — | — | — | — |
| 1001–1300 | — | — | — | — | — | — | — | — | — | — | — | — |
| Below 1000 | — | — | — | — | — | — | — | — | — | — | — | — |
| **Based on Data for Hurarie** | | | | | | | | | | | | |
| *Elevation (mm)* | | | | | | | | | | | | |
| Above 2500 | — | — | 85.2 | 98.5 | 155.5 | 119.1 | 107.6 | 47.5 | — | — | — | — |
| 2201–2500 | — | — | 67.4 | 98.5 | 155.5 | 119.1 | 107.6 | 36.6 | — | — | — | — |
| 1901–2200 | — | — | 48.8 | 98.5 | 155.5 | 119.1 | 84.3 | 25.0 | — | — | — | — |
| | | | | | | | | | | | | 0°C Elevation[a] (snow line) |
| 1601–1900 | — | — | 30.2 | 54.0 | 155.5 | 78.5 | 61.1 | 13.4 | — | — | — | — |
| 1301–1600 | — | — | 11.6 | 9.2 | 66.5 | 37.0 | 37.9 | 1.8 | — | — | — | — |
| 1001–1300 | — | — | — | — | — | — | — | — | — | — | — | — |
| Below 1000 | — | — | — | — | — | — | — | — | — | — | — | — |

[a]Above the snow line, monthly precipitation should be mostly as snow or ice, but is shown as meltwater (mm).

**Table 8.4.4 Estimated Dew Point and Snow Line Based on Data for Bloudan (1560 m) and Hurarie (1540 m), September 1971-August 1978, Hydrologic Years**

| Month | S | O | N | D | J | F | M | A | M | J | J | A |
|---|---|---|---|---|---|---|---|---|---|---|---|---|
| **A. Bloudan** | | | | | | | | | | | | |
| Relative humidity (%) | 34.90 | 42.79 | 61.13 | 74.80 | 77.40 | 70.17 | 62.59 | 53.79 | 44.70 | 37.13 | 34.21 | 36.67 |
| Monthly temperature (°C) | 20.30 | 15.77 | 9.16 | 3.50 | 1.86 | 3.67 | 6.40 | 10.68 | 15.17 | 18.92 | 21.16 | 21.74 |
| Approximate dew point (°C) | 3.0 | 2.3 | 2.4 | –5.0 | –2.1 | –2.1 | –2.0 | +2.0 | +2.4 | +4.0 | 4.0 | 5.0 |
| Subtotal[a] | 17.3 | 13.5 | 6.8 | 8.5 | 4.0 | 5.8 | 8.4 | 8.7 | 12.8 | 14.9 | 17.2 | 16.7 |
| Elevation of dew point (m) | 3723 | 3248 | 2410 | 2623 | 2060 | 22.85 | 2610 | 2648 | 3160 | 3423 | 3710 | 3647 |
| Elevation at 0°C | 4097[b] | 3535[b] | 2710[b] | 1997 | 1792 | 2019 | 2360 | 2895[b] | 3498[b] | 3923[b] | 4210[b] | 4273[b] |
| **B. Hurarie** | | | | | | | | | | | | |
| Relative humidity (%) | 50.80 | 57.91 | 70.70 | 84.76 | 84.86 | 79.96 | 75.24 | 70.51 | 63.01 | 56.71 | 49.87 | 47.37 |
| Monthly temperature (°C) | 19.8 | 15.88 | 9.56 | 3.58 | 2.00 | 4.01 | 6.50 | 10.81 | 15.78 | 19.60 | 22.30 | 21.90 |
| Approximate dew point (°C) | 8.0 | 7.5 | 2.5 | 1.6 | –1.8 | 1.6 | 2.0 | 6.0 | 9.0 | 10.0 | 10.9 | 7.5 |
| Subtotal[a] | 11.0 | 11.4 | 7.1 | 2.0 | 3.8 | 2.4 | 4.5 | 4.8 | 6.8 | 9.6 | 11.4 | 14.4 |
| Elevation of dew point (m) | 2915 | 2965 | 2426 | 1790 | 2015 | 1840 | 2103 | 2140 | 2390 | 2740 | 2965 | 3340 |
| Elevation at 0°C | 4015[b] | 3528[b] | 2735[b] | 1990 | 1790 | 2040 | 2353 | 28901 | 3515[b] | 3990[b] | 4328[b] | 4278[b] |

[a] Total number of degrees (C) air temperature must be lowered in order for dew point to be reached.

[b] These elevations are higher than and not included in the recharge area.

Assumptions: That winter precipitation results mostly from orographic conditions and that an adiabatic rate of 0.8°C per 100 m adequately represents climatic conditions upslope from the weather stations to the elevations of the dew point and snow line.

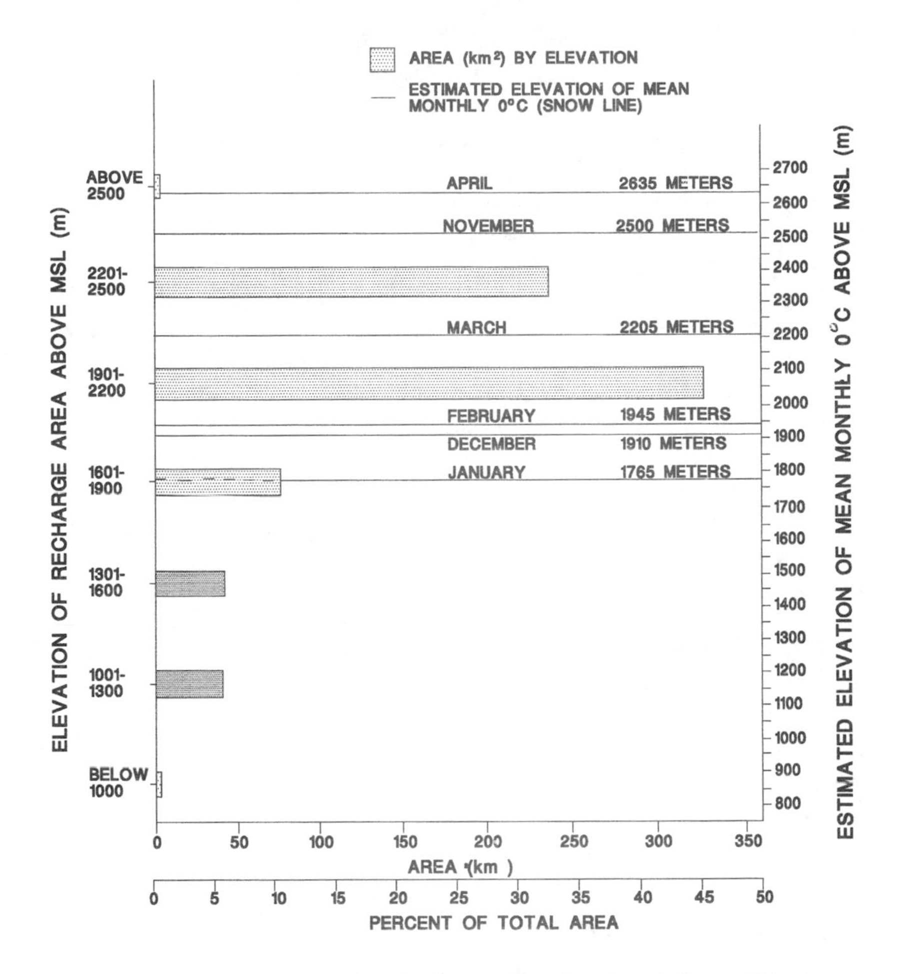


**Figure 8.4.13 Elevations of recharge area and estimated monthly means of 0°C (or snow line).**

years from 1971 through 1978. Much of this snow represents a snowpack being stored over the period and it appears to be released as melt water over a snowmelt period from mid-March through April.

The length of time during which the release of snowmelt becomes available for infiltration (plus any precipitation which might occur simultaneously) is approximately 45 days. The total amount of melt water was 398 mm over the 45 days; the amount available for infiltration in the model, after evaporation, was 239 mm, or 5.3 mm per day. This was calculated as follows:

$$Y = \frac{[(P_{4+5+6+7})Sp] + R}{T} \cdot Ir$$

where y = moisture available for infiltration

$P_{2+3+4+5}$ = precipitation for December, January, February and March (471 mm)

Sp = snowfall as a percent of total precipitation (74%)

R = rainfall in April (49 mm)

T = 45-day snow melt period

Ir = infiltration rate (0.6)

Consequently, an impulse equivalent to 239 mm of moisture, or 38% of the annual precipitation, should enter the recharge system of Figeh during the snowmelt period in March and April. This impulse should cause a response in the discharge at Figeh in a predictable manner.

The above calculations are approximations used to estimate the combined effect of five climatological variables on annual groundwater recharge within the recharge basin. The calculations are based on established examples but are no more than estimates because the existing climatological data represent only two stations, both of which are at similar elevations. These estimates, however, have scientific merit and must await better and more geographically dispersed data bases being refined.

The size and scope of the snowmelt impulse on the recharge area of Figeh is critical to understanding the dimensions and durations of not only high flows but also low flows within the Figeh system. The importance of accurately measuring the climatological variables and using these in determining recharge impulses and time windows of recharge cannot be overstressed as a long-term water management tool, particularly as the basis for safe and efficient low-flow augmentation.

### 8.4.6 DISCHARGE GROUNDWATER TO THE BARADA RIVER

Increase in discharge of groundwater to the Barada River has been recognized by Burdon[6,7] and subsequently by Dr. Meir,[8,9] Department of Irrigation and Hydraulic Work, in his studies made during the Fall of 1982. The interpretation of a conductivity traverse of the Barada indicated changes in chemical quality due to mixing that may be from spring discharge not discernible in the river bed. Ain Harouch and several springs are reported to emerge in the bottom of the Barada River between Ain Harouch and Side Spring. A reconnaissance survey of springs along the Barada River indicates several springs occur between the villages of Takeye and Khadra.

The discharge of the Barada River was measured at six locations by The Department of Irrigation and Hydraulic Power. The measurements were performed during the period of October 13 to October 30, 1982, at Ain Habib, Takeye Hydroelectric Power Plant, Kfer el Aquamid, Deir Moukarren, Ain Khadra, and Hemeh.

The study was designed to provide data to determine whether the Barada River gains or looses water through communication with the groundwater

**Table 8.4.5 Discharge of Barada River Data from Dr. Meir, 1982**

| | Discharge of Barada River ($m^3/s$) | | | | | |
|---|---|---|---|---|---|---|
| | 10/13 | 10/16 | 10/20 | 10/23 | 10/27 | 10/30 |
| Dam at Takeye | — | — | — | — | — | — |
| Habib Spring | 0.165 | 0.131 | 0.131 | 0.135 | 0.172 | 0.152 |
| Takeye Discharge Hydroplant | 1.968 | 2.362 | 2.120 | 1.706 | 1.929 | 1.70 |
| Kfer el Aquamid | 1.740 | — | 2.045 | 1.613 | 1.663 | 1.432 |
| Deir Moukarren | 1.108 | 1.568 | 1.532 | — | 1.325 | 1.373 |
| Ain Khadra | — | 2.303 | 2.249 | — | 1.994 | 1.603 |
| El Hama | — | 1.383 | 1.505 | — | 1.474 | 1.602 |

system and to allow determination of specific river reaches in which water is gained or lost. The study provided data indicating groundwater discharge from the karst system.

Data on Table 8.4.5 indicates losses of water from the Barada River in the two reaches between the Takeye Power Plant and Deir Moukarren (Figure 8.4.14). These measurements indicate loss of large quantities of groundwater to the Barada River that would be potentially available for the water supply of Damascus. The water losses are due to the recharge to the groundwater system.

During the period of measurement, the Barada gained 0.4 to 0.9 $m^3$ per second flow between Deir Moukarren and Ain Khadra. Much of this significant water gain must be a result of recharge from groundwater of the Figeh karst system in the area where Turonian strata are exposed in proximity of the river.

Of even greater significance is that discharge of the Barada River and from the Figeh karst system are hydrologically connected. The future long-term development of water from Figeh karst system and particularly any extensive augmentation of low flow from the Figeh karst system must take this into account from an environmental as well as a water supply aspect.

Data in Table 8.4.5 indicate loss of water from the Barada River in the reach between Ain Takeye and Hama. Fourteen water samples were collected from the Barada River and six water samples from springs adjacent to the river. The pH, specific conductance and temperature of each sample was measured in the field.

Temperature and conductivity for each of the 14 water samples from the Barada River and each of the 6 water samples from springs were obtained along the Barada River. In addition, ten locations where pH was measured are shown in Figure 8.4.14. A deflection of the plots for temperature, conductivity, and/or pH occurs at each location where significant additions of water from springs enter the Barada River, for example, at Ain Habib, at the spring at Souk Wadi Barada, at the spring at Kfer el Aquamid, at Ain Harouch, at Side Spring, and at Figeh Spring. Based on this information, PELA has inferred

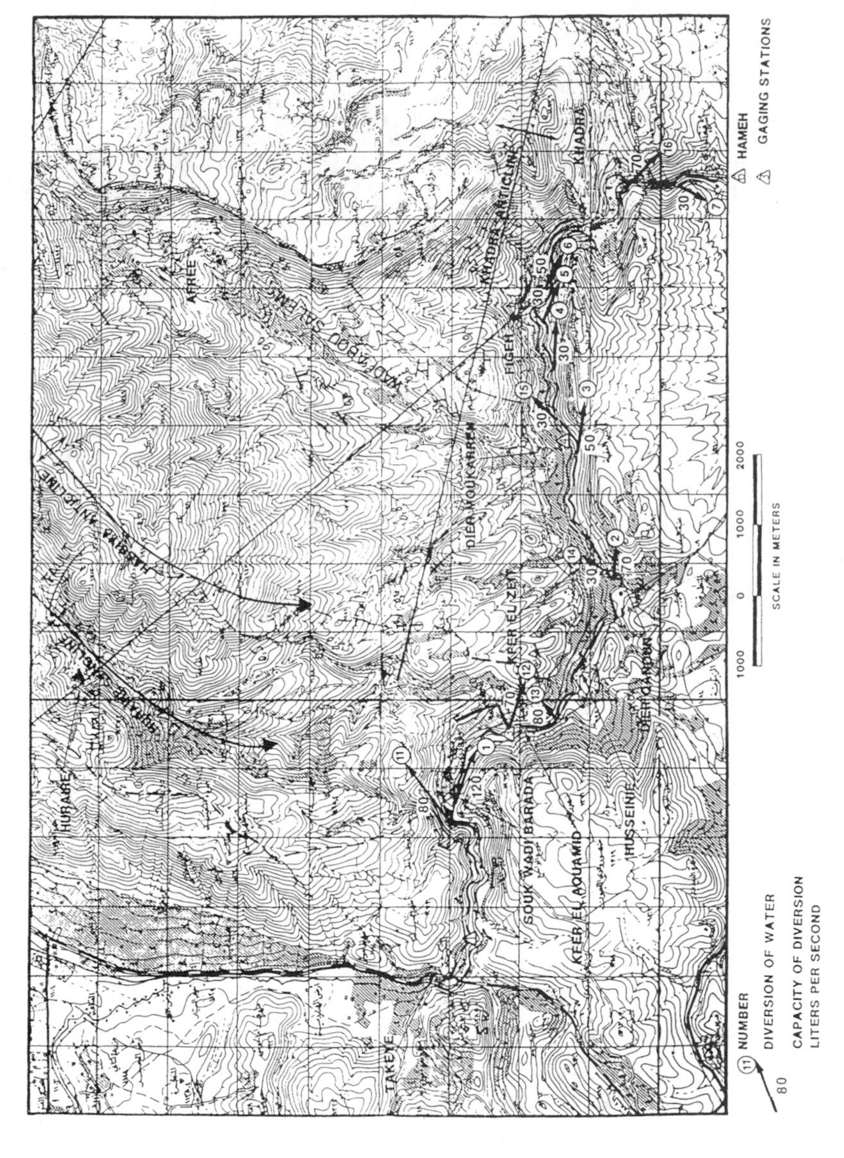


Figure 8.4.14 Sites for which measurements of discharge of the Barada River.

that additional springs are present in the reach of the Barada River between Kfer el Aquamid and Deir Qanoun. More detailed studies are needed on stream flow. The studies should include correlation with changes in water quality to determine relationships, as a basis for management and development of these water supplies.

#### 8.4.6.1 Pumping Test Studies

Extensive stratigraphic and structural geologic studies were a prelude to a series of pumping tests at critical points in the Figeh Spring area. Interpretation of pumping tests must be based on a knowledge of the stratigraphic sequence of rocks, their lithology, fracturing, karstification, and the recharge area, as well as their storage and discharge characteristics to properly interpret quantitative results from pumping tests.

Geologic mapping during the summer of 1981 led to the opinion that karstification of the Cenomanian strata acts as the principal aquifer to the Figeh system. Groundwater movement is controlled by dip of strata, plunge of folds, location of faults, and intensity and orientation of fractures. The presence of marls in the Cenomanian rocks retard the vertical movement of water.

Intersection of the anticlinal axes of the Hassiya and Khadra folds created an interference structure, Figeh Dome. These three structures control the outcrop pattern of Turonian rocks and, along with faults and fractures, the location of Figeh Spring.

The structural and stratigraphic setting dictates that groundwater in the system must be divided into two different, but interrelated, flow systems. Both systems flow southward, but part of the water is diverted around the western side of Figeh Dome. This water rises at Ain Harouch and some discharges directly to the Barada or is lost as underflow beneath the Barada River. The second major system allows water to move around the eastern side of the Figeh Dome and to rise at Ain Figeh. A portion of this water moves southeastward along the Khadra axis and may underflow the Barada River. The Side Spring at Figeh receives water from both flow systems, as well as groundwater that moves downgradient in Pleistocene alluvial deposits along the Barada River.

The hydrologic system of Figeh[10] represents a system in which numerous losses occur and have been observed. Losses include seasonal evapotranspiration and perennial springs, seeps, evaporation, underflow to the Barada River, and possibly underflow to other systems as underflow outside the area.

Another major loss from the system occurs from the withdrawals of groundwater from wells used for irrigation or urban use from within the Figeh karst system. One example is at Deir Moukarren, 3 km west of Figeh Spring where a large well and pumping station have been constructed on the southwest side of the Figeh Dome.

The series of carefully planned pumping tests were also performed to quantitatively delineate the aerial extent and the boundaries of the karst system supplying water to the Figeh Spring. It was recognized that this series of pumping tests would also be required to properly assess the interrelationship between Figeh Spring, Side Spring, the Pilot Development Project, Ain Harouch, and groundwater movement in the Deir Moukarren area.

Pumping tests[11] had to be carefully planned with representative monitor wells in all possible impacted water sources. Pumping tests were performed as follows:

1. From Figeh cavity opening;
2. Pilot Development Project;
3. Side Spring;
4. Ain Harouch, PELA test wells: PL-4; PL-5C; PL-5D; and
5. Wells in the vicinity of Deir Moukarren.

The pumping tests and monitoring carried out during the project at Figeh documented the following:

1. Extent of cone of depression (dewatering) in the Quaternary deposits above bedrock.
2. Changes in turbidity of samples from pilot development wells.
3. Changes in turbidity from samples from Figeh discharge.
4. Rapidity of changes in water level in Quaternary deposits.
5. Changes in water level in the underlying cavities (limestone) in the aquifer system.
6. Changes in turbidity of select samples from observation wells tapping the cavity system and at the spring discharge points.
7. Record of barometric pressure.
8. Change in land surface elevation at selected points around Figeh Spring.
9. Change in groundwater temperature from Main Spring, Side Spring, and selected wells in the surficial aquifer and the cavity system.
10. Change in discharge from Figeh and Side Spring.
11. Change in discharge from springs discharging to the Barada River.
12. Change in water level in 5 piezometers along the south and north sides of the Grout Curtain.
13. Area of impact, as related to the potential for pollution of the water from the Spring.

### 8.4.6.2 Main Cavity — Figeh Pumping Test

A monitoring network, with systematic collection of data using a combination of water level recorders, manual measurements, and stream gauging, was an important part of test results during pumping and recovery (Figure 8.4.15). The following summarizes the results of these tests reported in Table 8.4.6:

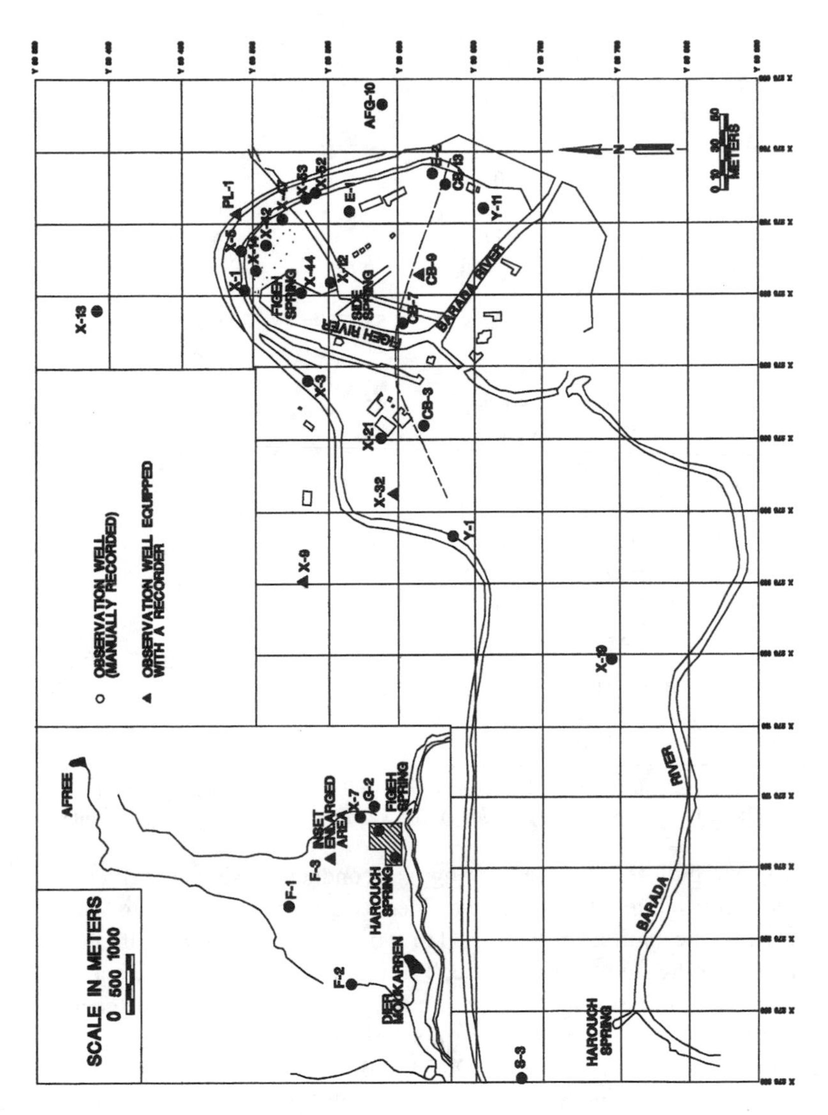


Figure 8.4.15 Location of observation wells for the cavity pumping test.

**Table 8.4.6 Results of Main Cavity Pumping Test**

| Observation well no. | r (m) | Q ($m^3$/day) | T ($m^2$/day) | Leakance ((m/day)/m) | S | Graphical plot |
|---|---|---|---|---|---|---|
| Cavity | | 21772.8 | 140935 | — | — | t-d |
| | | 29946.2 | 119213 | — | — | t-d |
| | | 51321.6 | 145933 | — | — | t-d |
| | | 51321.6 | 181605 | — | — | t-R |
| X-32 | 168 | 29946.2 | 851518 | $1.207 \times 10^{-4}$ | $9.22 \times 10^{-5}$ | t-d |
| | | 51321.6 | 68102 | $6.03 \times 10^{-5}$ | $4 \times 10^{-5}$ | t-R |
| X-42 | 28.5 | 21772.8 | 31518 | $3.88 \times 10^{-1}$ | $1.185 \times 10^{-1}$ | t-d |
| | | 29946.2 | 170304 | $6.58 \times 10^{-1}$ | $5.24 \times 10^{-3}$ | t-d |
| X-44 | 19 | 21772.8 | 99057 | $6.86 \times 10^{-1}$ | $1.29 \times 10^{-1}$ | t-d |
| | | 51321.6 | 81722 | $4.5 \times 10^{-2}$ | $4.4 \times 10^{-4}$ | t-d |
| X-53 | 64 | 21722.8 | 222244 | $1.356 \times 10^{-3}$ | $2.86 \times 10^{-3}$ | t-d |
| | | 51321.6 | 302675 | | $2.05 \times 10^{-4}$ | t-d |
| X-55 | 50 | 21772.8 | 22224 | $3.55 \times 10^{-1}$ | $8.89 \times 10^{-3}$ | t-d |
| | | 29946.2 | 50729 | $8.11 \times 10^{-3}$ | $2.87 \times 10^{-2}$ | t-d |
| | | 51321.6 | 56752 | — | $4.73 \times 10^{-3}$ | t-d |
| | | 34732.8 | 691337 | $2.765 \times 10^{-4}$ | $2 \times 10^{-4}$ | t-R |
| E-2 | 131 | 51321.6 | 120180 | — | $1.17 \times 10^{-1}$ | t-R |
| | | Scattered data | | — | — | t-d |
| CB-9 | 113 | 29946.2 | 72250 | — | $9.7 \times 10^{-1}$ | t-d |
| | | 51321.6 | 329525 | $2.58 \times 10^{-3}$ | $1.7 \times 10^{-1}$ | t-d |
| PL-1 | 53 | 21772.8 | 101971 | — | $3.4 \times 10^{-1}$ | t-d |
| | | 29946.2 | 70125 | — | $4.2 \times 10^{-1}$ | t-d |
| | | 31622.4 | 62943 | $8.96 \times 10^{-3}$ | $3.1 \times 10^{-1}$ | t-d |
| | | 51321.6 | 74293 | $1.05 \times 10^{-3}$ | $8.2 \times 10^{-1}$ | t-d |
| | | 51321.6 | 37147 | $5.29 \times 10^{-1}$ | $6.6 \times 10^{-1}$ | t-R |
| X-1 | 15 | 38102.4 | 195718 | — | $9.7 \times 10^{-1}$ | t-R |
| X-13 | 120 | 34732.8 | 74739 | $8.3 \times 10^{-1}$ | $9.4 \times 10^{-1}$ | t-R |

*Note:* r = Distance from pumping well to observation well (m). Q = Discharge rate ($m^3$/day). T = Coefficient of transmissivity ($m^2$/day). S = Coefficient of storage (dimensionless).

1. The pumping test at Figeh Spring was conducted using step-drawdown procedures. Discharge was increased in three steps, in order to stress the aquifer gradually and to avoid any potential for collapse at the Spring due to subsidence. The pumping test was performed for a period of seven days and recovery data were collected for a period of 3 days, until the system stabilized.
2. Observation wells X-9, X-32, X-53, CB-9, X-35, PELA-1, and F-3 were equipped with recorders, and Figeh Spring was monitored by reading an in-place staff gauge. Wells X-1, X-3, X-5, X-12, X-13, X-14, X-21, X-42, X-44, X-52, X-55, X-47, AFG-10, Y-11, E-1, E-2, CB-3, CB-13, and S-3 were manually monitored (Figure 8.4.15).

3. The measured drawdown data, after adjusting for change in barometric pressure, were plotted against time on double logarithmic paper of the same scale as type curve (time was plotted on the abscissa and drawdown on the ordinate). The time-drawdown plots of data from the wells showed steps or segments (a, b, c, and d) due to gradual increase in pumping rate.
4. Some irregular variation in time-drawdown may have been caused by an irregular distribution of fractures which act as recharge or barrier boundaries. The fractures, filled with water, contribute water and thus reduce the rate of drawdown, while empty fractures act as a barrier that increases the rate of drawdown.
5. Characteristics of the drawdown curve indicate that pumping at the cavity, during the early period, had a point impact on the system and the cone of depression deepened relatively more than lateral expansion. After pumping for a longer period with higher rate of withdrawal, the cone began to gradually expand outward. Pumping at a rate of 51,321.6 $m^3$/day did not affect the regular flow of Figeh Spring. The spring continued to flow at a rate expected for that particular time of the year.

The time-drawdown and time-recovery plots (Figures 8.4.16, 17, 18 and 19) show a decrease in time-rate of drawdown and recovery, which, based on

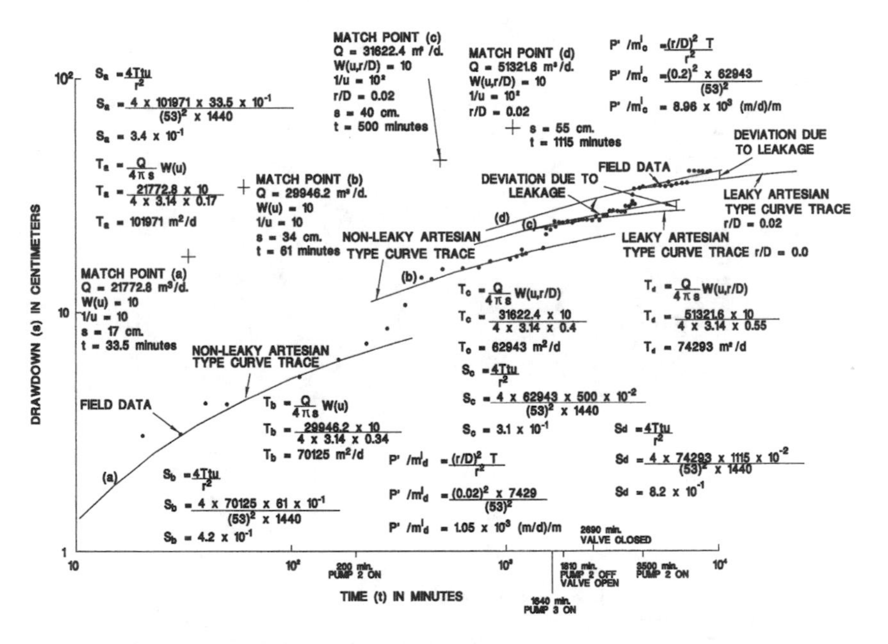


**Figure 8.4.16 Time-drawdown plot for observation well PELA-1 (cavity pumping test).**

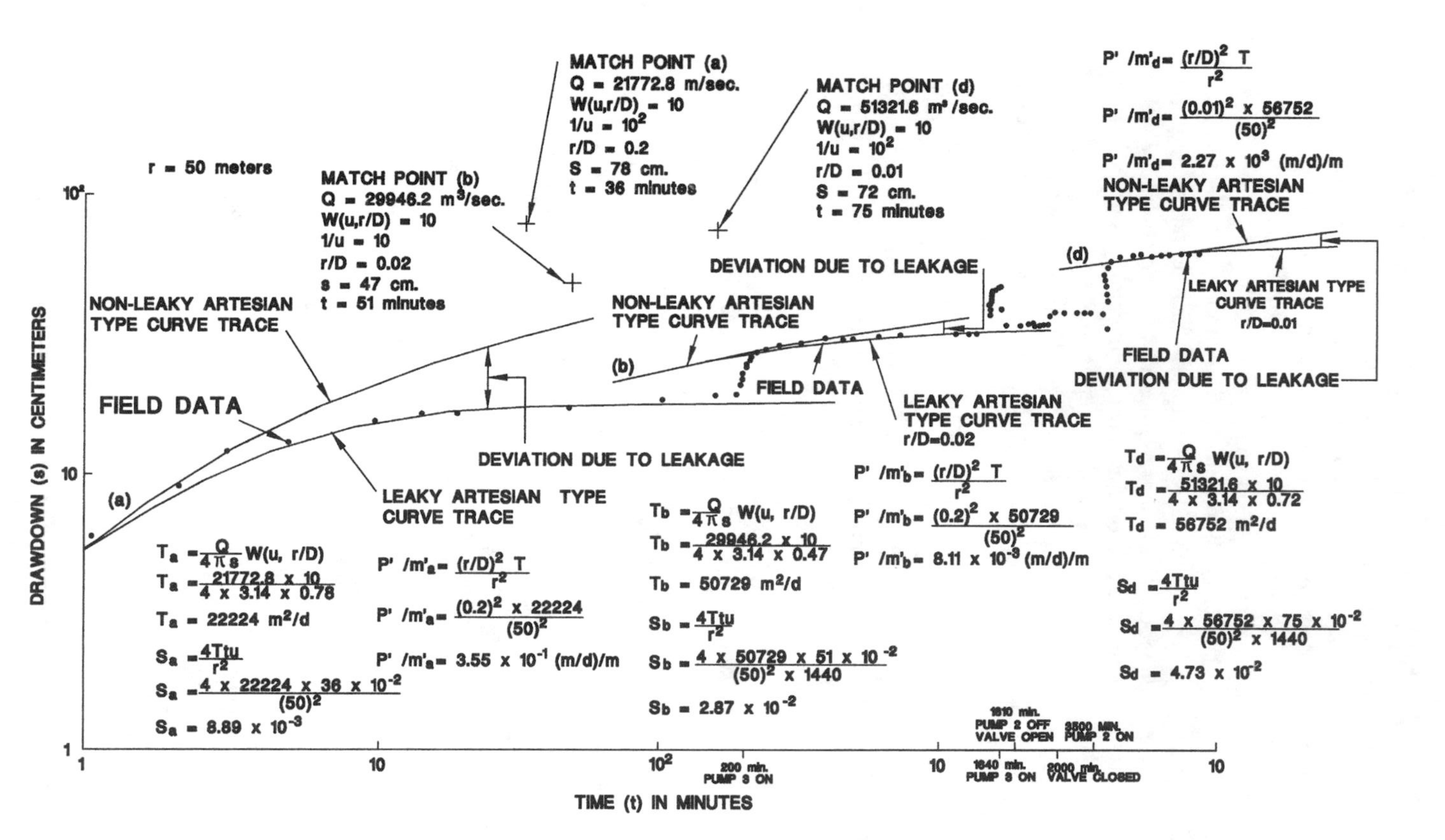


Figure 8.4.17 Time-drawdown plot for observation well X-55 (cavity pumping test).

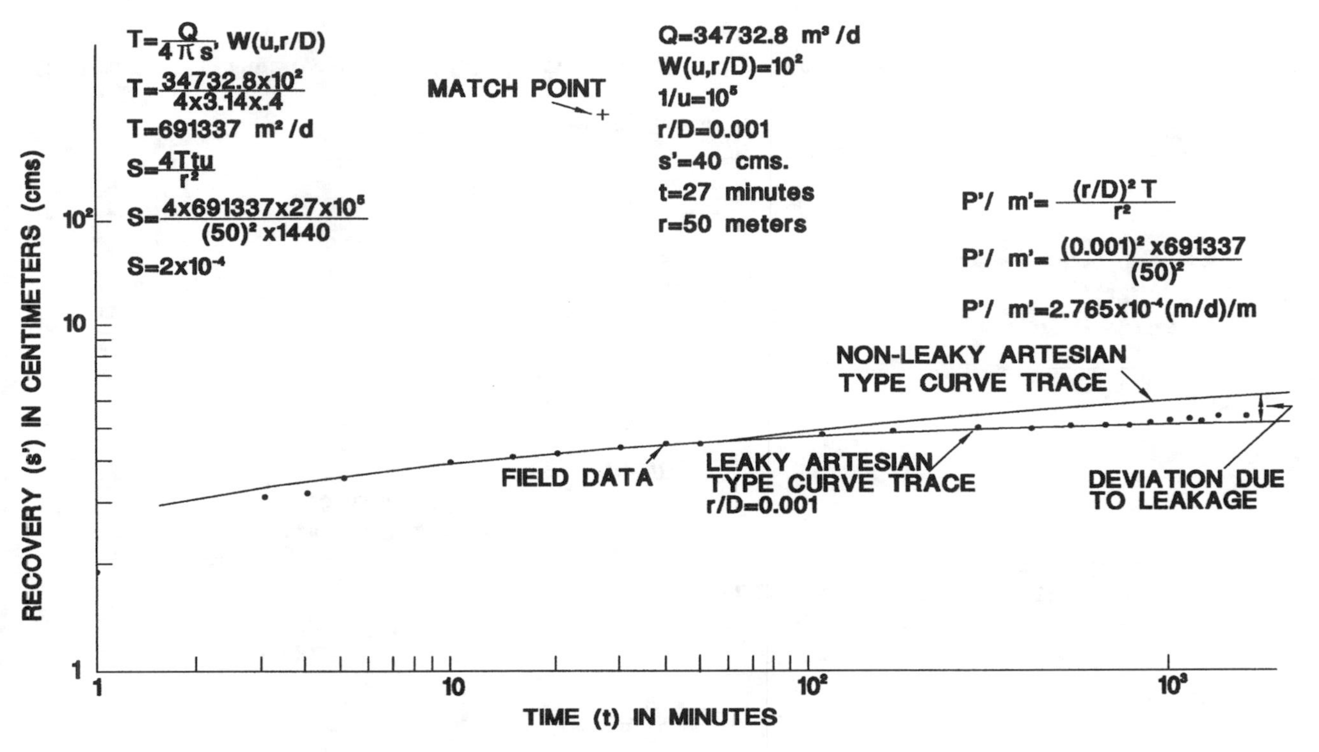


Figure 8.4.18 Time-recovery plot for observation well X-55 (cavity pumping test).

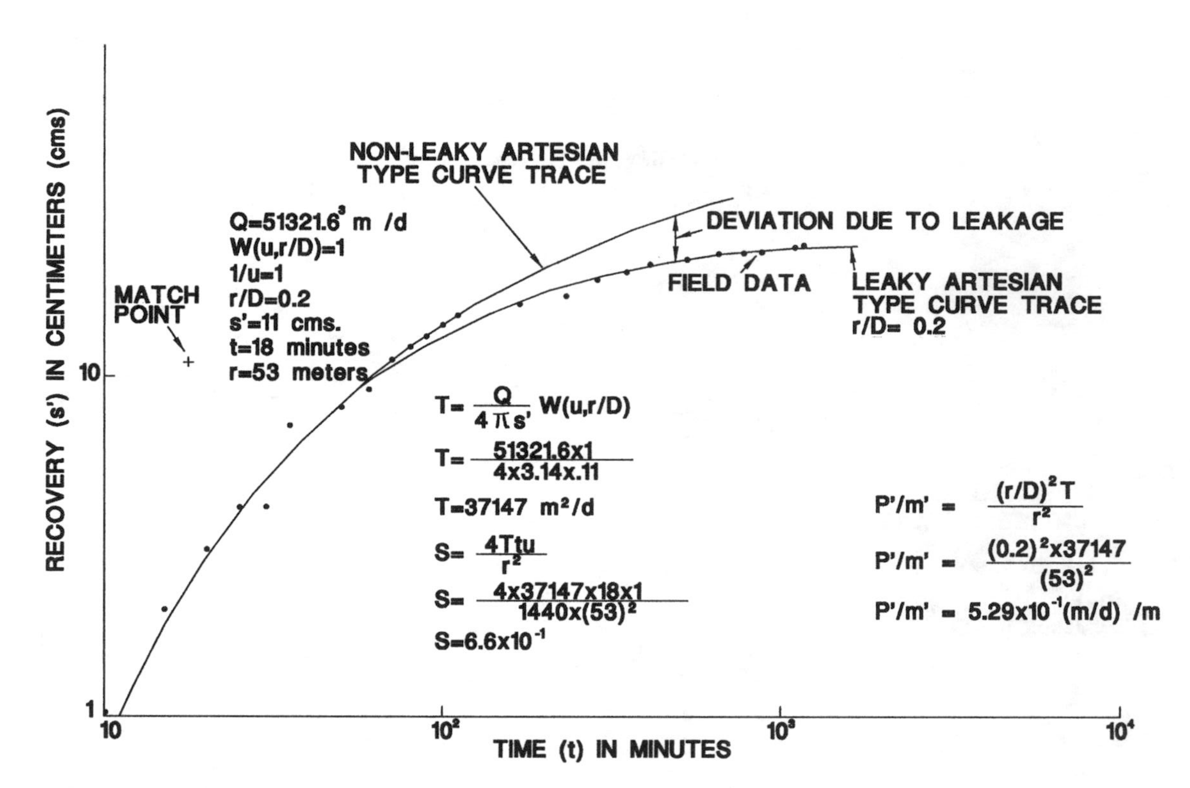


Figure 8.4.19 Time-recovery plot for observation well PELA-1 (cavity pumping test).

available hydrogeological data, can be attributed to the effects of leakage. The deviation of time-drawdown plots from the non-leaky artesian-type curve (Theis curve) indicates that the hydrologic system at Figeh is a semi-confined system.

### 8.4.6.3 Side Spring Pumping Test

A pumping test on Side Spring of ten days duration was performed during the period of September 5 to September 16, 1982. Well Z-1 was also used as one of the points of discharge while wells X-1, X-2, X-3, X-5, X-13, X-19, X-20, X-21, X-26, X-32, X-53, X-54, X-55, E-1, E-2, AFG-10, CB-1, CB-2, CB-3, CB-6, CB-7, CB-9, CB-11, CB-13, Y-1, Y-2, Y-3, Y-6, F-1, F-3, PL-4, F-2, Figeh Spring, and Side Spring were used as observation points. Observation wells X-9, X-32, X-52, X-55, CB-9, F-3, and PL-4 were equipped with recorders. The staff gauges in Figeh and Side Springs were read throughout the test and the remainder of the observation wells were manually measured (Figure 8.4.20).

Owing to difficulties in determining instantaneous discharge, the total average discharge rate during different periods of the test was calculated to be 1.546 $m^3$/sec (133,574.4 $m^3$/d) for the first 200 min of pumping; then the rate was increased to 2.04 $m^3$/sec (176,256 $m^3$/d) until 2,000 min of pumping.

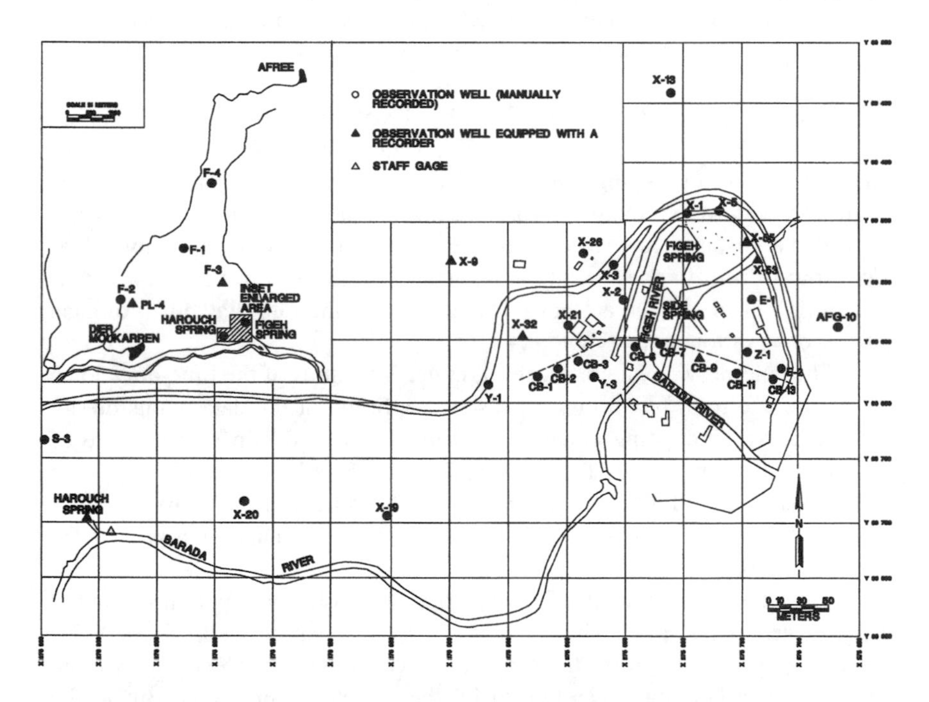


**Figure 8.4.20 Location of observation wells for the Side Spring pumping test.**

The discharge rate was increased to 2.218 $m^3$/sec (191,695.7 $m^3$/d), and it remained approximately the same until 10,000 min of pumping. During the final stage of pumping, the rate of discharge was increased to 2.227 $m^3$/sec (192,412.8 $m^3$/d) until termination of pumping. Discharge rates had to be calculated precisely because of the following factors:

1. There was no method for the direct instantaneous measurements of discharge rates from the Side Spring.
2. Failures of the pumps during test period.
3. Power surges due to uncontrollable voltage and power failure.
4. Extracted water was not fully contributed by the aquifer, but came from three different sources:
   a. *Source I* — The contribution of Figeh Spring increased exponentially with time, as the hydraulic gradient reversed due to pumping.
   b. *Source II* — A volume of water was contributed by the subsurface drainage (alluvial water) to the Side Spring. This contribution occurred because of lowering of water level in Side Spring due to pumping.
   c. *Source III* — Water was contributed from fractures in the limestone at a higher elevation than the Side Spring. These fractures initially contributed to Figeh Spring but pumping diverted the flow of groundwater from Figeh Spring toward Side Spring.

The quantity of water contributed by these various sources to pumping appears to be greater than the amount of water directly from the aquifer so that the impact of pumping from Side Spring on the aquifer system during the earlier period of pumping is from localized storage as the impact is not recorded at distant observation wells (r = +100 m). The transmissivity values determined from this test must be evaluated carefully, and the computations of coefficient of storage are not included in the time-rate drawdown of each of the observation wells (Table 8.4.7).

The time-drawdown graphs show, in general, steps of the time-drawdown curve are due to different rates of discharge. Some of the data points do not fall on curves indicating either sudden increases or declines in the rate of discharge thereby causing the variation in time-rate drawdown. Some of the variations in time-drawdown are the result of a single or collective effect of various factors previously described and must be evaluated as a part of the interpretive process.

The effects of pumping of Side Spring were observed within the initial ten minutes of pumping. The water levels in observation wells X-19, X-21, Y-1, Z-1, and CB-6, show direct hydraulic connection between Side Spring and the water-bearing zones tapped by these observation wells. But in the remaining observation wells X-1, X-2, X-5, X-13, X-20, X-26, X-32, X-53, CB-2, CB-3, CB-7, CB-9, CB-11, CB-13, E-1 and E-2, the impact of pumping

**Table 8.4.7 Results of Side Spring Pumping Test**

| Observation well no. | r (m) | Q ($m^3$/day) | T ($m^2$/day) | Leakance (m/day)/m | Graphical plot |
|---|---|---|---|---|---|
| X-1 | 84 | 176256 | 44550 | — | t-d |
| | | 191695.7 | 141318 | — | t-d |
| | | 192412.8 | 38299 | — | t-d |
| X-2 | 30 | 133574.4 | 42534 | $4.72 \times 10^{-1}$ | t-d |
| | | 176256 | 44550 | $1.23 \times 10^{-1}$ | t-d |
| | | 191695.7 | 30573 | — | t-d |
| X-19 | 286 | 133574.4 | 295414 | $144 \times 10^{-1}$ | t-d |
| | | 176256 | 25059 | $2.75 \times 10^{-2}$ | t-d |
| | | 191695.7 | 4489 | — | t-d |
| X-21 | 78 | 133574.4 | 42202 | $6.24 \times 10^{-1}$ | t-d |
| | | 176256 | 66823 | $2.74 \times 10^{-2}$ | t-d |
| | | 192412.8 | 56739 | — | t-d |
| X-26 | 78 | 133574.4 | 59082.8 | $2.4 \times 10^{-2}$ | t-d |
| | | 176256 | 444086 | $1.82 \times 10^{-1}$ | t-d |
| Y-1 | 153 | 133574.4 | 129694 | $2.216 \times 10^{-1}$ | t-d |
| | | 176256 | 55032 | $2.35 \times 10^{-2}$ | t-d |
| | | 191695.7 | 20082 | — | t-d |
| | | 192412.8 | 16297 | — | t-d |
| Z-1 | | 133574.4 | 128131 | — | t-d |
| | | 192412.8 | 64367 | — | t-d |
| CB-1 | | 133574.4 | 66468 | — | t-d |
| | | 176256 | 48390 | — | t-d |
| | | 192412.8 | 19149 | — | t-d |
| CB-2 | 104 | 133574.4 | 50642 | $1.87 \times 10^{-1}$ | t-d |
| | | 176256 | 45524 | $4.208 \times 10^{-2}$ | t-d |
| | | 192412.8 | 22528 | — | t-d |
| CB-3 | 84 | 133574.4 | 46239 | $2.62 \times 10^{-1}$ | t-d |
| | | 176256 | 43179 | $6.12 \times 10^{-2}$ | t-d |
| | | 192412.8 | 55707 | — | t-d |
| E-1 | 79 | 133574.4 | 393885 | $6.31 \times 10^{-1}$ | t-d |

*Note:* r = Discharge from pumping well to observation well (m). Q = Discharge rate ($m^3$/day). T = Coefficient of transmissivity ($m^2$/day). S = Coefficient of storage (dimensionless).

was not recorded or measured until after 100 min of pumping, indicating a more indirect hydraulic connection of Side Spring with the water-bearing zones tapped by these observation wells. The aquifer at these observation wells is either independent of main cavity system, or the cone of depression did not extend deep enough, until after 100 min of pumpage, to affect these wells and establish a hydraulically connected system between the Side Spring and water-bearing zones tapped by these observation wells. A similar situation was observed during the cavity pumping test (Table 8.4.7).

The computation and analyses of time-drawdown graphs for the one observation well is given in Figure 8.4.21. The summary of results are given in Table 8.4.7.

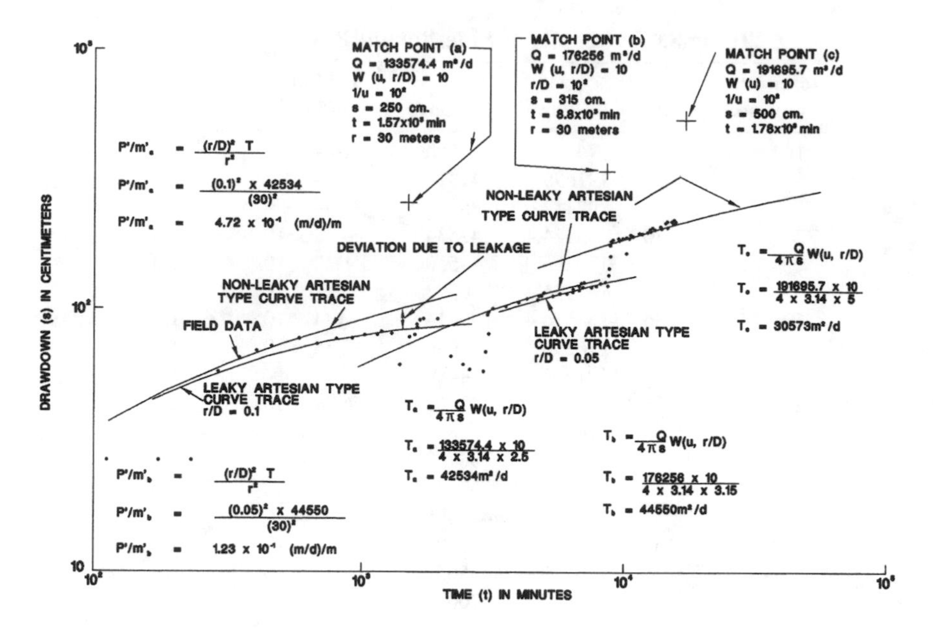


**Figure 8.4.21 Time-drawdown plot for observation well X-2 (Side Spring pumping test).**

## 8.4.6.4 Pilot Pumping Test

The design of the pilot pump test involved construction of a caisson containing 20 wells of 12 1/4" diameter penetrating the main cavity at Figeh Spring (see Figure 8.4.22). The caisson was connected to the pump intakes in the central control center via a tunnel of 1.5 m in diameter. The total discharge from the Spring and caisson was about 4.0 $m^3$/sec.

The objective of the test was to determine feasibility of augmentation of the flow from Source Figeh during periods of stable low flow without seriously impacting the flow from Figeh or adversely affecting the hydrologic system supplying Figeh Spring.

Specific objectives were to:

1. Augment flow from Figeh during periods of low flow.
2. Determine impact of pumping on the discharge from Figeh.
3. Determine the impact on the hydrologic system supplying water to Figeh.
4. Determine areal extent of cone of depression caused by pumping the pilot development facility.
5. Determine the most effective rate of pumping from the system.
6. Determine any changes in land surface (subsidence features) caused by pumping.

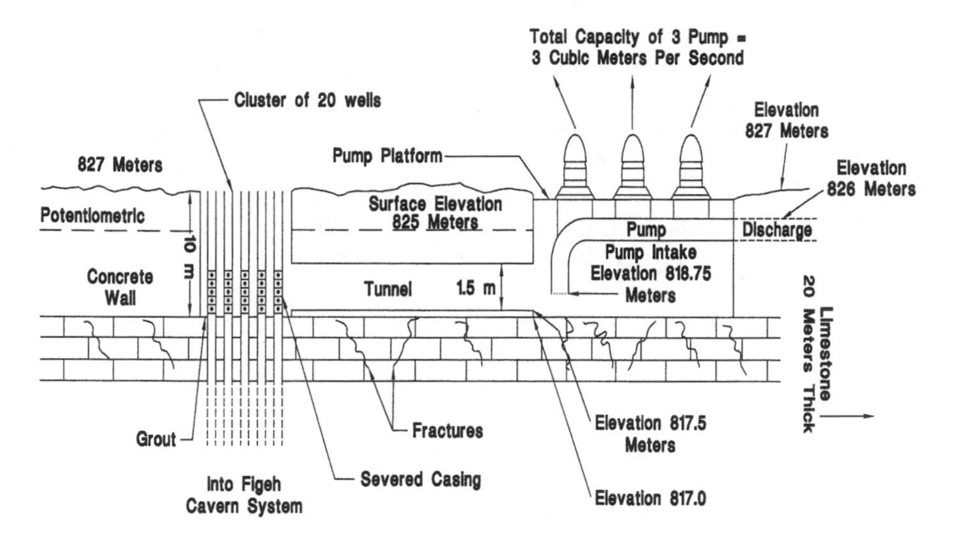


**Figure 8.4.22 Schematic pilot pumping installation.**

Pumping from the Pilot Project was from flow discharged to the pumps through tunnels 1.5 m in diameter. The head on the system, at the time of the test, was at an elevation of 825 m. The minimum elevation in head at which the pumping was performed was 818.75 m. The elevation of the pump intake provided for a maximum drawdown in potentiometric head of 6 meters. A schematic diagram of the pilot pumping installation is shown in Figure 8.4.22. The geologic setting of the well is shown on the geologic cross section.

The limestone underlying the site is at a depth of 10 m and forms a semi-confining bed for the underlying artesian system. The semi-confining bed of limestone is extensively fractured, and these fractures serve as avenues for limited hydraulic connection and flow of water from the Figeh aquifer system upward to the unconsolidated Quaternary alluvial deposits. Water was observed entering the well during drilling operations at the site of test hole X-37 after 1 m of limestone had been penetrated.

The geologic setting at the Pilot Development test site was of concern, because the limestone on which the Pilot Development caisson was set was fractured and the bearing strength was unknown. During test drilling and pumping there was substantial potential for collapse caused by drilling activities and vibration from pumping. A collapse of any portion of the underlying limestone would have an adverse effect on the hydrologic system supplying water to Figeh. As a part of the pumping test, therefore, a detailed monitoring program was implemented to detect changes in the surface features and the hydrologic system (e.g., muddying, change in water temperature or water quality) that would be indicative of changes within the system during withdrawal.

The monitoring program during pumping from the Pilot Development project at Figeh included the determination of the following items:

1. Close monitoring of the extent of cone of depression (dewatering) in the alluvial deposits above bedrock.
2. Changes in turbidity of samples from Pilot Development wells.
3. Changes in turbidity from samples from Figeh discharge.
4. Rapidity of changes in water level in Quaternary deposits.
5. Changes in water level in the underlying limestone aquifer system.
6. Changes in turbidity of selected examples from observation wells tapping the cavity system and at the spring discharge points.
8. Change in land surface elevation at selected points around Figeh Spring.
9. Change in groundwater temperature from Figeh Spring, Side Spring, and selected wells in the surficial aquifer and the cavity system.
10. Change in discharge from Side Spring.
11. Change in discharge from springs discharging to the Barada River.

The Pilot pump test was performed using step-drawdown procedure (Table 8.4.8). Three pumps were installed near the caisson to pump at a rate of 1 $m^3$/sec each. Four pumps were installed in the Main Spring. The collective discharge rate from these four pumps was 1.00 $m^3$/sec. During the test, the rate of discharge was increased in four steps at a rate of 1 $m^3$/sec.

The test was initiated at 10:00 a.m. on October 12, 1983, by starting pump number 1 at a rate of 1 $m^3$/sec. The pumping continued for 60 min, and then a second pump was started. The total discharge rate was 2 $m^3$/sec. The third pump was turned on after 120 min of pumping. The total discharge rate was 3 $m^3$/sec for a period of 180 min then the four cavity pumps were started one by one. The total pumping rate after 240 min of pumping was 4.00 $m^3$/sec and remained stable until the termination of the test at 10:45 a.m. on October 17, 1983. Recovery data were collected until 10:00 a.m. on October 21, 1983. The test was performed for a period of 5 days and recovery data collected for a period of 4 days until the system stabilized.

The observation wells, X-55, PL-7, PL-9, X-32, X-26, X-9, F-2, and F-3 were equipped with recorders and Main Spring, Side Spring, and Ain Harouch were maintained by reading in-place staff gauges. Two shallow wells, Sh-1 and Sh-2, were penetrating unconsolidated upper aquifer together with wells, X-1, X-2, X-5, X-7, X-8, X-11, X-13, X-14, X-18, X-20, X-21, X-26, X-43, X-44, X-46, X-50, X-51, X-52, X-54, X-55, Y-1, Y-2, Y-6, Y-7, Y-9, Y-11, CB-1, CB-5, CB-6, CB-10, CB-11, CB-13, E-1, E-2, F-1, F-3, Z-1, PL-2A, PL-2C, PL-3, PL-4, PL-5B, PL-5D, and PL-6 were manually monitored.

The time-drawdown or time-recovery plots of water levels for wells were obtained from location map recorders or by carefully monitored manual measurements to show current changes in spring flow and water levels due to

**Table 8.4.8 Results of Pumping Test Analysis (Pilot Test) Time-Drawdown**

| Well no. | r (m) Distance from cavity (Main Spring) | Transmissivity $T_1$ ($m^2$/day) | Transmissivity $T_2$ ($m^2$/day) | Transmissivity $T_3$ ($m^2$/day) | Transmissivity $T_4$ ($m^2$/day) | Leakance (m/day/m) |
|---|---|---|---|---|---|---|
| PL-2A | 83 | | | 295,000 | 980,000 | $1 \times 10^{-1}$ |
| Y-7 | 155 | | 84,000 | 299,000 | | |
| Y-9 | 185 | 79,000 | 785,700 | | | |
| Y-11 | 165 | | | | | |
| Ain Harrouch | 570 | | 206,200 | 190,000 | | |
| X-5 | 35 | | | 245,550 | | $8 \times 10^{-2}$ |
| PL-5D | 580 | | | 199,000 | | |
| PL-5B | 582 | | | 305,500 | | |
| S-3 | 540 | | | 172,000 | | |
| X-18 | 38 | | | 283,000 | | |
| X-13 | 120 | | | 81,000 | | |
| X-11 | 78 | | | 134,000 | | |
| X-8 | 23 | | | 149,000 | | $1.12 \times 10^{-1}$ |
| X-14 | 14 | | | 150,000 | | |
| X-55 | 48 | | | 150,000 | | |
| PL-7 | 90 | | | 225,000 | | |
| PL-9 | 138 | | | 335,000 | | |
| F-2 | | | | 150,000 | | |
| X-20 | 445 | | | 166,000 | | |
| Caisson (Ca-2) | 27 | | 153,000 | 172,000 | | |
| Sh-3 | | | 15,000 | 275,000 | | |
| CB-13 | 148 | | | 270,000 | | |
| X-21 | 135 | | | 550,000 | | |

**Table 8.4.8 Results of Pumping Test Analysis (Pilot Test) Time-Drawdown** *(Continued)*

| Well no. | r (m) Distance from cavity (Main Spring) | Transmissivity $T_1$ ($m^2$/day) | Transmissivity $T_2$ ($m^2$/day) | Transmissivity $T_3$ ($m^2$/day) | Transmissivity $T_4$ ($m^2$/day) | Leakance (m/day/m) |
|---|---|---|---|---|---|---|
| X-46 | 47 | | | 230,000 | | |
| PL-2C | 60 | | | 190,000 | 196,440 | |
| PL-1 | 55 | | | 180,000 | | $1.51 \times 10^{-1}$ |
| E-2 | 142 | | | 40,000 | | |
| X-26 | 93 | | | 392,000 | | |
| F-3 | | | 86,000 | 38,000 | | |
| Y-6 | 200 | | 86,000 | 60,000 | | |
| E-1 | 78 | | 120,000 | 190,000 | | |
| Main Spring | | | | 180,000 | 200,000 | |
| PL-3 | | | 37,000 | 200,000 | | |
| X-52 | 75 | | 10,000 | 50,000 | | |
| X-50 | 52 | | 170,000 | 180,000 | | |
| X-44 | 35 | | 90,000 | 150,000 | | |
| X-9 | 205 | | | 230,000 | | $5.5 \times 10^{-2}$ |
| Z-1 | 114 | | 355,000 | 480,000 | | |
| Cassion (Ca-1) | 25 | | | 67,000 | 140,000 | |
| X-1 | 20 | | | 145,000 | | $9 \times 10^{-1}$ |
| X-2 | 80 | | | 625,000 | | |
| X-32 | 170 | | | 720,000 | 500,000 | |
| X-43 | 36 | | | 197,000 | | $3.7 \times 10^{-3}$ |

*Note:* Q = 3 $m^3$/day, or 259200 $m^3$/day, $Q_2$ = 4 $m^3$/sec or 345600 $m^3$/day.

gradual increase in pumping rate. The results of the pumping tests are summarized in Tables 8.4.8 and 8.4.9.

The water level in shallow wells Sh-1 and Sh-2 did not show significant response to the pumping indicating minimum direct hydraulic connection with the underlying karst system. The responses measured in the observation wells during the test demonstrate that the karst system at Figeh responds heterogeneously and that the spring is supplied by interconnected or preferential flow paths within the area of observation wells. Barada River which flows south of Figeh Spring acts as a hydrologic boundary. The surficial and alluvial deposits occurring on the top of the bedrock constitute a water table aquifer which is in a delicate hydraulic balance with the underlying Figeh karst aquifer.

The karst system at Deir Moukarren is in poor hydraulic connection with the part of the karst system supplying Figeh Spring and an additional volume of 2 $m^3$/sec could be developed from the Deir Moukarren Basin in the general vicinity of PL-5 wells.

The water temperature data collected during the test indicates the presence of a deeper karst fissure system which contains water at a higher temperature than the water discharging from the upper part of the Figeh system.

In conclusion, the pumping of the cavity at the mouth of Figeh and from the Pilot Development Project at a total rate of 4 $m^3$/sec increased the yield from the system about 0.6 $m^3$/sec (600 l/sec). This pumpage could be safely increased to a rate of 1 $m^3$/sec and would result in an additional drawdown of 2 cm at Main Spring during low-season flows. The maximum drawdown at the cavity below the caisson, Side Spring, and Ain Harouch with pumping at a rate of 1 $m^3$/sec from the spring and 3 $m^3$/sec from pumping caisson, was 1.17, 1.83, 0.27, and 0.29, respectively.

The discharge at Ain Harouch is, in part, hydraulically connected with the Barada River and the overlying alluvial aquifer. It is also hydraulically connected with the irrigation ditches which recharge the karst aquifer in the vicinity of Ain Harouch (well X-19).

Pumping test results indicate that, with proper control and managed modifications at Figeh Spring, flow augmentation in the amount of 4 $m^3$/sec is available to support the needs of Damascus during the low-flow season. The reduction in storage will be replaced by the rains during the early part of the recharge period.

## 8.4.7 ENVIRONMENTAL CONSTRAINTS TO FUTURE USE OF FIGEH SYSTEM

1. The Cenomanian and Turonian strata comprise a karst aquifer system through which water is conveyed from recharge to storage and ultimately to discharge at the springs at Ain Figeh and Ain Harouch. There are many additional springs along the Barada River that are

**Table 8.4.9 Results of Pumping Test Analysis (Pilot Pumping Test) Time-Recovery**

| Well no. | r (m) Distance from cavity (Main Spring) | Transmissivity $T_1$ ($m^2$/day) | Transmissivity $T_2$ ($m^2$/day) | Transmissivity $T_3$ ($m^2$/day) | Leakance (m/day/m) |
|---|---|---|---|---|---|
| Sh-3 | | 213,000 | | | |
| X-9 | 205 | 327,400 | | | |
| PL-5B | 582 | 370,000 | 1,145,900 | | |
| PL-3 | | 92,000 | | | |
| Main Spring | | 1,833,500 | 4,297,000 | | |
| E-2 | 142 | 30,000 | | | |
| Z-1 | 114 | 226,000 | 598,000 | 1,262,000 | |
| Caisson (Ca-2) | 27 | 125,000 | | | |
| Caisson (Ca-1) | | 125,000 | | | $8 \times 10^{-2}$ |
| Y-6 | 200 | 1,300,000 | | | |
| X-52 | 75 | 110,000 | 125,000 | 610,000 | |
| X-50 | 52 | 275,000 | | | $1.02 \times 10^{-4}$ |
| X-43 | 36 | 240,000 | | | |
| X-44 | 35 | 250,000 | | | $8.16 \times 10^{-4}$ |

| | | | | | |
|---|---|---|---|---|---|
| X-20 | 445 | 197,000 | | | |
| Ain Harrouch | | 810,000 | | | |
| PL-7 | 90 | 382,000 | | | $1.2 \times 10^{-5}$ |
| Cb-3 | | | 55,000 | | |
| Y-7 | 155 | 130,000 | 292,000 | | |
| PL-2C | 60 | 371,000 | | | $2.6 \times 10^{-5}$ |
| X-1 | 20 | 370,000 | 170,000 | 420,000 | |
| E-1 | 78 | 3,055,775 | | | $1.25 \times 10^{-3}$ |
| X-2 | 80 | 177,500 | 723,000 | 1,146,000 | |
| X-8 | 23 | 180,000 | | | $1.3 \times 10^{-1}$ |
| X-5 | 35 | 275,000 | | | |
| PL-1 | 55 | 275,000 | 352,500 | | |
| X-14 | 14 | 83,000 | | | $1.7 \times 10^{-1}$ |
| X-18 | 38 | 177,400 | | | $3.07 \times 10^{-1}$ |
| X-11 | 78 | 125,000 | | | $2 \times 10^{-1}$ |
| X-13 | 120 | 88,700 | | | |
| X-46 | 47 | 289,500 | | | $1.31 \times 10^{-4}$ |
| PL-9 | 138 | 491,100 | | | $2.57 \times 10^{-1}$ |

*Note:* Q = 4 $m^3$/sec or 345600 $m^3$/day.

points of discharge. There is likewise a substantial upward flow from the Turonian rocks through fractures and solution cavities into the Barada which maintains its base flow.

2. The overlying Senonian marls act as aquitards that restrict discharge to land surface where they occur.
3. Detailed photogeologic and ground-truth studies have established that virtually the entire recharge area for the Figeh Spring system is underlain by rocks of Cenomanian Age, that are fractured and karstified in this area, and that a major source of recharge is from the snow fields and snow accumulation at altitudes above 1,500 meters upgradient from the springs.
4. Geologic mapping and hydrogeologic studies during the summer of 1981 and 1982 verified that the Cenomanian strata also contains a complex karstified aquifer system that supplies large quantities of water to the Figeh system. Groundwater movement is from the outcrop of these rocks into interconnected openings along bedding and fractures and locally in solution cavities to the underlying storage system. The specific movement of water is controlled by the dip of bedding, the plunge of folds, the location of faults, and the system and intensity of fracturing.
5. Major structural features influence the direction of groundwater movement from recharge to storage to discharge. Intersection of the anticlinal axis of the Hassiya and Khadra folds create an interference structure (the Figeh Dome). These three structures control the outcrop pattern of the Turonian rocks in the area and along with the faults and fractures the present location of Figeh and the other springs that discharge into the Barada River.
6. Interpretation of the structural and stratigraphic setting indicated that groundwater in the area should be divided into two different flow paths. Both regimes flow southward, but part of the water is diverted around the western flank of the Figeh Dome. Some of this water rises at Ain Harouch, and some is believed to be lost as underflow beneath the Barada River. The second major flow regime consists of the water that moves around the eastern side of the Figeh Dome, and rises at Ain Figeh. A portion of this water moves southeastward along the Khadra fold axis, and some may underflow the Barada River. The Side Spring at Figeh receives water from both flow regimes, plus flow through the alluvial beds of an ancient Barada River channel that now parallels and is adjacent to the Barada River.
7. Flow directions of groundwater are shown in the map. Points of discharge are shown by the location of the springs. (See Figure 8.4.14 and Table 8.4.5.) Figeh Spring discharges near a Pleistocene

meander of the Barada River. Figeh Spring, during Roman occupation (63 BC to 633 AD), discharged at a lower elevation than it does today. Figeh Spring at present discharges through an opening in the Turonian rocks. At this point, a reservoir system has been developed diverting the water into the channels that supply the city of Figeh. There is also withdrawal of groundwater from sump pumps that have been installed at the Spring. This water is under semi-artesian pressure.

8. The importance of understanding the reservoir system from which the water flows, the flow direction, and the rate of water withdrawn allows determination of the area around the Spring in which environmental restraints should be imposed on development. These are (as shown in Figure 8.4.23):

   Area A: In the closest semi-circular area around Ain Harouch, Side Spring, and Figeh, two karst flow systems are involved. Maximum security must be imposed to limit development that could cause physical (discharge), chemical change, or cause pollution to access near surface solution cavities and rock fractures that are connected to the water in the Spring. This area has been titled "Maximum Security Area." No development other than those activities associated with the production of water from the spring should be undertaken.

   Area B: A second semi-circle delineates an upgradient area in the spring-flow system in which development such as housing, agriculture, grazing, and other activities should be restrained. There should be no placement of landfills or hazardous, toxic, or radioactive waste, and all construction or other development should be carefully reviewed and approved by representatives of Figeh Spring.

   Area C: The third outlying semi-circular area shown on the map is an area that would be safe and proper for development of light agriculture and/or light grazing with minimum use of pesticides, insecticides, and fertilizers. Minimum housing, urban development, or construction should be allowed by approval of Figeh. This area should be restricted from the assignment or placement of hazardous, toxic, and/or radioactive waste.

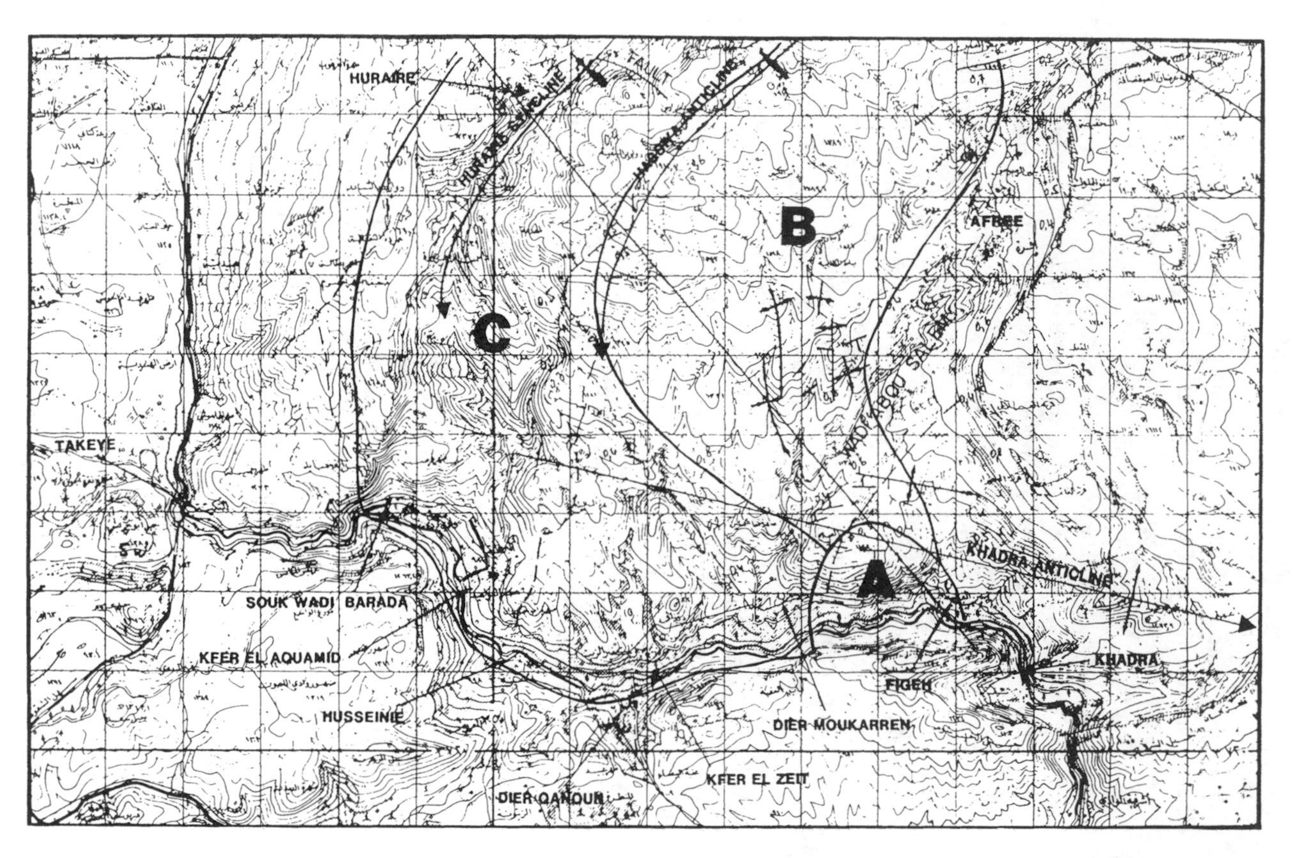


Figure 8.4.23 Map of environmental protection zones: (A) total protection — maximum security; (B) permitted — housing, other development; (C) permit proper development — light agriculture.

## REFERENCES

1. Dubertret, L., Carte Geologique au 50,000 du Liban. Feuille de Zebdani et Notice Explicative, Beyrouth, 1949.
2. Ponikarov, V. P., Ed., Vinogradskey, A. V., Trans, *The Geological Map of Syria, Scale 1:50,000, with Explanatory Notes*, Minestry of Industry, Syriani Arab Republic, 1968, 120 pp.
3. Bazin, F. (SOGREAH), Etude hydrologique et hydrogeologique de la Source Figeh, rapport final, A. Cartes; R. 11.422, 1973.
4. Bazin, F. (SOGREAH), Etude hydrologique et hydrogique de la Source Figeh; R. 11.343, 1973.
5. Bazin, F. (SOGREAH), 1973, and others, Etude hydrologique et hydrogeologique de la Source Figeh, rapport final; R. 11.422, 1973.
6. Burdon, D. J., Geological features of the Barada Valley in relation to the proposed storage reservoirs, Technical FAO United Nations Report No. 337 to the Government of Syria, 1954.
7. Burdon, D. J., The Groundwater Resources of the Damascus Basin — A Preliminary Report, Department of Irrigation and Hydraulic Power, Ministry of Public Works, Government of Syria, 1959, 19 pp.
8. Burdon, D. J., Groundwater Development and Conservation in Syria, Expanded Technical Assistance Plan Rept. No. 1270, FAO, United Nations, Rome, 1961, 84 pp.
9. Burdon, D. J. and Safadi, C., Ras-El-Ain (the great karst spring of Mesopotamia), *J. Hydrology*, 7, 1, 1963.
10. LaMoreaux, P. E., Remote-sensing techniques and the detection of karst, *Assoc. Eng. Geol.*, 16, 3, 1979.
11. LaMoreaux, P. E. and Associates, Assessment of pilot development project report, 28 July–8 August, 1982, with a section on Calculation of stability of boreholes at the project Al-Fije Spring, by Walid Kanaan, *Establissement Public des Eaux de Figeh*, P. E. LaMoreaux and Associates, Tuscaloosa, AL, 1982, 22 pp.

# APPENDIX A
# GLOSSARY

The terms included in this glossary originated from the following references:

*Glossary of Geology and Related Sciences*, American Geological Institute, Washington, DC, 1996.

*Hydrology and Water Resources for Sustainable Development in a Changing Environment*, UNESCO, New York, 1990.

**Acid precipitation** Any atmospheric precipitation which has an acid reaction through the absorption of acid producing substances such as sulfur dioxide.

**Aggradation** The general building up of the land by deposition processes.

**Air instability** This state exists in a body of air that is marked by a strong vertical temperature decrease and high moisture content. Unstable air tends to rise.

**Albedo** The amount of light reflected by a given surface compared to the amount received.

**Alluvium** Material deposited by running water (gravel, sand, silt, clay).

**Anaerobic condition** Characterized by absence of air or free oxygen.

**Andesitic basalt** A fine-grained extrusive igneous rock composed of plagioclase feldspars and ferromegnesian silicates.

**Anticline** A fold in which the rocks are bent convex upward.

**Aquiclude** A body of relatively impermeable rock that is capable of absorbing water slowly, but functions as an upper or lower boundary of an aquifer and does not transmit groundwater rapidly enough to supply a well or spring.

**Aquifer** A porous, permeable, water-bearing geologic body of rock. Generally restricted to materials capable of yielding an appreciable amount of water.

**Aquifuge** A rock which contains no interconnected openings or interstices and therefore neither absorbs nor transmits water.

**Aquitard** A confining bed that retards but does not prevent the flow of water to or from an adjacent aquifer; a leaky confining bed. It does not readily yield water to wells or springs, but may serve as a storage unit for groundwater.

**Artesian** An adjective referring to groundwater confined under sufficient hydrostatic pressure to rise above the upper surface of the aquifer.

**Artesian aquifer** Confined aquifer.

**Artesian head** The level to which water from a well will rise when confined in a standing pipe.

**Artesian well** A well in which water from a confined aquifer rises above the top of the aquifer. Some wells may flow without the aid of pumping.

**Auger mining** A method of extracting ore by boring horizontally into a seam, much like a drill bores a hole into wood.

**Avalanche** A large mass of either snow, rock debris, soil, or ice which detaches and slides down a mountain slope.

**Barometer** An instrument which measures atmospheric pressure. The first liquid barometer was designed by Torricelli in 1644.

**Basalt** A fine-grained, dark-colored igneous rock composed of ferromagnesian minerals.

**Bedding plane** A plane which separates or delineates layers of sedimentary rock.

**Biosphere** That part of the earth system that supports life.

**BOD (biochemical oxygen demand)** The oxygen used in meeting the metabolic needs of aquatic aerobic microorganisms. A high BOD correlates with accelerated eutrophication.

**Brackish water** Water with a salinity intermediate between that of freshwater and seawater.

**Brine** Concentrated salt solution remaining after removal of distilled product; also, concentrated brackish saline or sea waters containing more than 100,000 mg/l of total dissolved solids.

**Canopy fire** A forest fire that involves the crowns of trees. It is also called a crown fire.

**Carbon dioxide ($CO_2$)** A gaseous product of combustion about 1.5 times as heavy as air. A rise in $CO_2$ in the atmosphere increases the greenhouse effect.

**Carbon monoxide (CO)** A product of incomplete combustion. CO is colorless, and has no odor, and combines with hemoglobin in the blood leading to suffocation caused by oxygen deficiency.

**Cement** Chemically precipitated mineral material that occurs in the spaces among the individual grains of a consolidated sedimentary rock, thereby binding the grains together as a rigid coherent mass; it may be derived from the sediment or its entrapped waters, or it may be brought in by solution from outside sources. The most common cements are silica (quartz, opal, chalcedony), carbonates (calcite, dolomite, siderite), and various iron oxides; others include barite, gypsum, anhydrite, and pyrite.

Clay minerals and other fine clastic particles should not be considered as cements.

**Centipoise** A unit of viscosity based on the standard of water at 20°C (which has a viscosity of 1.005 centipoises).

**Chain reaction** A self-sustaining nuclear reaction which, once started, passes from one atom to another (see also, fission).

**Chemical treatment** Any process involving the addition of chemicals to obtain a desired result.

**Clay minerals** One of a complex and loosely defined group of finely crystalline, meta colloidal, or amorphous hydrous silicates essentially of aluminum with a monoclinic crystal lattice of the two or three layer type in which silicon and aluminum ions have tetrahedral coordination in respect to oxygen. Clay minerals are formed chiefly by chemical alteration or weathering of primary silicate minerals such as feldspars, pyroxenes, and amphiboles and are found in clay deposits, soils, shales, and mixed with sand grains in many sandstones. They are characterized by small particle size, ability to adsorb substantial amounts of water, and ions on the surfaces of the particles. The most common clay minerals belong to the kaolin, montmorrilionite, and illite groups.

**Climate** The statistical sum total of meteorological conditions (averages and extremes) for a given point or area over a long period of time.

**Cloud seeding** The artificial introduction of condensation nuclei (dry ice or silver iodide) into clouds to force precipitation.

**Cold front** The boundary on the earth's surface, or aloft, along which warm air is displaced by cold air.

**Colloidal dispersion** The process of extremely small particles (colloids) being dispersed and suspended in a medium of liquids or gases.

**Colluvium** An accumulation of soil and rock fragments at the foot of a cliff or slope under the direct influence of gravity.

**Compressibility** The reciprocal of bulk modules of elasticity. Its symbol is $\beta$. *Syn*: modulus of compression.

**Concentration** (a) The amount of a given substance dissolved in a unit volume of solution. (b) The process of increasing the dissolved solids per unit volume of solution, usually by evaporation of the liquid.

**Concentration tank** A settling tank of relatively short detention period in which sludge is concentrated by sedimentation of floatation before treatment, dewatering, or disposal.

**Cone volcano** A steep-sided and cone-shaped volcano which is composed of both lava flows and layers of pyroclastic materials. This type is also called a stratovolcano.

**Confined aquifer** An aquifer bounded above and below by impermeable beds or beds of distinctly lower permeability than that of the aquifer itself; an aquifer containing confined groundwater.

**Confined groundwater** A body of groundwater overlain by material sufficiently impervious to sever free hydraulic connection with overlying groundwater except at the intake. Confined water moves in conduits under the pressure due to difference in head between intake and discharge areas of the confined water body.

**Confining bed** A body of impermeable or distinctly less permeable material stratigraphically adjacent to one or more aquifers. *Cf:* aquitard; aquifuge; aquiclude.

**Convection** Mass motion within gases and liquids caused by differences in density brought about by cooling or heating.

**Core barrel** (a) A hollow tube or cylinder above the bit of a core drill, used to receive and preserve a continuous section or core of the material penetrated during drilling. The core is recovered from the core barrel. (b) The tubular section of a corer, in which ocean-bottom sediments are collected either directly in the tube or in a plastic liner placed inside the tube.

**Core drill** (a) A drill (usually a rotary drill, rarely a cable-tool drill) that cuts, removes, and brings to the surface a cylindrical rock sample (core) from the drill hole. It is equipped with a core bit and a core barrel. (b) A lightweight, usually mobile drill that uses drill tubing instead of drill pipe and that can (but need not) core down from grass roots.

**Corrasion** Wearing away of the earth's surface forming sinkholes and caves and widening them due to running water.

**Corrosion** The gradual deterioration or destruction of a substance or material by chemical action, frequently induced by electrochemical processes. The action proceeds inward from the surface.

**Creep** A slow movement of unconsolidated surface materials (soil, rock fragments) under the influence of water, strong wind, or gravity.

**Crustal plates** In the theory of plate tectonics it is stated that the earth's crust is not continuous but is composed of many large and small plate units that are in relative motion to one another.

**Cuttings** Rock chips or fragments produced by drilling and brought to the surface. The term does not include the core recovered from core drilling. *Also*: well cuttings; sludge; drillings. *Syn*: drill cuttings.

**Darcy** A standard unit of permeability, equivalent to the passage of one cubic centimeter of fluid of one centipoise viscosity flowing in one second under a pressure differential of one atmosphere through a porous medium having an area of cross section of one square centimeter and a length of one centimeter. A millidarcy is one one-thousandth of a darcy.

**Darcy's law** A derived formula for the flow of fluids on the assumption that the flow is laminar and that inertia can be neglected. The numerical formulation of this law is used generally in studies of gas, oil, and water production from underground formations.

**DDT (dichloro diphenyl trichloroethane)** An insecticide, one of several chlorinated hydrocarbons.

**Debris slide** A sudden downslope movement of unconsolidated earth materials or mine waste particularly once it becomes water saturated.

**Deep-well injection** A technique for disposal of liquid waste materials by pressurized infusion into porous bedrock formations or cavities.

**Degradation** The general lowering of the land by erosional processes.

**Desalination** Any process capable of converting saline water to potable water.

**Desertification** The creation of desert-like conditions, or the expansion of deserts as a result of man's actions which include overgrazing, excessive extraction of water, and deforestation.

**Desertization** A relatively new term that denotes the natural growth of deserts in response to climatic change.

**Deserts** Permanently arid regions of the world where annual evaporation by far exceeds annual precipitation. They cover about 16% of the earth.

**Detrital** Relates to deposits formed of minerals and rock fragments transported to the place of deposition.

**Diamond dust** This phenomenon occurs in arctic haze and produces bright, shimmering lights intermixed with rainbow colors that result in confusing images and reduced visibility.

**Dip slope** Topographic slope conforming with the dip of the underlying bedrock.

**Discharge** The volume of water passing a given point within a given period of time.

**Doppler radar** This instrument emits a radar frequency which appears to be changing as the wave is bounced back from a moving object. The frequency lengthens when the distance between transmitter and object increases, and it shortens as the distance decreases.

**Downdrafts** Downward and sometimes violent cold air currents frequently associated with cumulonimbus clouds and thunderstorms.

**Downwind effect** Severe turbulence can develop on the downwind [leeward] side of large buildings and mountains. This turbulence could be called the "snow fence" effect; it can be dangerous to aircraft.

**Drainage basin** The area which is drained by a river and its tributaries.

**Drilling fluid** A heavy suspension, usually in water but sometimes in oil, used in rotary drilling, consisting of various substances in a finely divided state (commonly bentonitic clays and chemical additives such as barite), introduced continuously down the drill pipe under hydrostatic pressure, out through openings in the drill bit, and back up in the annular space between the pipe and the borehole walls and to a surface pit where cuttings are removed. The fluid is then reintroduced into the pipe. It is used to lubricate and cool the bit, to carry the cuttings up from the bottom, and to prevent sloughing and cave-ins by plastering and consolidating the walls with a clay lining, thereby making casing unnecessary during drilling, and also offsetting pressures of fluid and gas that may exist in the subsurface. *Syn*: drilling mud.

**Drill-stem test** A procedure for determining productivity of an oil or gas well by measuring reservoir pressures and flow capacities while the drill pipe is in the hole and the well is full of drilling mud. A drill stem test may be done in a cased or uncased hole.

**Drought** An extended period of below-normal precipitation especially in regions of sparse precipitation. Prolonged droughts can lead to crop failures, famines, and sharply declining water resources.

**Dust storm** A severe weather system, usually in dry area, which is characterized by high winds and dust-laden air. Major dust storms were observed during the 1930s in the Dust Bowl region of the United States.

**Earthquake** A sudden movement and tremors within the earth's crust caused by fault slippage or subsurface volcanic activity.

**Ecosystem** A functional system based on the interaction between all living organisms and the physical components of a given area.

**Effective porosity** The measure of the total volume of interconnected void space of a rock, soil or other substance. Effective porosity is usually expressed as a percentage of the bulk volume of material occupied by the interconnected void space.

**Effective stress** The average normal force per unit area transmitted directly from particle to particle of a soil or rock mass. It is the stress that is effective in mobilizing internal friction. In a saturated soil, in equilibrium, the effective stress is the difference between the total stress and the neutral stress of the water in the voids; it attains a maximum value at complete consolidation of the soil.

**Ejecta** Solid material thrown out of a volcano. It includes volcanic ash, lapilli, and bombs.

**Elastic rebound** This concept implies that rocks, after breaking in response to prolonged strain, rebound back to their previous position or one similar to it. This sudden breaking and rebound may cause earthquakes.

**Entomologist** A scientist who studies insects.

**Environmentalism** This concept, also called environmental determinism, proposes that the total environment is the most influential control factor in the development of individuals or cultures.

**Epicenter** The point on the earth's surface which is located directly above the focus on an earthquake.

**Evapotranspiration** The sum of evapotranspiration from wetted surfaces and of transpiration by vegetation.

**Extension fault** A branch rupture extending from a major fault line.

**Eye** (of a hurricane) The mostly cloudless, calm center area of a hurricane. This center is surrounded by near-vertical cloud walls.

**Facies** A term used to refer to a distinguished part or parts of a single geologic entity, differing from other parts in some general aspect; e.g., any two or more significantly different parts of a recognized body of rock or stratigraphic composition. The term implies physical closeness and genetic relation or connection between the parts.

**Facies change** A lateral or vertical variation in the lithologic or paleontologic characteristics of contemporaneous sedimentary deposits. It is caused by, or reflects, a change in the depositional environment. *Cf*: facies evolution.

**Facies map** A broad term for a stratigraphic map showing the gross areal variation or distribution (in total or relative content) of observable attributes or aspects of different rock types occurring within a designated stratigraphic unit, without regard to the position or thickness of individual beds in the vertical succession; specifically a lithofacies map. Conventional facies maps are prepared by drawing lines of equal magnitude through a field of numbers representing the observed values of the measured rock attributes. *Cf*: vertical-variability map.

**Fault** A surface or zone of rock fracture along which there has been displacement, from a few centimeters to a few kilometers.

**Filtrate** The liquid which has passed through a filter.

**Filtration** The process of passing a liquid through a filtering medium (which may consist of granular material, such as sand, magnetite, or diatomaceous earth, or may be finely woven cloth, unglazed porcelain, or specially prepared paper) for the removal of suspended or colloidal matter.

**Firestorm** A violent and nearly stationary mass fire which develops its own inblowing wind system. It develops mostly in the absence of preexisting ground wind.

**Fishery** The commercial extraction of fish in a given region.

**Fission** The splitting of an atom into nuclei of lighter atoms through bombardment with neutrons. Enormous amounts of energy are released in this process which is used in the development of nuclear power and weapons.

**Fissure eruption** A type of volcanic eruption which takes place along a ground fracture instead of through a crater.

**Flank eruption** A type of volcanic eruption that takes place on the side of a volcano instead of from the crater. This typically occurs when the crater is blocked by previous lava eruptions.

**Flash flood** A local, very sudden flood that typically occurs in usually dry river beds and narrow canyons as a result of heavy precipitation generated by mountain thunderstorms.

**Floc** Small masses, commonly gelatinous, formed in a liquid by the reaction of a coagulant, through biochemical processes, or by agglomeration.

**Flood crest** The peak of a flood event, also called a flood wave, which moves downstream and shows as a curve crest on a hydrograph.

**Floodplain** A stretch of relatively level land bordering a stream. This plain is composed of river sediments and is subject to flooding.

**Flood stage** The stage at which overflow of the natural banks of a stream begins to cause damage in the reach in which the elevation is measured.

**Flow rate** The volume per time given to the flow of water or other liquid substance, which emerges from an orifice, pump, or turbine or passes along a conduit or channel, usually expressed as cubic feet per second (cfs), gallons per minute (gpm) or million gallons per day (mgd).

**Focus** The point of earthquake origin in the earth's crust from where earthquake waves travel in all directions.

**Foliation** A textural term referring to the planar arrangement of mineral grains in metamorphic rock.

**Formation** A body of rock characterized by a degree of lithologic homogeneity; it is prevailingly, but not necessarily, tabular and is mappable on the earth's surface or traceable in the subsurface.

**Formation water** Water present in a water-bearing formation under natural conditions as opposed to introduced fluids, such as drilling mud.

**Fossil fuel** Fuels such as natural gas, petroleum, and coal that developed from ancient deposits of organic deposition and subsequent decomposition.

**Fuel load** The total mass of combustible materials available to a fire.

**Geophysical logs** The records of a variety of logging tools which measure the geophysical properties of geologic formations penetrated and their contained fluids. These properties include electrical conductivity and resistivity, the ability to transmit and reflect sonic energy, natural radioactivity, hydrogen ion content, temperature, and gravity. These geophysical properties are then interpreted in terms of lithology, porosity, fluid content, and chemistry.

**Geothermal gradient** The rate of increase of temperature in the earth with depth. The gradient near the surface of the earth varies from place to place depending upon the heat flow in the region and on the thermal conductivity of the rocks. The approximate geothermal gradient in the earth's crust is about 25°C/km.

**Glacial drift** A general term applied to sedimentary material transported and deposited by glacial ice.

**Glacis** A protective earthen bank that slopes away from the outer walls of a fortification.

**Graben** A down-faulted block; may be bounded by upthrown blocks (horsts).

**Gradation** The leveling of the land through erosion, transportation, and deposition.

**Granite** A light-colored, or reddish, coarse-grained intrusive igneous rock that forms the typical base rock of continental shields.

**Greenhouse effect** The trapping and reradiation of the earth's infrared radiation by atmospheric water vapor, carbon dioxide, and ozone. The atmosphere acts like the glass cover of a greenhouse.

**Ground avalanche** An avalanche type which involves the entire thickness of the snowpack and usually includes soil and rock fragments.

**Ground fire** A type of fire which occurs beneath the surface and burns rootwork and peaty materials.

**Group** (general) An association of any kind based upon some feature of similarity or relationship. (stratigraphy) Lithostratigraphic unit consisting of two or more formations; more or less informally recognized succession of strata too thick or inclusive to be considered a formation; subdivisions of a series.

**Grout** A cementitious component of high water content, fluid enough to be poured or injected into spaces such as fissures surrounding a well bore and thereby filling or sealing them. Specifically a pumpable slurry of portland cement, sand, and water forced under pressure into a borehole during well drilling to seal crevices and prevent the mixing of groundwater from different aquifers.

**Horst** An up-faulted block; may be bounded by downthrown blocks (grabens).

**Hot spot** (geol.) Excessively hot magma centers in the asthenosphere that usually lead to the formation of volcanoes.

**Humus** The partially or fully decomposed organic matter in soils. It is generally dark in color and partly of colloidal size.

**Hurricane** A tropical low-pressure storm (also called baguio, tropical cyclone, typhoon, willy willy). Hurricanes may have a diameter of up to 400 miles (640 km) and a calm center (the eye) and must have wind velocities higher than 75 mph (120 k/mh). Some storms attained wind velocities of 200 mph (320 km/h).

**Hydrate** Refers to those compounds containing chemically combined water.

**Hydraulic** [eng.] Conveyed, operated, effected, or moved by means of water or other fluids, such as a "hydraulic dredge," using a centrifugal pump to draw sediments from a river channel.

**Hydraulic** [hydraul.] Pertaining to a fluid in motion, or to movement or action caused by water.

**Hydraulic action** The mechanical loosening and removal of weakly resistant material solely by the pressure and hydraulic force of flowing water, as by a stream surging into rock cracks or impinging against the bank on the outside of a bend, or by ocean waves and currents pounding the base of a cliff.

**Hydraulic conductivity** Ratio of flow velocity to driving force for viscous flow under saturated conditions of a specified liquid in a porous medium.

**Hydraulic gradient** In an aquifer, the rate of change of total head per unit of distance of flow at a given point and in a given direction.

**Hydraulic head** (a) The height of the free surface of a body of water above a given subsurface point. (b) The water level at a point upstream from a given point downstream. (c) The elevation of the hydraulic grade line at a given point above a given point of a pressure pipe.

**Hydraulics** The aspect of engineering that deals with the flow of water or other liquids; the practical application of hydromechanics.

**Hydrocarbon** Organic compounds containing only carbon and hydrogen. Commonly found in petroleum, natural gas, and coal.

**Hydrodynamics** The aspect of hydromechanics that deals with forces that produce motion.

**Hydrogeology** The science that deals with subsurface waters and with related geologic aspects of surface waters. Also used in the more restricted sense of groundwater geology only. The term was defined by Mead (1919) as the study of the laws of the occurrence and movement of subterranean waters. More recently it has been used interchangeably with geohydrology.

**Hydrograph** A graph which shows the rate of river discharge over a given time period.

**Hydrography** (a) The science that deals with the physical aspects of all waters on the Earth's surface, esp. the compilation of navigational charts of bodies of water. (b) The body of facts encompassed by hydrography.

**Hydrologic cycle** The constant circulation of water from the sea, through the atmosphere, to the land, and its eventual return to the atmosphere by way of transpiration and evaporation from the sea and the land surfaces.

**Hydrologic system** A complex of related parts — physical, conceptual, or both — forming an orderly working body of hydrologic units and their man-related aspects such as the use, treatment, reuse, and disposal of water and the costs and benefits thereof, and the interaction of hydrologic factors with those of sociology, economics, and ecology.

**Hydrology** (a) The science that deals with global water (both liquid and solid), its properties, circulation, and distribution, on and under the Earth's surface and in the atmosphere, from the moment of its precipitation until it is returned to the atmosphere through evapotranspiration or is discharged into the ocean. In recent years the scope of hydrology has been expanded to include environmental and economic aspects. At one time there was a tendency in the U.S. (as well as in Germany) to restrict the term *hydrology* to the study of subsurface waters (DeWeist, 1965). (b) The sum of the factors studied in hydrology; the hydrology of an area or district.

**Hydrosphere** The waters of the Earth, as distinguished from the rocks (lithosphere), living things (biosphere), and the air (atmosphere). Includes the waters of the oceans, rivers, lakes, and other bodies of surface water in liquid form on the continents; snow, ice, and glaciers; and liquid water, ice, and water vapor in both the unsaturated and saturated zones below the land surface. Included by some, but excluded by others, is water in the atmosphere, which includes water vapor, clouds, and all forms of precipitation while still in the atmosphere.

**Hydrothermal** Of or pertaining to hot water, to the action of hot water, or to the products of this action, such as a mineral deposit precipitated from a hot aqueous solution, with or without demonstrable association with igneous processes, also said of the solution itself. *Hydrothermal* is generally used for any hot water but has been restricted by some to water of magmatic origin.

**Hydrothermal processes** Those processes associated with igneous activity that involve heated or superheated water, esp. alteration, space filling, and replacement.

**Hygroscopic particles** Condensation nuclei in the atmosphere that attract water molecules (carbon, sulfur, salt, dust, ice particles).

**Impermeable** Impervious to the natural movement of fluids.

**Induction** The creation of an electric charge in a body by a neighboring body without having physical contact.

**Injection well** (a) A recharge well. (b) A well into which water or a gas is pumped for the purpose of increasing the yield of other wells in the area. (c) A well used to dispose of fluids in the subsurface environment by allowing them to enter by gravity flow or injection under pressure.

**Intensity** (earthquake) A measurement of the effects of an earthquake on the environment expressed by the Mercalli scale in stages from I to XII.

**Ion** An electrically charged molecule or atom that lost or gained electrons and therefore has a smaller or greater number of electrons than the originally neutral molecule or atom.

**Ionization** The process of creating ions.

**Iron Age** The period that followed the Bronze Age when mankind began the use of iron for making implements and weapons around 800 BC. The earliest use of iron may go back to 2500 BC.

**Ironstone** A term sometimes used to describe a hardened plinthite layer in tropical soils. It is primarily composed of iron oxides bonded to kaolinitic clays.

**Isopach** A line drawn on a map through points of equal thickness of a designated stratigraphic unit or group of stratigraphic units.

**Isopach map** A map that shows the thickness of a bed, formation, or other tabular body throughout a geographic area; a map that shows the varying true thickness of a designated stratigraphic unit or group of stratigraphic units by means of isopachs plotted normal to the bedding or other bounding surface at regular intervals.

**Isotopes** Atoms of a given element having the same atomic number but differ in atomic weight because of variations in the number of neutrons.

**Jet stream** A high-velocity, high-altitude (25,000 to 40,000 feet or 7,700 to 12,200 m) wind that moves within a relatively narrow oscillating band within the upper westerly winds.

**Joint** (geol.) A natural fissure in a rock formation along which no movement has taken place.

**Karst** A type of topography characterized by closed depressions (sinkholes), caves, and subsurface streams.

**Landslide** A general term that denotes a rapid downslope movement of soil or rock masses.

**Land-subsidence** A gradual or sudden lowering of the land surface caused by natural or man-induced factors such as solution (see karst) or the extraction of water or oil.

**Lapse rate** Expresses the rate of change (temperature or pressure) of atmospheric values with a change in elevation.

**Latent energy** The heat energy that produces changes of state in a substance without increasing the temperature of such substance. An example would be the melting of ice into liquid water and the subsequent evaporation to vapor. Latent energy is released when the processes are reversed.

**Leachate** The solution obtained by the leaching action of water as it percolates through soil or other materials such as wastes containing soluble substances.

**Lithification** The conversion of unconsolidated material into rock.

**Lithology** (a) The description of rocks on the basis of such characteristics as color, structures, mineralogic composition, and grain size. (b) The physical character of a rock.

**Lithosphere** The outer solid layer of the earth which rests on the nonsolid asthenosphere. The lithosphere averages about 60 miles (100 km) in thickness.

**Loess** Fine silt-like soil particles which have been transported and deposited by wind action. Some loess deposits may be hundreds of feet thick.

**Magma** Naturally occurring molten rock which may also contain variable amounts of volcanic gases. It issues at the earth's surface as lava.

**Magma chambers** Underground reservoirs of molten rock (magma) that are usually found beneath volcanic areas.

**Mantle** (geol.) The intermediate zone of the earth found beneath the crust and resting on the core. The mantle is believed to be about 1,800 miles (2,900 km) thick.

**Member** A division of a formation, generally of distinct lithologic character or of only local extent. A specially developed part of a varied formation is called a member, if it has considerable geographic extent. Members are commonly, though not necessarily, named.

**Mercalli scale** Used to describe the effects of an earthquake's intensity on a scale of I to XII ranging from "imperceptible" to "major catastrophe." It is not a quantified scale.

**Metamorphism** The process which induces physical or compositional changes in rocks caused by heat, pressure, or chemically active fluids.

**Meteorology** The scientific study of weather and atmospheric physics.

**Millidarcy** The customary unit of fluid permeability, equivalent to 0.001 darcy. Abbrev: md.

**Moho discontinuity** A zone between the earth's crust and mantle which shows a marked change in the travel velocity of seismic waves caused by density changes between these layers. Named after the seismologist Mohorovicic who discovered this discontinuity in 1909.

**Mudflow** A downslope movement of water-saturated earth materials such as soil, rock fragments, or volcanic ash.

**Mud logs** The record of continuous analysis of a drilling mud or fluid for oil and gas content.

**Neutralization** Reaction of acid or alkali with the opposite reagent until the concentrations of hydrogen and hydroxyl ions in the solution are approximately equal.

**Non-renewable resources** Resources (coal, oil, ores, etc.) that cannot be renewed once they have been used up. In contrast, wood, air, and water are renewable resources.

**Overburden** (spoil) Barren bedrock or surficial material which must be removed before the underlying mineral deposit can be mined.

**Oxidation** The addition of oxygen to a compound. More generally, any reaction which involves the loss of electrons from an atom.

**pH** The negative logarithm of the hydrogen-ion concentration. The concentration is the weight of hydrogen ions, in grams, per liter or solution. Neutral water, for example, has a pH value of 7, a hydrogen ion concentration of 10.

**Packer** In well drilling, a device lowered in the lining tubes which swells automatically or can be expanded by manipulation from the surface at the correct time to produce a water-tight joint against the sides of the borehole or the casing, thus entirely excluding water from different horizons.

**Percentage map** A facies map that depicts the relative amount (thickness) of a single rock type in a given stratigraphic unit.

**Perched aquifer** A water body that is not hydraulically connected to the main zone of saturation.

**Permafrost** Permanently frozen ground.

**Permeability** The property of capacity of a porous rock, sediment, or soil for transmitting a fluid without impairment of the structure of the medium; it is a measure of the relative ease of fluid flow under unequal pressure. The customary unit of measurement is the millidarcy.

**Pesticide** Any chemical used for killing noxious organisms.

**Plugging** The act or process of stopping the flow of water, oil, or gas in strata penetrated by a borehole or well so that fluid from one stratum will not escape into another or to the surface; especially the sealing up of a well that is tube abandoned. It is usually accomplished by inserting a plug into the hole, by sealing off cracks and openings in the sidewalls of the hole, or by cementing a block inside the casing. Capping the hole with a metal plate should never be considered as an adequate method of plugging a well.

**Porosity** The property of a rock, soil, or other material of containing interstices. It is commonly expressed as a percentage of the bulk volume of material occupied by interstices, whether isolated or connected.

**Potentiometric surface** An imaginary surface representing the static head of groundwater and defined by the level to which water will rise in a well. The water table is a particular potentiometric surface.

**Pressure** (a) The total load or force acting on a surface. (b) In hydraulics, without qualifications, usually the pressure per unit area or intensity of pressure above local atmospheric pressure expressed, for example, in pounds per square inch, kilograms per square centimeter.

**Primary porosity** The porosity that develops during the final stages of sedimentation or that was present within sedimentary particles at the time of deposition. It includes all depositional porosity of the sediments or the rock.

**Resistivity** Refers to the resistance of material to electrical current. The reciprocal of conductivity.

**Refusal** During drilling, the maximum depth beyond which augers (drill bits) can not be advanced, usually the top of bedrock.

**Resource** A concentration of naturally occurring solid, liquid, or gaseous materials in or on the earth's crust in such form that economic extraction of a commodity is currently or potentially feasible.

**Rotary drilling** A common method of drilling, being a hydraulic process consisting of a rotating drill pipe at the bottom of which is attached a hard-toothed drill bit. The rotary motion is transmitted through the pipe from a rotary table at the surface, i.e., as the pipe turns, the bit loosens or grinds a hole in the bottom material. During drilling, a stream of drilling mud is in constant circulation down the pipe and out through the bit from where it and the cuttings from the bit are forced back up the hole outside the pipe and into pits where the cuttings are removed and the mud is picked up by pumps and forced back down the pipe.

**Runoff** That part of precipitation which flows over the surface of the land as sheet wash and stream flow.

**Salinization** The excessive build-up of soluble salts in soils or in water. This often is a serious problem in crop irrigation system.

**Saltation** A form of wind erosion where small particles are picked up by wind and fall back to the surface in a "leap and bound" fashion. The impact of the particles loosens other soil particles rendering them prone to further erosion.

**Sanitary landfill** A land site where solid waste is dumped, compacted, and covered with soil in order to minimize environmental degradation.

**Sea level** An imaginary average level of the ocean as it exists over a long period of time. It is also used to establish a common reference for standard atmospheric pressure at this level.

**Secondary porosity** The porosity developed in a rock formation subsequent to its deposition or emplacement, either through natural processes of dissolution or stress distortion or artificially through acidization or the mechanical injection of coarse sand.

**Secondary wave** (S) A body earthquake which travels more slowly than a primary wave (P). The wave energy moves earth materials at a right angle to the direction of wave travel. This type of shear wave cannot pass through liquids.

**Sedimentation** The process of removal of solids from water by gravitational settling.

**Seismic activity** Earth vibrations or disturbances produced by earthquakes.

**Seismic survey** The gathering of seismic data from an area; the initial phase of seismic prospecting.

**Seismograph** A device that measures and records the magnitude of earthquakes and other shock waves such as underground nuclear explosions.

**Seismology** The science that is concerned with earthquake phenomena.

**Seismometer** An instrument, often portable, designed to detect earthquakes and other types of shock waves.

**Semi-arid regions** Transition zones with very unreliable precipitation that are located between true deserts and subhumid climates. The vegetation consists usually of scattered short grasses and drought-resistant shrubs.

**Septic tank system** An onsite disposal system consisting of an underground tank and a soil absorption field. Untreated sewage enters the tank where solids undergo decomposition. Liquid effluent moves from the tank to the absorption field via perforated pipe.

**Shear** The movement of one part of a mass relative to another leading to lateral deformation without resulting in a change in volume.

**Shear strength** The internal resistance of a mass to lateral deformation (see shear). Shear strength is mostly determined by internal friction and the cohesive forces between particles.

**Sinkhole** A topographic depression developed by the solution of limestone, rock salt, or gypsum bedrock.

**Sludge** (a) Mud obtained from a drill hole in boring; mud from drill cuttings. The term has also been used for the cuttings produced by drilling. (b) A semi-fluid, slushy, and murky mass or sediment of solid matter resulting from treatment of water, sewage, or industrial and mining wastes, and often appearing as local bottom deposits in polluted bodies of water.

**Slurry** A very wet, highly mobile, semiviscous mixture or suspension of finely divided, insoluble matter.

**Soil failure** Slippage or shearing within a soil mass because of some stress force that exceeds the shear strength of the soil.

**Soil liquefaction** The liquefying of clayey soils that lose their cohesion when they become saturated with water and are subjected to stress or vibrations.

**Soil salinization** The process of accumulation of soluble salts (mostly chlorides and sulfates) in soils caused by the rise of mineralized groundwater or the lack of adequate drainage when irrigation is practiced.

**Soil structure** The arrangement of soil particles into aggregates which can be classified according to their shapes and sizes.

**Soil texture** The relative proportions of various particle sizes (clay, silt, sand) in soils.

**Solution** A process of chemical weathering by which rock material passes into calcium carbonate in limestone or chalk by carbonic acid derived from rainwater containing carbon dioxide acquired during its passage through the atmosphere.

**Sorting** A dynamic gradational process which segregates sedimentary particles by size or shape. Well-sorted material has a limited size range whereas poorly sorted material has a large size range.

**Specific conductance** The electrical conductivity of a water sample at 25°C (77°F), expressed in micro-ohms per centimeter.

**Specific gravity** The ratio of the mass of a body to the mass of an equal volume of water.

**Spontaneous combustion** Type of fire started by the accumulation of the heat of oxidation until the kindling temperature of the material is reached.

**Stage** Refers to the height of a water surface above an established datum plane.

**Standing wave** An oscillating type of wave on the surface of an enclosed body of water. The wave acts similarly to water sloshing back and forth in an open dish.

**Stock** An irregularly shaped discordant pluton that is less than 100 $km^2$ in surface exposure.

**Storage coefficient** In an aquifer, the volume of water released from storage in a vertical column of 1 square foot when the water table or other potentiometric surface declines 1 foot. In an unconfined aquifer, it is approximately equal to the specific yield.

**Stratification** The structure produced by a series of sedimentary layers or beds (strata).

**Stratigraphy** The study of rock strata including their age relations, geographic distribution, composition, and history.

**Stratosphere** The part of the upper atmosphere that shows little change in temperature with altitude. Its base begins at about 7 miles (11 km) and its upper limits reach to about 22 miles (35 km).

**Stream terraces** Elevated remainders of previous floodplains; they generally parallel the stream channel.

**Stress** Compressional, tensional, or torsional forces that act to change the geometry of a body.

**Structure-contour map** A map that portrays subsurface configuration by means of structure contour lines; contour map; tectonic map. *Syn*: structural map, structure map.

**Summit aridity** Dry conditions that may develop on convex hills as a result of excessive drainage and thin soil layers.

**Surface casing** The first string of a well casing to be installed in the well. The length will vary according to the surface conditions and the type of well.

**Surficial deposit** Unconsolidated transported or residual materials such as soil, alluvial, or glacial deposits.

**Surge** A momentary increase in flow in an open conduit or pressure in a closed conduit that passes longitudinally along the conduit, usually due to sudden changes in velocity.

**Swab** A piston-like device equipped with an upward-opening check valve and provided with flexible rubber suction caps, lowered into a borehole or casing by means of a wire line for the purpose of cleaning out drilling mud or lifting oil.

**Talus debris** Unconsolidated rock fragments which form a slope at the base of a steep surface.

**Tectonic** Said of or pertaining to the forces involved in, or the resulting structures or features of, tectonics. *Syn*: geotectonic.

**Till** Unstratified and unsorted sediments deposited by glacial ice.

**Topsoil** The surface layer of a soil that is rich in organic materials.

**Tornado** A highly destructive and violently rotating vortex storm that frequently forms from cumulonimbus clouds. It is also referred to as a twister.

**Total porosity** The measure of all void space of a rock, soil, or other substance. Total porosity is usually expressed as a percentage of the bulk volume of material occupied by the void space.

**Toxin** A colloidal, proteinaceous, poisonous substance that is a specific product of the metabolic activities of a living organism and is usually very unstable, notably toxic when introduced into the tissues, and typically capable of inducing antibody formation.

**Transmissivity** In an aquifer, the rate at which water of the prevailing kinematic viscosity is transmitted through a unit width under a unit hydraulic gradient. Though spoken of as a property of the aquifer, it embodies also the saturated thickness and the properties of the contained liquid.

**Transpiration** The process by which water absorbed by plants is evaporated into the atmosphere from the plant surface.

**Triangulation** A survey technique used to determine the location of the third point of a triangle by measuring the angles from the known end points of a base line to the third point.

**Tsunami** A Japanese term that refers to a seismic sea wave which can be generated by severe submarine fault slippages or volcanic eruptions. The tsunami reaches great heights when it enters shallow waters, but it is unnoticeable on the high seas.

**Turbulence** (meteorol.) Any irregular or disturbed wind motion in the air.

**Twister** An American term used for a tornado.

**Unconfined aquifer** A groundwater body that is under water table conditions.

**Unconsolidated material** A sediment that is loosely arranged, or whose particles are not cemented together, occurring either at the surface or at depth.

**Urbanization** The transformation of rural areas into urban areas. Also referred to as urban sprawl.

**Vapor pressure** That part of the total atmospheric pressure which is contributed by water vapor. It is usually expressed in inches of mercury or in millibars.

**Vesicular** A textural term indicating the presence of many small cavities in a rock.

**Viscosity** The property of a substance to offer internal resistance to flow; its internal friction. The ratio of the rate of shear stress to the rate of shear strain is known as the coefficient of viscosity.

**Vorticity** (meteorol.) Any rotary flow of air such as in tornadoes, mid-latitude cyclones, and hurricanes.

**Wastewater** Spent water. According to the source, it may be a combination of the liquid and water-carried wastes from residence, commercial buildings, industrial plants, and institutions, together with any groundwater, surface water, and storm water which may be present. In recent years, the term *wastewater* has taken precedence over the term *sewage*.

**Water quality** The chemical, physical, and biological characteristics of water with respect to its suitability for a particular purpose.

**Water table** The surface marking the boundary between the zone of saturation and the zone of aeration. It approximates the surface topography.

**Weather** The physical state of the atmosphere (wind, precipitation, temperature, pressure, cloudiness, etc.) at a given time and location.

**Well log** A log obtained from a well, showing such information as resistivity, radioactivity, spontaneous potential, and acoustic velocity as a function of depth; esp. a lithologic record of the rocks penetrated.

**Well monitoring** The measurement, by on-site instruments or laboratory methods, of the water quality of a water well. Monitoring may be periodic or continuous.

**Well plug** A water-tight and gas-tight seal installed in a borehole or well to prevent movement of fluids. The plug can be a block cemented inside the casing.

**Well record** A concise statement of the available data regarding a well, such as a scout ticket; a full history or day-by-day account of a well, from the day the well was surveyed to the day production ceased.

**Well stimulation** Term used to describe several processes used to clean the well bore, enlarge channels, and increase pore space in the interval to be injected, thus making it possible for wastewater to move more readily into the formation. Well stimulation techniques include: surging, jetting, blasting, acidizing, and hydraulic fracturing.

**Windbreak** Natural or planted groups or rows of trees that slow down the wind velocity and protect against soil erosion.

**Zone of aeration** The zone in which the pore spaces in permeable materials are not filled (except temporarily) with water. Also referred to as unsaturated zone or vadose zone.

**Zone of saturation** The zone in which pore spaces are filled with water. Also referred to as phreatic zone.

# APPENDIX B
# CONVERSION TABLES

**English-SI Conversion Table**

| | | | | | |
|---|---|---|---|---|---|
| **Length** | 1 inch | = 2.54 cm | **Mass** | 1 oz | = 28.35 g |
| | 1 ft | = 0.3048 m | | 1 $lb_m$ | = 0.4536 kg |
| | 1 mi | = 1.609 km | | 1 s. ton | = 907 kg |
| | | | | 1 l. ton | = 1016 kg |
| **Area** | 1 $inch^2$ | = 6.4516 $cm^2$ | | 1 $lb_m/ft^3$ | = 16.02 $kg/m^3$ |
| | 1 $ft^2$ | = 0.0929 $m^2$ | | | |
| | 1 acre | = 0.4047 ha | **Force** | 1 $lb_f$ | = 4.448 N |
| | | = $0.4047 \cdot 10^4$ $cm^2$ | **Temperature** | x°F | = (9/5)x°C + 32 |
| | 1 $mi^2$ | = 2.590 $km^2$ | | | |
| | | | **Stress and pressure** | 1 $lb_f/foot^2$ | = 47.88 Pa |
| **Volume** | 1 US ft oz | = 29.54 $cm^3$ | | 1 psi | = $6.895 \cdot 10^3$ Pa |
| | 1 $ft^3$ | = $2.832 \cdot 10^{-2}$ $m^3$ | | 1 atm | = $1.013 \cdot 10^5$ Pa |
| | | = 28.32 liter | | 1 bar | = $10^5$ Pa |
| | 1 US gal | = $3.785 \cdot 10^{-3}$ $m^3$ | | | = 0.1 MPa |
| | | = 3.785 liter | | | |
| | 1 UK gal | = $4.546 \cdot 10^{-3}$ $m^3$ | **Work or energy** | 1 ft $lb_f$ | = 1.356 J |
| | | = 4.546 liter | | 1 calorie | = 4.185 J |
| | 1 US bushel | = $3.524 \cdot 10^{-2}$ $m^3$ | | 1 BTU | = $1.055 \cdot 10^3$ J |
| | | = 35.24 liter | | | |
| | 1 oil barrel | = 0.156 $m^3$ | **Hydraulic conductivity** | 1 ft/s | = 0.3048 m/s |
| | | = 156 liter | | 1 US gal/day $ft^2$ | = $4.720 \cdot 10^{-7}$ m/s |
| **Fluid flux** | 1 cubic ft/s | = $2.832 \cdot 10^{-2}$ $m^3/s$ | **Transmissivity** | 1 $ft^2/s$ | = $9.290 \cdot 10^{-2}$ $m^2/s$ |
| | | = 28.32 liter/s | | 1 US gal/day ft | = $1.438 \cdot 10^{-7}$ $m^2/s$ |
| | 1 US gal/min | = $6.309 \cdot 10^{-5}$ $m^3/s$ | | | |
| | | = $6.309 \cdot 10^{-2}$ liter/s | **Intrinsic permeability** | 1 $ft^2$ | = $9.290 \cdot 10^{-2}$ $m^2$ |
| | 1 UK gal/min | = $7.576 \cdot 10^{-5}$ $m^3/s$ | | | = $9.412 \cdot 10^{10}$ darcy |
| | | = $7.576 \cdot 10^{-2}$ liter/s | | 1 darcy | = $0.987 \cdot 10^{-12}$ $m^2$ |

**Prefixes for Multiplying and Dividing of SI Units**

| | | | | | | | |
|---|---|---|---|---|---|---|---|
| $10^{-1}$ | tenth | deci | d | $10^{1}$ | ten | deca | da |
| $10^{-2}$ | hundredth | centi | c (%) | $10^{2}$ | hundred | hecto | h |
| $10^{-3}$ | thousandth | milli | m (‰) | $10^{3}$ | thousand | kilo | k |
| $10^{-6}$ | millionth | micro | μ (ppm) | $10^{6}$ | million | mega | M |
| $10^{-9}$ | billionth | nano | n | $10^{9}$ | billion | giga | G |
| $10^{-12}$ | trillionth | pico | p (ppb) | $10^{12}$ | trillion | tera | T |

*Note:* ppm (parts per million) — particles per million particles; ppb (parts per billion) — particles per billion particles.

**Chemical Symbols**

| | | | |
|---|---|---|---|
| Mass number | | Ion charge | Examples |
| | SYMBOL | | $^{12}_{6}C$, $Ca^{2+}$, $O_2$ |
| Number of protons | | Number of atoms | |

# APPENDIX C
# MATH MODELING AND USEFUL PROGRAMS

Groundwater governing 2-D equation:

$$\frac{\delta^2 h}{\delta x^2} + \frac{\delta^2 h}{\delta y^2} = \frac{S}{T}\frac{\delta h}{\delta r} \tag{9.1}$$

a. Finite difference method (backward-difference simulation), see Figure 9.1

$$\frac{h(i-1,J),n + h(i+1,J),n + h(i,J-1),n + h(i,J+1),n - 4h(i,j),n}{a^2} = \frac{S}{T}\frac{h(i,J),n - h(i,J),(n-1)}{\Delta t} \tag{9.2}$$

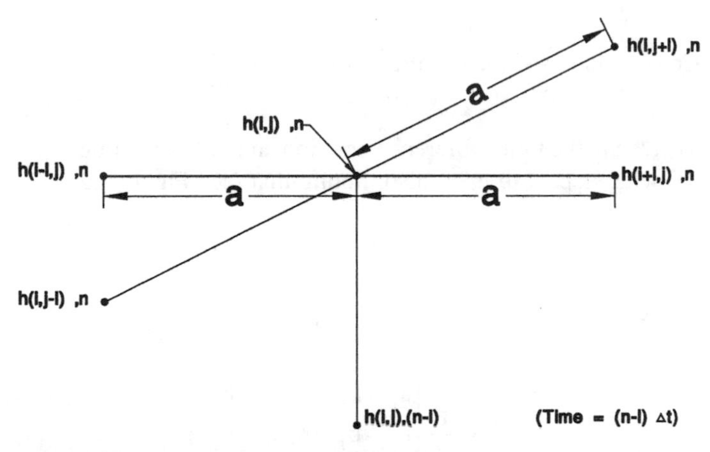


**Figure 9.1 Finite difference notation.**

where

$$a = \Delta x = \Delta y$$

b. Finite element method

Equation 9.1 may take the following form including recharging (+) or discharging (–) well Q,

$$\nabla \cdot [T_{(x,y)} \nabla h] \pm Q_{(x,y,t)} = S_{(x,y)} \frac{\delta h}{\delta t} \tag{9.3}$$

Using variational methods, Equation 9.3 may take the following form:

$$U = \iint \tfrac{1}{2} \left[ T_x \frac{\delta h^2}{\delta x} + T_y \frac{\delta h^2}{\delta y} + S \frac{\delta h}{\delta t} \mp Q \right) h \Bigg] dxdy \tag{9.3}$$

minimizing Equation 9.3

$$\frac{\delta U}{\delta h_i} = \int_s e \left[ T_x \frac{\delta h}{\delta x} \left( \frac{\delta}{\delta h_i} \frac{\delta h}{\delta x} \right) + T_y \frac{\delta h}{\delta y} \left( \frac{\delta}{\delta h_i} \frac{\delta h}{\delta y} \right) + \right. \tag{9.4}$$

$$\left. s \frac{\delta h}{\delta t} \pm Q \frac{\delta h}{\delta h_i} \right] ds$$

Considering $\frac{\delta h}{\delta t}$ for simplicity to be $\frac{hn - hn - 1}{\Delta t}$

as given in the finite difference notation and differentiating Equation 9.3 with respect to $h_J, h_k$, also, noting that (see Figure 9.2)

$$hp = a_1 + a_2 x + a_3 y \tag{9.5}$$

where

$$\begin{bmatrix} a_1 \\ a_2 \\ a_3 \end{bmatrix} = \frac{1}{2\Delta} \begin{bmatrix} a_i & a_j & a_k \\ b_i & b_J & b_k \\ c_i & c_J & c_k \end{bmatrix} \begin{bmatrix} h_{pi} \\ h_{pJ} \\ h_{pk} \end{bmatrix} \tag{9.6}$$

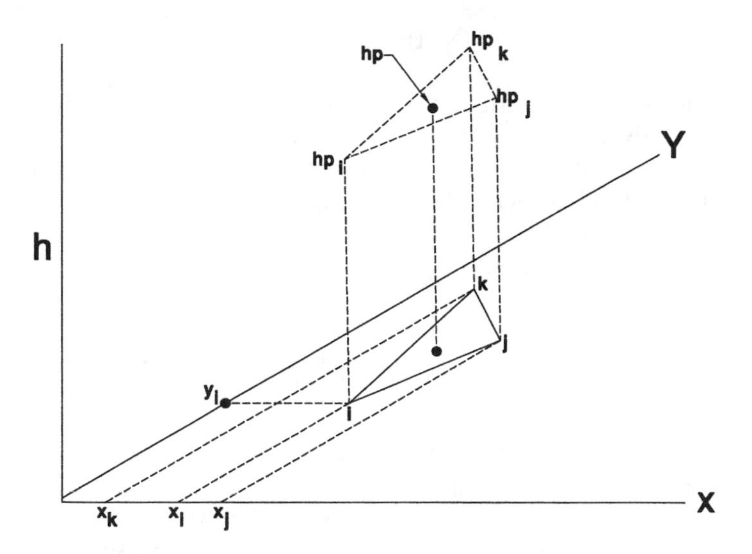


**Figure 9.2 Finite element (for a triangular element).**

$a_{i,J,k}$, $b_{i,J,k}$ and $c_{i,J,k}$ can be given by substitution in Equation 9.5 in terms of $x_{i,J,k}$ and $y_{i,J,k}$ $2\Delta = 2$ area of the triangle = determinate of the triangle.

Equation 9.5 then becomes

$$hp = \frac{1}{2\Delta}[(a_i + b_i x + c_i y)h_{pi} + (a_J + b_J x + c_J y)h_{pJ} + (a_k + b_k x + c_k y)h_{pk}] \tag{9.7}$$

having $N_i = (a_i + b_i x + c_i y)/2\Delta$,
$N_J = (a_J + b_J x + c_J y)/2\Delta$,
$N_K = (a_K + b_K x + c_h y)/2\Delta$

Equation 9.7 then becomes

$$hp = [N_i \quad N_J \quad N_k] \begin{bmatrix} h_{pi} \\ h_{pJ} \\ h_{pk} \end{bmatrix} = [N](h) \tag{9.8}$$

After assembling the whole set of minimizing equations for the whole region,

$$[N][H] + [F] = 0 \tag{9.9}$$

Equation 9.9 describes the steady state condition and for the unsteady state conditions $\Delta t$ can be set for reasonable time length and the finite difference approximation for $h_n - h_{n-1}/\Delta t$ can be included in Equation 9.9.

For the solute transport equation:

$$\frac{\delta c}{\delta t} + u\frac{\delta c}{\delta x} + w\frac{\delta c}{\delta y} - \frac{\delta}{\delta x}\left(D_L \frac{\delta c}{\delta x}\right) + \frac{\delta}{\delta y}\left(D_T \frac{\delta c}{\delta y}\right) + kc' = 0 \tag{9.10}$$

Equation 9.4 is identical with Equation 7.7 and can be simulated the same way as the groundwater Equation 9.3 when finite element method is used. The difference between the two equations is that Equation 9.9 contains symmetrical matrices whereas solute transport finite element equations do not have this symmetry. For this reason, for any numerical model of a solute transport problem a memory of at least 1 MB in the computer is required for more than 500 elements model.

# APPENDIX D
# SOFTWARE MANUAL OF DRAWDOWN AROUND MULTIPLE WELLS

*Prof. Dr. Mostafa M. Soliman*

This program can be used to calculate the drawdown of the water table around multiple well systems for dewatering purposes or for any other environmental problem. Modified Theis equation is used to develop the drawdown contours around the well system for both confined and unconfined aquifers. The aquifers are considered isotropic for simplicity, having uniform saturated thickness. Two-dimensional solution is adopted in this respect. Three-dimensional solution is tried for solute transport problems which will be demonstrated in another book by the author.

The executable files are written in TPASCAL designed by Dr. A. A. Hassan and modified by the author to suit some environmental problems. The name of this file is MULTIP4.EXE. The data file name can be any name the user can select. For the example given here the data file name is MULTIP4.DAT. This file should include all the boundary conditions, the aquifer characteristics, the well positions, the complete network of the area, etc. The following steps can be followed to create your data file.

**Step 1:**

Assign the name of your data file and write the data needed for each of the following step on a separate line.

**Step 2:** (line 1)

Specify the kind of aquifer, 0 for confined, and 1 for unconfined. This should be stated on the first line of your data file.

**Step 3:** (line 2)

Assign the maximum and minimum co-ordinates of the problem boundaries with their scales as given in the following line:

xm1 xm2 ym1 ym2 delta(x) delta(y)

**Step 4:** (line 3)

Try to adjust the groundwater flow direction parallel to one of the axes and then type the values of transmissivity, T, storativity, S, water table slope, gi, and the saturated thickness, M, as:

T S gi M

**Step 5:** (line 4)

Write the number of the pumping wells, n.

**Step 6:**

Type the x, y co-ordinates of the pumping well and discharge (Q) of each well. For example, if we have three wells those values should be arranged as:

(line 5) X1 Y1 Q1
(line 6) X2 Y2 Q2
(line 7) X3 Y3 Q3

**Step 7:** (line 8)

On the next line type the number of time NT selected: NT

**Step 8:** (line 9)

Type the value of the selected times (it) for the drawdown values to be calculated. *Note:* keep in mind the time units to be the same in the problem; this means that if T is m2/day, the time unit is a day. Thus, line 9 contains

t1 t2 ............ tr

**Step 9:** (line 10)

Type the number of points in the domain (np), the number of rows (nrows), and the number of vertical lines (nl) as follows:

np nrows nl

**Step 10:** (line 11)

For irregular boundary problem assign an integer 0 or 1 for regular boundary (regular boundaries are either square or rectangular in shape).

0 or 1

**Step 11:** (line 12)

On line 12, type the number of points on x-axis (nx) and the number of points on y-axis (ny) as:

nx ny

**Step 12:** (line 13)

For drawing the drawdown contours assign the element dimensions delta x1 and delta y1 as:

delta x1 delta y1

**Step 13:** (line 14)

Write the number of the corners: nc

**Step 14:** (line 15)

Write the corner numbers nc(i) of the boundary problem as:

i ii iii iv (supposing we have 4 corners)

**Step 15:** (line 16)

Write the number of contours nhl you want to draw as:

nhl

**Step 16:** (line 17)

Write the contour values, hl(i) (note that you must have the exact number of contour values as specified in step 15):

hl(1) hl(2) ............ hl(i)

**Step 17:** (line 18)

In the last line of the data file write the contour information, considering a = contour interval, b = the lowest value of the contours, and c = the highest contour needed for demonstration; then line 18 should include

contour intervals = a (from b to c)

**Step 18:**

After preparing the data file, you can start the program using MULTIP4.EXE . Write the data file name upon request, then press (enter ←). The first figure appears on the screen showing the well locations. Press (enter ←) to give the drawdown contours as specified after the first period. Press (enter ←) again to show the drawdown contours for the second period and so on until you execute the number of time periods required in the problem.

As an example, the following problem is given:

For an unconfined aquifer with a saturated thickness, 50 m, three wells were drilled to discharge 100 cum/h /well. Find the draw down contours around the well group after three successive time periods, 0.5, 1, and 1.5 days, respectively, if you are given the following data:

$$T = 750 \text{ sq. m/day}, \quad S = 0.01, \quad ig = 0.0001$$

if the aquifer occupies an area of 20 × 10 km, and the well co-ordinates are (3500,4500), (3250,4250), and (3750, 4000) m, respectively.

Notice that the data file for this problem is given in MULTIP4.DAT, which follows Steps 1 to 17.

# INDEX

## A

## D



## E

## F



## G

## H

# I



# J



# K



# L

# R

## S

## T

## Z